FOURTH EDITION

FUNDAMENTALS OF
PHYSICS

EXTENDED CHAPTERS 43 - 49

DAVID HALLIDAY
University of Pittsburgh

ROBERT RESNICK
Rensselaer Polytechnic Institute

JEARL WALKER
Cleveland State University

JOHN WILEY & SONS, INC.
New York Chichester Brisbane Toronto Singapore

ACQUISITIONS EDITOR Cliff Mills
DEVELOPMENTAL EDITOR Barbara Heaney
MARKETING MANAGER Catherine Faduska
PRODUCTION SUPERVISOR Lucille Buonocore
INTERIOR DESIGN Dawn L. Stanley
COVER DESIGN Jeanette Jacobs Design
MANUFACTURING MANAGER Andrea Price
COPY EDITING SUPERVISOR Deborah Herbert
PHOTO RESEARCH DIRECTOR Stella Kupferberg
PHOTO RESEARCHERS Charles Hamilton
Hilary Newman
Pat Cadley
ILLUSTRATION COORDINATOR Edward Starr
ILLUSTRATIONS Precision Graphics
COVER PHOTO Courtesy FPG International

ISBN 0-471-59571-3 (paper)

Printed in the United States of America

10 9 8 7 6 5 4 3 2 1

Recognizing the importance of preserving what has been
written, it is a policy of John Wiley & Sons, Inc. to have books
of enduring value published in the United States printed on acid-
free paper, and we exert our best efforts to that end.

This book was set in 10/12 New Baskerville by Progressive Typog-
raphers and printed and bound by Von Hoffman Press. The cover
was printed by Phoenix Color.

PREFACE

Tremendous advances have taken place over the last few years in understanding the needs and preparation of physics students for their careers in science and engineering. In writing this fourth edition of *Fundamentals of Physics,* we have been guided by this ferment of activity. With the insights provided by a new coauthor, Jearl Walker, we have completely reexamined our approaches and coverage, and we hope that this new edition will contribute to the enhancement of physics education.

CHANGES IN THE FOURTH EDITION

Although we have retained the organizational framework of the third edition, we have rewritten many chapters and many sections of other chapters. Each chapter has been scrutinized to ensure clarity and currency of coverage, reflecting the needs of science and engineering students. In particular, changes have been made in the coverage of friction, work and energy, electrostatics, and optics.

We have completely reexamined current concepts and derivations to see whether there are better or clearer ways to treat them. In many instances, we have added more explanations or intermediate steps. We have also added more Sample Problems within each chapter, with the goal not only of providing more examples to students but also of tying these Sample Problems more closely to the end-of-chapter Exercises and Problems.

We have also analyzed all of the end-of-chapter Questions, Exercises, and Problems, adding many to their number and editing them for even greater clarity and interest. At the ends of most chapters we have added a new section of problems, entitled "Additional Problems," which are unreferenced to chapter sections.

We have devoted considerable attention to illustrating real-world applications of physics topics. A prime example is the "puzzler" that opens each chapter. These examples of curious phenomena, many of which are common, were chosen so as to intrigue a student. Explanations of the puzzlers are given within the chapters, either in a text discussion or within Sample Problems. Because students will likely see these or related phenomena well after the physics course is completed, the puzzlers should provide long-term reinforcement of the associated physics.

Because the diagrams that accompany discussions of physics are vital to understanding, we have reviewed every diagram in the book for its clarity and usefulness. Nearly all the diagrams in this edition have been changed in some way, and many new diagrams have been added.

In addition, we have used full color for all the diagrams and most of the photographs. In the diagrams, color allowed us to show the various parts more clearly, to emphasize the important aspects, and to give a sense of depth to three-dimensional situations.

CHAPTER FEATURES

The features of each chapter were carefully planned to motivate students and guide their reasoning processes.

Opening Puzzlers

Each chapter opens with a physics "puzzler," describing a curious phenomenon that is intended to entice a student. These puzzlers are carefully linked to the physics of the associated chapters, and the memorable photographs of the puzzlers have been chosen in order that the relevant physics also be memorable. The explanations are given within text, for qualitative explanations, or within Sample Problems, for quantitative explanations. When answered in the latter form, the puzzler is designed to prepare the students for some of the more challenging end-of-chapter problems.

Sample Problems

We have increased the number of Sample Problems in this edition to over 400, so as to provide problem-solving models for all aspects of the chapter. We have modified many of the Sample Problems from the previous edition to tie them more closely to the end-of-chapter Exercises and Problems. All Sample Problems have been carefully edited to make them even more helpful to students. Thus, more than 50% of the Sample Problems are in some way new.

These Sample Problems offer a student the chance to work through a problem with the authors, to see how to begin with a question and end with an answer. Thus the Sample Problems provide a bridge from the physics of the text to the end-of-chapter problems. They also provide an opportunity to sort out concepts, terminology, and symbolization, to strengthen mathematical skills, and to sharpen the ability to spot "dead-end" solution strategies.

Problem Solving Tactics

Careful attention to developing a student's problem-solving skills has been a hallmark of previous editions of this book. This feature is continued in the present edition with more than 70 sections titled Problem Solving Tactics, an increase of 50% over the third edition. In these sections, we emphasize techniques of skilled problem solvers, review the logic of Sample Problems, and discuss common misunderstandings of terminology and physics concepts. As in the third edition, most of these guideposts to learning fall within the first half of the book where students need the most help, but many now appear in the second half of the book when especially tricky situations arise.

Questions, Exercises, and Problems

The sets of end-of-chapter Questions, Exercises, and Problems are by far the largest and most varied of any introductory physics text. We have edited the highly praised sets of the earlier editions to achieve even greater clarity and interest and have added a substantial number of new applied and conceptual Questions, Exercises, and Problems. We have been careful to maintain the variation in level and breadth of scope that have characterized our texts.

At the same time, we have been careful not to discard the many tried-and-true problems that have survived the test of the classroom for many years. Long-time users of our text will not find their favorites missing.

A more generous use of figures and photographs serves better to illustrate the Questions, Exercises, and Problems than before.

Questions. These thought questions have always been a special feature of our books. They are used as sources of classroom discussion and for clarification of homework concepts. Now numbering approximately 1150, they relate even more to everyday phenomena, serve to arouse curiosity and interest, and stress conceptual aspects of physics.

Exercises and Problems. The total number of Exercises and Problems has been increased to over 3400 in this edition, up from 3160 in the third edition. Exercises, identified by an "E" after their number, typically involve one step or formula or represent a single application. They thus build student confidence for the problem sets. Problems are identified by a "P" after their number; among them are a small number of advanced problems, identified by asterisks (*) next to their number.

In addition to labeling Exercises "E" and Problems "P", we have organized them by difficulty for each section of the chapter. Our goal was to simplify the selection process for instructors in the face of the voluminous material now available. Hence, instructors can vary the content emphasis and the level of difficulty to suit their tastes and the preparation of the student body while still putting aside a significant number of Exercises and Problems for many years of instruction.

Additional Problems. At the request of many instructors, we have added a new section, entitled "Additional Problems," to the ends of most chapters. While working these problems, which are unreferenced to chapter sections, students must identify for themselves the relevant physics principles.

Applications and Guest Essays

To emphasize the relevance of what physicists do and to motivate the students further, we have included within the chapters numerous applications of physics in engineering, technology, medicine, and familiar everyday phenomena.

In addition, we feature 17 essays, written by distinguished scientists and distributed at appropriate locations within the text, on the applications of physics to special topics of student interest, such as dance, sports, the greenhouse effect, lasers, holography, and many more. (See the table of contents.)

Four of the essays are new, and the authors of the 13 essays carried over from the third edition have revised and updated their material. And most of the essays, although self-contained, make references to the material covered in the immediately preceding chapter(s) and contain questions to engage the student's thought process.

MODERN PHYSICS

Like the third edition, this edition is available in a single volume of 42 chapters, ending with relativity, and in an Extended Version of 49 chapters, containing in addition a development of quantum physics and its applications to atoms, solids, nuclei, and particles. The former is meant for introductory courses that treat quantum physics in a subsequent separate course or semester.

In the early chapters, we have sought to pave the way for the systematic study of quantum physics. We have done this in three ways. (1) In appropriate places we have called attention—by specific example—to the impact of quantum ideas on our daily lives. (2) We have stressed those concepts (conservation principles, symmetry arguments, reference frames, role of aesthetics, similarity of methods, use of models, field concepts, wave concepts, etc.) that are common to both classical and quantum physics. (3) Finally, we have included a number of short, optional sections in which selected quantum (and relativistic) ideas are presented in ways that lay the foundation for the detailed and systematic treatments of relativity, atomic, nuclear, solid state, and particle physics given in later chapters.

FLEXIBILITY

In addition to the quantum physics chapters and the optional sections on quantum topics, we have included numerous optional sections throughout the text that are of an advanced, historical, general, or specialized nature.

Thus, we have consciously made available much more material than any one course or instructor is expected to "cover." Just as a textbook alone is not a course, so a course does not include the entire textbook. Indeed, more can be "uncovered" by doing less. The process of physics and its essential unity can be revealed by judicious selective coverage of many

fewer chapters than are contained here and by coverage of only portions of many included chapters. Rather than give numerous examples of such coherent selections, we urge instructors to be guided by their own interests and circumstances and to plan ahead so that some topics in relativistic and quantum physics are always included.

SUPPLEMENTS

- *A Student's Companion* by J. RICHARD CHRISTMAN, U.S. Coast Guard Academy. Much more than a traditional study guide, this student manual is designed to be used in close conjunction with the text. The Student's Companion is divided into four parts, each of which corresponds to a major section of the text, beginning with an overview "chapter." These overviews are designed to help students understand how the important topics are integrated and how the text is organized. For each chapter of the text, the corresponding Companion chapter offers: Basic Concepts, Problem Solving, Notes, Mathematical Skills, Computer Projects and Notes.
- *Solutions Manual* by J. RICHARD CHRISTMAN, U.S. Coast Guard Academy and EDWARD DERRINGH, Wentworth Institute. This manual provides students with complete worked-out solutions to 30% of the exercises and problems found at the end of each chapter within the text.
- *Interactive Learningware,* by JAMES TANNER, Georgia Institute of Technology, with the assistance of Gary Lewis, Kennesaw State College. This software contains 200 problems from the end-of-chapter exercises and problems, presented in an interactive format, providing detailed feedback for the student. Problems from Chapters 1 to 22 are included in Part 1, from Chapters 23 to 42 in Part 2. The accompanying workbooks allow the student to keep a record of the worked-out problems. The Learningware is available in IBM 3.5'' and Macintosh formats.
- *Instructor's Manual* by J. RICHARD CHRISTMAN, U.S. Coast Guard Academy. This manual contains lecture notes outlining the most important topics of each chapter, as well as demonstration experiments, laboratory and computer exercises; film and video sources are also included. Separate sections contain articles that have appeared recently

in the *American Journal of Physics* and *The Physics Teacher*.

- *Test Bank* by J. RICHARD CHRISTMAN, U.S. Coast Guard Academy. More than 2200 multiple-choice questions are included in the Test Bank for *Fundamentals of Physics*.
- *Computerized Test Bank.* IBM and Macintosh versions of the entire Test Bank are available with full editing features to help you customize tests.
- *Animated Illustrations.* Approximately 85 text illustrations are animated for enhanced lecture demonstrations.
- *Transparencies.* More than 200 four-color illustrations from the text are provided in a form suitable for projection in the classroom.

ACKNOWLEDGMENTS

A textbook contains far more contributions to the elucidation of a subject than those made by the authors alone. J. Richard Christman, of the U.S. Coast Guard Academy, has once again created many fine supplements for us; his knowledge of our book and his recommendations to students and faculty are invaluable. James Tanner, of Georgia Institute of Technology, has provided us with innovative software, closely tied to the text's exercises and problems. Albert Altman, of the University of Lowell, Massachusetts, and Harry Dulaney, of Georgia Institute of Technology, contributed many new problems to the text. We thank John Merrill, of Brigham Young University, and Edward Derringh, of the Wentworth Institute of Technology for their many contributions in the past.

Our guest essayists contributed their expertise in many areas of applied physics. We thank Charles Bean, Rensselaer Polytechnic Institute; Peter Brancazio, Brooklyn College of SUNY; Patricia Cladis, AT&T Bell Laboratories; Joseph Ford, Georgia Institute of Technology; Elsa Garmire, University of Southern California; Ivar Giaever, Rensselaer Polytechnic Institute; Tung H. Jeong, Lake Forest College; Barbara Levi, *Physics Today*; Kenneth Laws, Dickinson College; Peter Lindenfeld, State University of New Jersey–Rutgers; Suzanne Nagel, AT&T Laboratories; Sally K. Ride, University of California at San Diego; John Rigden, American Institute of Physics; Thomas D. Rossing, Northern Illinois University; and Raymond Turner, Clemson University.

A team of graduate students at Johns Hopkins University checked every exercise and problem, a truly formidable task. We thank Anton Andreev, Kevin Fournier, Jidong Jiang, John Kordomenos, Mark May, Jason McPhate, Patrick Morrissey, Mark Sincell, Olaf Vancura, John Q. Xiao, and Andrew Zwicker, our coordinator.

At John Wiley, publishers, we have been fortunate to receive strong coordination and support from Cliff Mills, our editor. He has guided our efforts and encouraged us along the way. Barbara Heaney has coordinated the developmental editing and multilayered preproduction process. Catherine Faduska, our marketing manager, has been tireless in her efforts on behalf of this edition, as well as the previous edition. Joan Kalkut has built a fine supporting package of ancillary materials. Anne Scargill edited the essays. Cathy Donovan and Julia Salsbury managed the review and administrative duties admirably.

We thank Lucille Buonocore, our able production manager, for pulling all the pieces together and guiding us through the complex production process. We also thank Dawn Stanley, for her design; Deborah Herbert, for supervising the detailed copy editing; Christina Della Bartolomea, for her copy editing; Edward Starr, for managing the line art program; Lilian Brady, for her proofreading; and all other members of the production team.

Stella Kupferberg and her team of photo researchers, particularly Charles Hamilton, Hilary Newman, and Pat Cadley, were inspired in their search for unusual and interesting photographs that communicate physics principles beautifully. We thank Edward Millman and Irene Nunes for their careful development of a full-color line art program, for which they scrutinized and suggested revisions of every piece. We also owe a debt of gratitude for the line art to the late John Balbalis, whose careful hand and understanding of physics can still be seen in every diagram.

Finally, we thank Edward Millman for his developmental work on the manuscript. With us, he has read every word, asking questions from the point of view of a student. Many of his questions and suggested changes have added to the clarity of this volume. Irene Nunes added a final, valuable developmental check in the last stages of the book.

Our external reviewers have been outstanding and we acknowledge here our debt to each member of that team:

Professor Maris A. Abolins
Michigan State University

Professor Barbara Andereck
Ohio Wesleyan University

Professor Albert Bartlett
University of Colorado

Professor Timothy J. Burns
Leeward Community College

Professor Joseph Buschi
Manhattan College

Professor Philip A. Casabella
Rensselaer Polytechnic Institute

Professor Randall Caton
Christopher Newport College

Professor Roger Clapp
University of South Florida

Professor W. R. Conkie
Queen's University

Professor Peter Crooker
University of Hawaii at Manoa

Professor William P. Crummett
Montana College of Mineral Science and Technology

Professor Robert Endorf
University of Cincinnati

Professor F. Paul Esposito
University of Cincinnati

Professor Jerry Finkelstein
San Jose State University

Professor Alexander Firestone
Iowa State University

Professor Alexander Gardner
Howard University

Professor Andrew L. Gardner
Brigham Young University

Professor John Gieniec
Central Missouri State University

Professor John B. Gruber
San Jose State University

Professor Ann Hanks
American River College

Professor Samuel Harris
Purdue University

Emily Haught
Georgia Institute of Technology

Professor Laurent Hodges
Iowa State University

Professor John Hubisz
College of the Mainland

Professor Joey Huston
Michigan State University

Professor Darrell Huwe
Ohio University

Professor Claude Kacser
University of Maryland

Professor Leonard Kleinman
University of Texas at Austin

Professor Arthur Z. Kovacs
Rochester Institute of Technology

Professor Kenneth Krane
Oregon State University

Professor Sol Krasner
University of Illinois at Chicago

Professor Robert R. Marchini
Memphis State University

Professor David Markowitz
University of Connecticut

Professor Howard C. McAllister
University of Hawaii at Manoa

Professor W. Scott McCullough
Oklahoma State University

Professor Roy Middleton
University of Pennsylvania

Professor Irvin A. Miller
Drexel University

Professor Eugene Mosca
United States Naval Academy

Professor Patrick Papin
San Diego State University

Professor Robert Pelcovits
Brown University

Professor Oren P. Quist
South Dakota State University

Professor Jonathan Reichert
SUNY–Buffalo

Professor Manuel Schwartz
University of Louisville

Professor John Spangler
St. Norbert College

Professor Ross L. Spencer
Brigham Young University

Professor Harold Stokes
Brigham Young University

Professor David Toot
Alfred University

Professor J. S. Turner
University of Texas at Austin

Professor T. S. Venkataraman
Drexel University

Professor Gianfranco Vidali
Syracuse University

Professor Fred Wang
Prairie View A & M

Professor George A. Williams
University of Utah

Professor David Wolfe
University of New Mexico

This new edition traces its origins to the text *Physics for Students of Science and Engineering* (John Wiley & Sons, Inc., 1960) by the authors of the third edition. Since that time it is estimated that well over 5 million students have been introduced to physics at the college or university level by this text and those that flowed from it, including among them translations into many languages. We dedicate this fourth edition to those students, and we hope that it will be as well received by those for whom it has been written.

DAVID HALLIDAY
5110 Kenilworth Place NE
Seattle WA 98105

ROBERT RESNICK
Rensselaer Polytechnic Institute
Troy NY 12181

JEARL WALKER
Cleveland State University
Cleveland OH 44115

CONTENTS

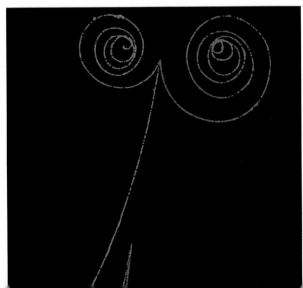

CHAPTER 46

THE CONDUCTION OF ELECTRICITY IN SOLIDS *1209*

What is a transistor and how does it function?

CHAPTER 47

NUCLEAR PHYSICS *1231*

Why and how do (some) nuclei undergo decay?

CHAPTER 48

ENERGY FROM THE NUCLEUS *1259*

What physics underlies the image that has horrified the world since World War II?

CHAPTER 49

QUARKS, LEPTONS, AND THE BIG BANG *1283*

How can a photograph of the universe 15×10^9 y ago be taken?

QUANTUM PHYSICS—I

First-order diffraction spectra are seen in this night scene photographed through a diffraction grating. All colors are emitted by the headlights on a car approaching at bottom left but only red shines from a traffic light and only individual bright colors come from the mercury lamps on the house at center right and at the top of a pole at left. Why?

43-1 A NEW DIRECTION

So far we have studied light—and by that word we now mean not only visible light but radiation throughout the entire electromagnetic spectrum—under the headings of reflection, refraction, polarization, interference, and diffraction. We can explain all these phenomena by treating light as an electromagnetic *wave,* governed by Maxwell's equations. The experimental support for this treatment is pretty overwhelming.

We now move off in an entirely new direction and consider experiments that can be understood only by making quite a different assumption about light, namely, that it behaves like a stream of *particles,* each with a specified energy and momentum.

You may well ask, "Well, which is it, wave or particle?" These concepts are so different that it is hard to see how light can model itself after both at the same time. We will face this question squarely in Section 44-10. Meanwhile, we will not worry about it but will simply look at the strong experimental evidence that light is particle-like. The path we are taking will open the door to the world of quantum physics and will allow us to begin to discuss how atoms are constructed.

43-2 EINSTEIN MAKES A PROPOSAL

In 1905, Einstein made the bold hypothesis—since convincingly confirmed by experiment—that light sometimes behaves as if its energy were concentrated in discrete bundles that he called *light quanta;* we now call them **photons.** He proposed that the energy of a single photon is

$$E = hf \quad \text{(photon energy)}, \quad (43\text{-}1)$$

in which f is the frequency of the light and h is the Planck constant. This constant, introduced into physics a few years earlier in another connection by Max Planck, has the value

$$h = 6.63 \times 10^{-34} \text{ J} \cdot \text{s}$$
$$= 4.14 \times 10^{-15} \text{ eV} \cdot \text{s}. \quad (43\text{-}2)$$

Photons carry not only energy but also linear momentum. We can find an expression for the momentum of a photon by starting with Eq. 42-41,

$$E^2 = (pc)^2 + (mc^2)^2. \quad (43\text{-}3)$$

This expression gives the relativistic relationship between the momentum p and the total energy E of a particle, such as an electron or a proton, of mass m.

We can apply Eq. 43-3 to a photon by putting $E = hf$ and $m = 0$ since a photon, traveling at the speed of light, must have zero mass. Equation 43-3 then becomes $hf = pc$; solving for p and using the relation $c = \lambda f$ (from Eq. 17-14) leads to

$$p = \frac{h}{\lambda} \quad \text{(photon momentum)}, \quad (43\text{-}4)$$

in which λ is the wavelength of the light.

Note how the wave and photon models are intimately connected. The energy E of the *photon* is related to the frequency f of the *wave* by Eq. 43-1. Similarly, the momentum p of the *photon* is related to the wavelength λ of the *wave* by Eq. 43-4. In each case, the factor of proportionality is the Planck constant h.

Equations 43-1 and 43-4 permit us to look at the electromagnetic spectrum in a new way. In Fig. 38-1 we displayed this spectrum as a range of wavelengths or, equivalently, of frequencies. We can now also display it as a range of photon energies or (if we wish) of photon momenta. Table 43-1 shows some corre-

TABLE 43-1

SOME CORRESPONDING WAVELENGTHS, FREQUENCIES, AND PHOTON ENERGIES

REGION OF THE ELECTROMAGNETIC SPECTRUM	WAVELENGTH	FREQUENCY (HZ)	PHOTON ENERGY
Gamma ray	50 fm	6×10^{21}	25 MeV
X ray	50 pm	6×10^{18}	25 keV
Ultraviolet	100 nm	3×10^{15}	12 eV
Visible	550 nm	5×10^{14}	2 eV
Infrared	10 μm	3×10^{13}	120 meV
Microwave	1 cm	3×10^{10}	120 μeV
Radio wave	1 km	3×10^{5}	1.2 neV

sponding wavelengths, frequencies, and photon energies for selected regions of the electromagnetic spectrum.

In 1905 most physicists were quite comfortable with the wave theory of light and did not look kindly upon Einstein's photon idea. Prominent among those who were slow to believe was Max Planck, the very person who introduced the constant h into physics. In recommending Einstein for membership in the Royal Prussian Academy of Sciences in 1913, for example, Planck wrote, "that he may sometimes have missed the target in his speculations, as for example in his theory of light quanta, cannot really be held against him." It is almost commonplace that radical ideas are accepted only slowly, even by such men of genius as Planck.

SAMPLE PROBLEM 43-1

Yellow light from a sodium vapor lamp has an effective wavelength of 589 nm. What is the energy, in electron-volts, of the corresponding photons?

SOLUTION From Eq. 43-1 we have, using the relation $c = \lambda f$,

$$E = hf = \frac{hc}{\lambda}$$

$$= \frac{(4.14 \times 10^{-15} \text{ eV}\cdot\text{s})(3.00 \times 10^8 \text{ m/s})}{589 \times 10^{-9} \text{ m}}$$

$$= 2.11 \text{ eV.} \qquad \text{(Answer)}$$

This is the energy that a single electron or proton would acquire if it were accelerated through a potential difference of 2.11 V.

SAMPLE PROBLEM 43-2

During radioactive decay, a certain nucleus emits a gamma ray whose photon energy is 1.35 MeV.

a. To what wavelength does this photon correspond?

SOLUTION From Eq. 43-1 and the relation $c = \lambda f$ we have

$$\lambda = \frac{c}{f} = \frac{hc}{hf} = \frac{hc}{E}$$

$$= \frac{(4.14 \times 10^{-15} \text{ eV}\cdot\text{s})(3.00 \times 10^8 \text{ m/s})}{1.35 \times 10^6 \text{ eV}}$$

$$= 9.20 \times 10^{-13} \text{ m} = 920 \text{ fm.} \qquad \text{(Answer)}$$

b. What is the momentum of this photon?

SOLUTION From Eq. 43-4 we can write (again using Eq. 43-1 and the relation $c = \lambda f$)

$$p = \frac{h}{\lambda} = \frac{hf}{\lambda f} = \frac{E}{c}, \qquad (43\text{-}5)$$

in which E is the photon energy. Substituting in a straightforward way, we find

$$p = \frac{E}{c} = \frac{(1.35 \text{ MeV})(1.60 \times 10^{-13} \text{ J/MeV})}{3.00 \times 10^8 \text{ m/s}}$$

$$= 7.20 \times 10^{-22} \text{ kg}\cdot\text{m/s.} \qquad \text{(Answer)}$$

Although this answer is correct, physicists in the field of high-energy particle physics do not ordinarily express the momenta of photons (or of particles such as electrons or protons) in SI units. Instead, they express a momentum as an energy divided by the speed of light, as discussed in Sample Problem 42-6. Thus, from Eq. 43-5,

$$p = \frac{E}{c} = \frac{1.35 \text{ MeV}}{c} = 1.35 \text{ MeV}/c. \qquad \text{(Answer)}$$

One advantage of this practice is that, given the energy of a photon, you at once know its momentum, and conversely. (This also holds for material particles if their total energies greatly exceed their rest energies, so that the last term in Eq. 43-3 can be neglected.)

43-3 THE PHOTOELECTRIC EFFECT

Here we consider the first of several experiments whose results cannot be interpreted in terms of a wave model for light, but that find a ready explanation if we assume that light is made up of photons.

If you shine a beam of light on a clean metal surface and if conditions are right, the light can knock electrons out of that surface. Most of us are familiar with applications of this **photoelectric effect,** as it is called, in automatic door openers or security alarm systems. When the photoelectric effect is studied carefully in the laboratory, we find that the experimental results cannot be explained at all in terms of the wave model of light. However, as Einstein pointed out, the explanation of the photoelectric effect is quite straightforward if we view it as a "collision" between an incident photon and an electron within the metal.

Figure 43-1 shows a typical apparatus for studying the photoelectric effect. Light of frequency f illuminates a metal plate P and knocks electrons out of the plate. A suitable potential difference V be-

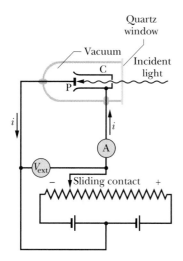

Quartz
window

Vacuum
Incident
light

FIGURE 43-1 An apparatus used to study the photoelectric effect. The incident light falls on plate P, ejecting photoelectrons, which are collected by collector cup C. The photoelectrons move in the circuit in a direction opposite that of the conventional current arrows. The batteries and variable resistor are used to produce and adjust the electric potential between P and C.

tween P and collector cup C sweeps up these *photoelectrons,* displaying them as a photoelectric current in ammeter A. The potential difference V is given by

$$V = V_{ext} + V_{cpd}, \qquad (43\text{-}6)$$

in which the first term on the right is the reading of the voltmeter in Fig. 43-1 and the second is the measured *contact potential difference* (a battery effect) introduced by the fact that the plate and the collector are usually made of different materials.

The essential data obtained in this experiment are displayed in Figs. 43-2 and 43-3. Figure 43-2 shows the photoelectric current i as a function of V, for incident light of two different intensities but the same wavelength. The *stopping potential* V_0 is the potential difference required to stop the fastest photoelectrons, thus bringing the photoelectric current to zero. Note that eV_0 measures the kinetic energy of the most energetic photoelectrons. That is,

$$K_m = eV_0. \qquad (43\text{-}7)$$

The central feature of Fig. 43-2 is that V_0 is the same for both curves. This observation may be generalized to the following: *the kinetic energy of the most energetic photoelectrons is independent of the intensity of the incident light.*

Figure 43-3 shows the stopping potential plotted against the frequency of the incident light, for several experiments like that of Fig. 43-2. We see by extrapolation that there is a certain *cutoff frequency* f_0 corresponding to a stopping potential of zero. For light with frequencies below f_0 the photoelectric effect simply does not happen.

Let us now see how the photon model succeeds —and the wave model fails—in explaining these experimental results:

1. *The Intensity Problem.* In wave theory, when you increase the intensity of a beam of light, you increase the magnitude of the oscillating electric field vector **E**. The force that the incident beam exerts on an electron is $e\mathbf{E}$. You might then expect that the more intense the light, the more energetic would be the

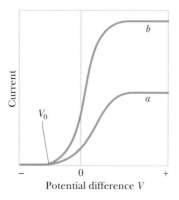

FIGURE 43-2 A plot (not to scale) of data taken with the apparatus of Fig. 43-1. The intensity of the incident light is twice as great for curve *b* as for curve *a*. The wavelength of the incident light is the same for both curves.

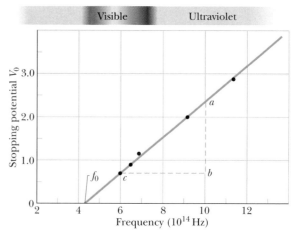

FIGURE 43-3 The stopping potential as a function of the frequency of the incident light when plate P in Fig. 43-1 is sodium. (Data reported by R. A. Millikan in 1916.)

ejected photoelectrons. However, as Fig. 43-2 shows, V_0 (and thus K_m, as given by Eq. 43-7) *does not* depend on the light intensity; this has been tested and verified experimentally over an intensity range of about 10^7.

For the photon model, however, the "intensity problem" is no problem. If we double the light intensity, we simply double the number of photons but we do not change the energy of the individual photons as given by Eq. 43-1. Thus K_m, the maximum kinetic energy that an electron can pick up from a photon during a collision, remains unchanged.

2. *The Frequency Problem.* According to wave theory, the photoelectric effect should occur at *any* frequency of the incident light, provided only that the light is intense enough. However, as Fig. 43-3 shows, there is a characteristic cutoff frequency below which there is no photoelectric effect, *no matter how intense the light.*

Again, the "frequency problem" is no problem if we think in terms of photons. The conduction electrons are held within the metal target by an electric field. Thus to be ejected, an electron must obtain a certain minimum energy ϕ, called the **work function** of the material. If the photon energy exceeds the work function (that is, if $hf > \phi$), the photoelectric effect can occur. If it does not (that is, if $hf < \phi$), the effect will not occur. This is exactly what Fig. 43-3 shows.

3. *The Time Delay Problem.* According to wave theory, the energy of an ejected photoelectron must be soaked up from the incident wave. The effective area from which the electron soaks up this energy cannot be much larger than the cross section of an atom. Thus if the light is feeble enough, there should be a measurable time delay between when the light strikes the surface and when the electron has accumulated sufficient energy to emerge from it. No such delay has ever been found. The time delay problem is that there is no time delay!

The "time delay problem" does not exist for the photon model because it postulates that the photon energy is delivered to the ejected photoelectron in a single collision event.

A Quantitative Analysis

Einstein wrote the principle of conservation of energy for the photoelectric effect as

$$hf = \phi + K_m \qquad \text{(photoelectric equation)}, \qquad (43\text{-}8)$$

where hf is the energy of the photon. Equation 43-8 tells us that a photon carries an energy hf into the surface, where the photon interacts with an electron. If the electron is to escape, an amount of energy ϕ (the **work function** of the material) must be provided to surmount the electric field that exists at the surface. The remaining energy ($= hf - \phi$) is equal to K_m, the *maximum* kinetic energy that the ejected electron can have.

Let us rewrite Eq. 43-8 by substituting for K_m from Eq. 43-7. After some rearrangement, we have

$$V_0 = (h/e)f - (\phi/e). \qquad (43\text{-}9)$$

Thus Einstein's photon theory predicts a linear relationship between the stopping potential V_0 and the frequency f, in complete agreement with Fig. 43-3. The slope of the experimental curve in that figure should be h/e, or

$$\frac{h}{e} = \frac{ab}{bc} = \frac{2.35 \text{ V} - 0.72 \text{ V}}{(10 \times 10^{14} - 6 \times 10^{14}) \text{ Hz}}$$

$$= 4.1 \times 10^{-15} \text{ V} \cdot \text{s}.$$

By multiplying this by the electronic charge e, we find

$$h = (4.1 \times 10^{-15} \text{ V} \cdot \text{s})(1.6 \times 10^{-19} \text{ C})$$

$$= 6.6 \times 10^{-34} \text{ J} \cdot \text{s},$$

which is in full agreement with the value given in Eq. 43-2.

SAMPLE PROBLEM 43-3

A potassium foil is a distance $r = 3.5$ m from a light source whose power P is 1.5 W. Assuming that the light incident on the foil from the source is a wave, how long would it take for the foil to soak up enough energy ($= 1.8$ eV) to eject a photoelectron? Assume that the electron collects its energy from a circular area of the foil whose radius is 5.3×10^{-11} m. (This value, called the *Bohr radius*, is roughly equal to the radius of an average atom. It is a (non-SI) length unit useful on the scale of atomic dimensions.)

SOLUTION   The target area A is $\pi(5.3 \times 10^{-11}$ m$)^2$ or 8.8×10^{-21} m^2. If the light source radiates uniformly in all directions, the intensity I at the foil is (see Sample Problem 38-1)

$$I = \frac{P}{4\pi r^2} = \frac{1.5 \text{ W}}{(4\pi)(3.5 \text{ m})^2} = 9.7 \times 10^{-3} \text{ W/m}^2.$$

The rate at which energy is intercepted by the target area is then

$$R = IA = (9.7 \times 10^{-3} \text{ W/m}^2)(8.8 \times 10^{-21} \text{ m}^2)$$

$$= 8.5 \times 10^{-23} \text{ W}.$$

If all this incoming energy is absorbed, the time required to accumulate enough energy for the electron to escape is

$$t = \left(\frac{1.8 \text{ eV}}{8.5 \times 10^{-23} \text{ J/s}}\right)\left(\frac{1.60 \times 10^{-19} \text{ J}}{1 \text{ eV}}\right)\left(\frac{1 \text{ min}}{60 \text{ s}}\right)$$

$$= 56 \text{ min!} \qquad\qquad \text{(Answer)}$$

However, no measurable time delay is observed.

SAMPLE PROBLEM 43-4

At what rate do photons strike the foil in Sample Problem 43-3? Assume a wavelength of 589 nm (yellow sodium light) and an area of 1.0 cm^2.

SOLUTION   Using results from Sample Problem 43-3, we can express the intensity at the foil as

$$I = (9.7 \times 10^{-3} \text{ W/m}^2)(1 \text{ eV}/1.6 \times 10^{-19} \text{ J})$$

$$= 6.1 \times 10^{16} \text{ eV/m}^2 \cdot \text{s}.$$

In Sample Problem 43-1, we found the energy of each photon of 589-nm yellow light to be 2.11 eV. The rate at which photons strike the plate is then

$$R = (6.1 \times 10^{16} \text{ eV/m}^2 \cdot \text{s})\left(\frac{1 \text{ photon}}{2.11 \text{ eV}}\right)(10^{-4} \text{ m}^2)$$

$$= 2.9 \times 10^{12} \text{ photons/s.} \qquad \text{(Answer)}$$

Even at this low intensity (about 1 μW/cm^2), the photon incidence rate is very great, about 10^8 photons falling every second on a patch the size of a period on this page. Small wonder that we do not ordinarily notice the granularity of light.

SAMPLE PROBLEM 43-5

Find the work function of sodium from the data plotted in Fig. 43-3.

SOLUTION   The straight line in Fig. 43-3 intersects the frequency axis at the cutoff frequency f_0. Putting the values $V_0 = 0$ and $f = f_0$ in Eq. 43-9 yields

$$0 = \frac{h}{e}f_0 - \frac{\phi}{e},$$

or

$$\phi = hf_0 = (6.63 \times 10^{-34} \text{ J} \cdot \text{s})(4.3 \times 10^{14} \text{ Hz})$$

$$= 2.9 \times 10^{-19} \text{ J} = 1.8 \text{ eV}, \qquad \text{(Answer)}$$

where we have taken the value of f_0 from the plot of Fig. 43-3.

We note from Eq. 43-9 that, to find the Planck constant h, you need know only the slope of the straight line in Fig. 43-3. To find the work function, you need know only the frequency intercept.

43-4 THE COMPTON EFFECT

Here we have another experiment that can be understood readily in terms of a photon model for light but that cannot be understood at all in terms of a wave model. Historically, this experiment proved to be a great "convincer" of the reality of photons because it introduced photon *momentum*, as well as photon energy, into an experimental situation. Furthermore, it showed that the photon model applies not only to visible and ultraviolet light—the domain of the photoelectric effect—but also to x rays.

In 1923, Arthur Holly Compton at Washington University in St. Louis arranged for a beam of x rays of wavelength λ to fall on a graphite target T, as in Fig. 43-4. He measured, as a function of wavelength, the intensity of the x rays scattered from the target in several selected directions. Figure 43-5 shows his results. We see that, although the incident beam contains only a single wavelength, the scattered x rays have intensity peaks at two wavelengths. One peak corresponds to the incident wavelength λ, the other to a wavelength λ' that is longer than λ by an amount $\Delta\lambda$. This **Compton shift,** as it is called, varies with the angle at which the scattered x rays are observed.

The scattered peak of wavelength λ' cannot be understood at all if you think of the incident x-ray beam as a wave. In this model, the incident wave, with frequency f, causes the electrons in the target to oscillate at that same frequency. These oscillating electrons, like charges surging back and forth in a small transmitting antenna, must radiate at this same

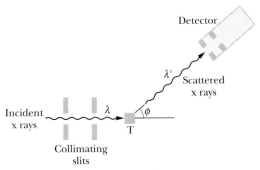

FIGURE 43-4 An apparatus used to study the Compton effect. A beam of x rays falls on a graphite target T. The x rays scattered from the target are observed at various angles ϕ to the incident direction. The detector measures both the intensity and the wavelength of these scattered x rays.

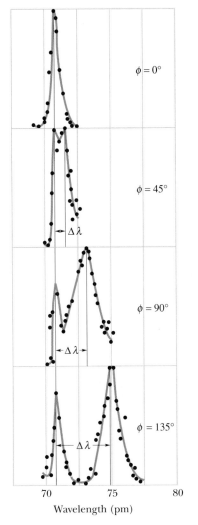

FIGURE 43-5 Compton's results for four values of the scattering angle ϕ. Note that the Compton shift $\Delta\lambda$ increases as the scattering angle increases.

frequency. Thus the scattered beam should have only the same frequency—and only the same wavelength—as the incident beam. But it doesn't.

Compton viewed the incident beam as a stream of photons of energy $E\ (=hf)$ and momentum p $(=h/\lambda)$ and assumed that some of these photons made billiard-ball-like collisions with individual free electrons in the target. Since an electron picks up some kinetic energy in such an encounter, the scattered photon must have a lower energy E' than the incident photon. It will therefore have a lower frequency f' and, correspondingly, a longer wavelength λ', exactly as we observe. Thus we account qualitatively for the Compton shift.

A Quantitative Analysis

Figure 43-6 suggests a collision between a photon and a free electron in the target. Let us apply the principle of conservation of energy. Because the electron emerges from the collision with a speed that may be comparable to the speed of light, we must use the relativistic expression for its kinetic energy (Eq. 42-38). Thus we have

$$hf = hf' + mc^2\left(\frac{1}{\sqrt{1 - (v/c)^2}} - 1\right),$$

in which the second term on the right is the kinetic energy of the recoiling electron. Substituting c/λ for

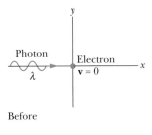

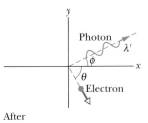

FIGURE 43-6 A photon of wavelength λ strikes a resting electron. The photon is scattered at angle ϕ with an increased wavelength λ'. The electron moves off with speed v at angle θ.

f and c/λ' for f' leads to

$$\frac{h}{\lambda} = \frac{h}{\lambda'} + mc\left(\frac{1}{\sqrt{1-(v/c)^2}} - 1\right)$$

(energy conservation).   (43-10)

Now let us apply the (vector) law of conservation of momentum to the collision of Fig. 43-6. The momentum of a photon is given by Eq. 43-4 ($p = h/\lambda$). For the electron, the relativistic expression for the momentum is given by Eq. 9-24,

$$p = \frac{mv}{\sqrt{1-(v/c)^2}}$$ (electron momentum).   (43-11)

We can then express the conservation of momentum for the photon–electron collision as

$$\frac{h}{\lambda} = \frac{h}{\lambda'}\cos\phi + \frac{mv}{\sqrt{1-(v/c)^2}}\cos\theta$$

(x component)   (43-12)

and

$$0 = \frac{h}{\lambda'}\sin\phi - \frac{mv}{\sqrt{1-(v/c)^2}}\sin\theta$$

(y component).   (43-13)

Our aim is to find $\Delta\lambda$ ($= \lambda' - \lambda$), the wavelength shift of the scattered photons. Of the five collision variables (λ, λ', v, ϕ, and θ) that appear in Eqs. 43-10, 43-12, and 43-13, we can eliminate two. We choose to eliminate v and θ, which deal only with the recoil electron.

Carrying out the necessary algebra leads to this simple result:

$$\Delta\lambda = \frac{h}{mc}(1 - \cos\phi)$$ (Compton shift),   (43-14)

in which the quantity h/mc (called the *Compton wavelength*) has the value of 2.43×10^{-12} m or 2.43 pm. Equation 43-14 agrees exactly with Compton's experimental results.

Equation 43-14 tells us that the Compton shift depends only on the scattering angle ϕ and not on the initial photon energy. The predicted shift varies from zero (for $\phi = 0$, a grazing collision, the incident photon being scarcely deflected) to $2h/mc$ (for $\phi = 180°$, a head-on collision, the incident photon being reversed in direction).

It remains to explain the peak in Fig. 43-5 in which the wavelength does *not* change. This peak results from scattering from electrons that are not free—as we have assumed so far—but are tightly bound to the atoms of the target. For a carbon target, the effective mass of such electrons is that of carbon atoms, or about $22{,}000\,m$, m being the electron mass. If we replace m in Eq. 43-14 by $22{,}000\,m$, we see that the Compton shift for bound electrons is immeasurably small, just as we observe.

The Compton effect is responsible for the so-called *electromagnetic pulse* (EMP) caused by thermonuclear explosions high in the atmosphere. The x rays and gamma rays generated in such explosions have Compton collisions with electrons in the upper atmosphere, knocking them sharply forward. This sudden, enormous surge of charge sets up electromagnetic fields that can play havoc with unshielded electric circuits on the Earth's surface. The effect was first noticed when power and communication circuits in Hawaii failed at the time of a thermonuclear airburst test in the Pacific Ocean, many miles away.

SAMPLE PROBLEM 43-6

X rays of wavelength 22 pm (photon energy = 56 keV) are scattered from a carbon target, the scattered radiation being viewed at 85° to the incident beam.

a. What is the Compton shift?

SOLUTION   From Eq. 43-14 we have

$$\Delta\lambda = \frac{h}{mc}(1 - \cos\phi)$$

$$= \frac{(6.63 \times 10^{-34}\,\text{J·s})(1 - \cos 85°)}{(9.11 \times 10^{-31}\,\text{kg})(3.00 \times 10^8\,\text{m/s})}$$

$$= 2.21 \times 10^{-12}\,\text{m} = 2.21\,\text{pm}.$$ (Answer)

b. What percentage of its initial energy does an incident x-ray photon lose?

SOLUTION   The fraction energy loss *frac* is

$$frac = \frac{E - E'}{E} = \frac{hf - hf'}{hf} = \frac{(c/\lambda) - (c/\lambda')}{(c/\lambda)} = \frac{\lambda' - \lambda}{\lambda'}$$

$$= \frac{\Delta\lambda}{\lambda + \Delta\lambda}.$$ (43-15)

Substitution yields

$$frac = \frac{2.21 \text{ pm}}{22 \text{ pm} + 2.21 \text{ pm}} = 0.091 \text{ or } 9.1\%. \quad \text{(Answer)}$$

Equation 43-14 reminds us that the Compton shift $\Delta\lambda$ is independent of the wavelength λ of the incident photon. Equation 43-15 then tells us that the shorter this incident wavelength (that is, the more energetic the incoming photon), the larger will be the fractional energy loss. The Compton effect shows up more strongly for more energetic photons.

43-5 PLANCK AND HIS CONSTANT: HISTORICAL ASIDE

At any given time, there are always one or more "hot problems" that attract the attention of the most able physicists. The related problems of the fundamental nature of matter and the evolution of the universe rank high on today's list. At the turn of the century, however, the problem that attracted the attention of the best and the brightest was a hot problem in more ways than one. It was that of understanding the wavelength distribution of the radiation emitted by heated objects.

The radiation emitted by red hot pokers or bonfires depends on too many variables to be of fundamental significance. The physicists of 1900 turned instead to the study of the radiation emitted by an *ideal radiator,* that is, a radiator whose emitted radiation depends *only* on the temperature of the radiator and not on the material from which the radiator is made, the nature of its surface, or anything other than temperature.

We can make such an ideal radiator in the laboratory by forming a cavity within a body and holding the walls of the cavity at a uniform temperature. We must drill a small hole through the cavity wall so that a sample of the radiation inside the cavity can escape into the laboratory, where we can study it. Experiment shows that such **cavity radiation** has a very simple spectrum, determined indeed only by the temperature of the walls. Cavity radiation (photons in a box) helps us to understand radiation, just as the ideal gas (atoms in a box) helped us to understand matter.

Figure 43-7 shows a simple cavity radiator made of a thin-walled tungsten tube about a millimeter in diameter, heated to incandescence by passing a current through it. We see the bright cavity radiation

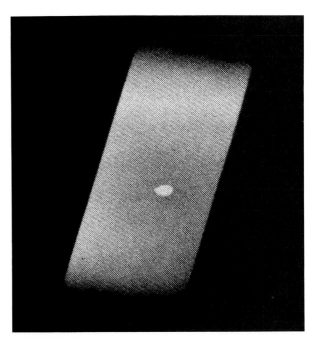

FIGURE 43-7 A thin-walled tungsten cylinder, heated to incandescence. Cavity radiation emerges from the small hole drilled through its wall.

emerging from a small hole in the wall. The cavity radiation is much brighter than the radiation from the outer wall of the cavity, even though the temperatures of the outer and inner walls are more or less equal.

The property of the cavity radiation that we seek to measure is its *spectral radiancy* $S(\lambda)$, defined so that $S(\lambda)\,d\lambda$ gives the radiated power per unit area of the cavity aperture that lies in the wavelength interval λ to $\lambda + d\lambda$. The solid curve in Fig. 43-8 shows the measured spectral radiancy for a cavity whose walls are held at 2000 K. Although such a radiator would glow brightly in a dark room, we see from the wavelength scale of the figure that only a small part of its radiated energy lies in the visible region of the spectrum. Most of it—by far—lies in the infrared. You do not have to linger too long near a bonfire to believe that it emits plenty of energy in the form of (warming) infrared rays.

The prediction of classical theory for the variation of the spectral radiancy with wavelength (at a given temperature) is

$$S(\lambda) = \frac{2\pi ckT}{\lambda^4} \quad \text{(classical radiation law).} \quad (43\text{-}16)$$

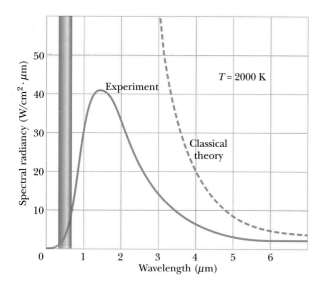

FIGURE 43-8 The solid curve shows the experimental spectral radiancy for a cavity at 2000 K. Note the failure of the classical theory, whose results are shown by the dashed curve. The range of visible wavelengths is indicated.

Here c is the speed of light and k is the *Boltzmann constant*, a quantity that we met in Section 21-5; its value is

$$k = 1.38 \times 10^{-23} \, \text{J/K}$$
$$= 8.62 \times 10^{-5} \, \text{eV/K}. \quad (43\text{-}17)$$

Equation 43-16 (with $T = 2000$ K) is plotted in Fig. 43-8. Although classical theory and experiment agree quite well at very long wavelengths (far beyond the scale shown in Fig. 43-8), the disagreement between theory and experiment at shorter wavelengths is total. The theoretical prediction does not even pass through a maximum. If the experiments are correct—and they are—then something is seriously wrong with the classical theory.

In 1900, Planck proposed a formula for the spectral radiancy that fitted the experimental data *perfectly* at all wavelengths and for all temperatures. His prediction is

$$S(\lambda) = \frac{2\pi c^2 h}{\lambda^5} \frac{1}{e^{hc/\lambda kT} - 1}$$

(Planck's radiation law). (43-18)

In deriving this formula, Planck introduced the important constant h (now called the *Planck constant*) into physics. By fitting Eq. 43-18 to experimental spectral radiancy data at a variety of temperatures,

Planck was able to arrive at a value for his constant that agreed within a few percent with the present accepted value. Modern quantum physics began with Planck's radiation law.

43-6 THE QUANTIZATION OF ENERGY

The assumptions that must be made to derive Planck's radiation law represent such a break with classical ideas that they were not at all clear to physicists of the day including—by his own admission—Planck himself. In 1917, however (17 years after Planck had advanced his radiation law), Einstein offered a remarkably straightforward derivation of Eq. 43-18 that made its underlying assumptions abundantly clear.*

The first of the assumptions underlying Eq. 43-18 is that the energy of the radiation in the cavity is quantized. That is, this radiation exists in the form of photons, of energy $E = hf$. The second assumption is that the energy of the atoms that form the cavity walls is quantized. That is,

The atoms that form the walls of the cavity can exist only in states corresponding to specific values of energy; states with intermediate energies are forbidden.

If you assume these quantization-of-energy principles for the radiation in the cavity and for the atoms of the cavity walls, you can derive Planck's law; if you don't, you can't.

We discussed energy quantization for atoms in a preliminary way in Section 8-9, which you may wish to reread at this time. Further developments showed that energy quantization is universal; it holds, not only for atoms, but for all kinds of systems—be they atoms, nuclei, molecules, or electrons in solids.

43-7 THE CORRESPONDENCE PRINCIPLE

We have seen that the equations of relativistic mechanics reduce to those of classical Newtonian mechanics under conditions (low particle speeds) in

---

*Einstein's derivation of Eq. 43-18 is given in Robert Resnick and David Halliday, *Basic Concepts in Relativity and Early Quantum Theory* (Macmillan, 1992), 2nd ed., Supplementary Topic E.

which the classical laws are known to agree with experiment.

A similar **correspondence principle** holds in quantum physics. That is,

> The equations of quantum physics must reduce to familiar classical laws under conditions in which the classical laws are known to agree with experiment.

Let us explore this principle for the case of Eq. 43-18, Planck's radiation law. The classical radiation law (Eq. 43-16) is known to agree with experiment at very large wavelengths. Let us see whether Eq. 43-18 reduces to Eq. 43-16 in this limiting case. We note, however, that if we simply substitute $\lambda = \infty$ in Eq. 43-18, an indeterminate value for $S(\lambda)$ results. We must adopt a more subtle approach.

To simplify the algebra, we write Eq. 43-18 in the form

$$S = \frac{2\pi c^2 h}{\lambda^5} \frac{1}{e^x - 1}, \qquad (43\text{-}19)$$

in which $x = hc/\lambda kT$. The limiting case of $\lambda \to \infty$ corresponds to $x \to 0$. For small enough values of x we can drop the squared and higher terms of the series

$$e^x = 1 + x + \frac{x^2}{2} + \frac{x^3}{6} + \cdots$$

and write

$$e^x - 1 = x.$$

Equation 43-19 then becomes

$$S = \frac{2\pi c^2 h}{\lambda^5} \left(\frac{1}{x}\right) = \frac{2\pi c^2 h}{\lambda^5} \left(\frac{\lambda kT}{hc}\right)$$
$$= \frac{2\pi ckT}{\lambda^4}.$$

This is exactly Eq. 43-16, the classical radiation law! Thus the correspondence principle holds. Note how the Planck constant h—that sure indicator of a quantum equation—has conveniently canceled out in the process of obtaining the classical (nonquantum) equation.

43-8 ATOMIC STRUCTURE

A question of ancient standing is, "What is the internal structure of an atom like?" It is appropriate that we begin to answer that question here, by examining one of the major clues to the structure of atoms, namely, the nature of the light that atoms emit.

In Fig. 43-8 we have an example of the light emitted by atoms when they are assembled to form the solid wall of a cavity radiator. We saw in Section 43-6 that what we can learn from the study of *this* radiation is the very important fact that the energies of the atoms that form the cavity walls are quantized. We can get no detailed information about specific atoms, however, because the cavity radiation does not depend on the nature of the atoms that make up the cavity walls.

To learn about the detailed structure of individual atoms (hydrogen, carbon, copper, and so on) we must study the light that they emit or absorb when they are alone, isolated from other atoms. To approximate this isolation, the atoms are usually put in a gaseous state. If they are then illuminated with light, they absorb at only sharply defined wavelengths, or **spectral lines,** of that light. Similarly, if the atoms are somehow given energy, such as via current in a discharge lamp, they emit light at only sharply defined wavelengths or spectral lines. In both cases, the specific set of spectral lines is a characteristic signature of the type of atom involved. Indeed, by measuring the spectral lines, a researcher can then identify the type of atom.

Such **line spectra** are characteristic, not only of isolated atoms, but also of isolated molecules or atomic nuclei. Figure 43-9 shows some selected line spectra, associated with the emission or absorption of radiation by such entities. These curves, which involve wavelengths from all over the electromagnetic spectrum, only begin to suggest the bewildering variety of such spectra that can be measured in the laboratory.

It is our plan to concentrate on the spectrum of the hydrogen atom. Hydrogen is the simplest atom and, not surprisingly, it has the simplest spectrum. In the remainder of this chapter we shall trace out the preliminary attempts by Niels Bohr to understand the structure of the hydrogen atom. In the next chapter we shall move on to a full quantum description of the hydrogen atom.

43-9 NIELS BOHR AND THE HYDROGEN ATOM

The wavelengths of the lines in the spectrum of atomic hydrogen (see Fig. 43-10) have been known with precision for many years. They stand as a testing

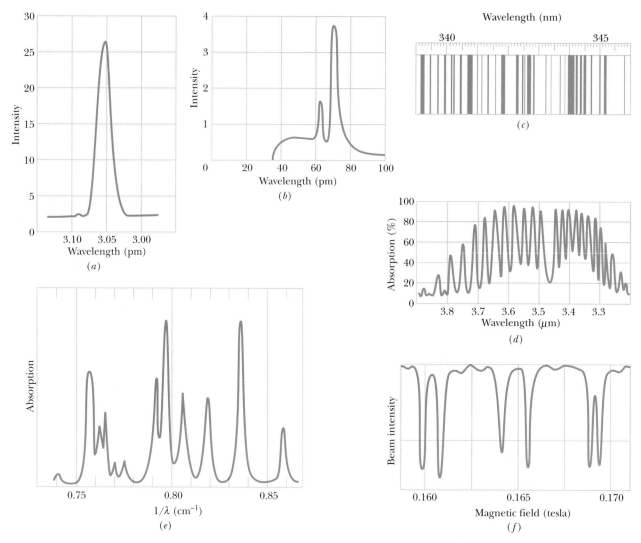

FIGURE 43-9 Selected spectral lines of emission or absorption, represented in a variety of ways. The intensity of (*a*) the gamma-ray emission line of ^{198}Hg nuclei and (*b*) two x-ray emission lines of Mo atoms, plotted versus wavelength. (*c*) Wavelengths and wavelength widths of ultraviolet emission lines of Fe atoms. (*d*) Infrared absorption by HCl molecules plotted versus wavelength. (*e*) Microwave absorption of NH$_3$ molecules plotted versus inverse wavelength (which is proportional to frequency). (*f*) Radiowave absorption by H$_2$ molecules versus the strength of the magnetic field to which the molecules are subjected.

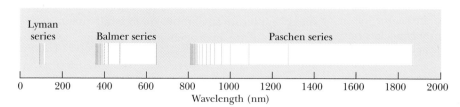

FIGURE 43-10 The Lyman series (ultraviolet), Balmer series (partly visible), and Paschen series (infrared) of spectral lines in the spectrum of atomic hydrogen. In each series, the lines bunch toward shorter wavelengths, approaching a series limit.

ground for any theory of the structure of the hydrogen atom that may be put forward.

Classical Theory

Let us first review the problems that arise when we try to determine the structure of the hydrogen atom by the methods of classical physics. We can imagine that the electron in the hydrogen atom revolves about the central nucleus (a proton) in a circular orbit of radius r, as in Fig. 43-11. We can then suppose that the frequency of the radiation that the atom emits is equal to the frequency at which the electron circulates in this orbit. Classical theory predicts that such an orbiting electron will indeed radiate, *and* at its orbital frequency. However, the theory has a fatal flaw. The orbiting electron will radiate its energy completely away, moving closer to the nucleus with each rotation and emitting a continuous spectrum of radiation as it spirals in toward the nucleus. In other words, the great classical theories of Newton and Maxwell stand helpless before the simplest atom. They cannot even account for the existence of the spectral lines, let alone predict their wavelengths. Indeed, they predict that atoms cannot exist!

Bohr's Theory

In 1913, just two years after English physicist Ernest Rutherford had put forward the idea that the atom has a nucleus, the great Danish physicist Niels Bohr (see Fig. 43-12) proposed a model for the hydrogen atom that not only accounted for the presence of the spectral lines but predicted their wavelengths to an accuracy of about 0.02%. Although Bohr's theory was successful for hydrogen, it proved less useful for more complex atoms. We now regard Bohr's theory as an inspired first step toward the more comprehensive quantum theory that followed it.

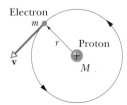

FIGURE 43-11 A classical model of the hydrogen atom, showing an electron of mass m circulating about a central nucleus of mass M. We assume that $M \gg m$.

FIGURE 43-12 Niels Bohr with Aage Bohr, one of his five sons. Both earned Nobel prizes in physics, Niels in 1922 and Aage in 1975.

Bohr, realizing that classical physics had come to a dead end with the structure of the hydrogen atom, put forward two bold postulates. Both turned out to be enduring features that carry over in full force to modern quantum physics. Moreover, both turned out to be quite general, applying not only to the hydrogen atom but to atomic, molecular, and nuclear systems of all kinds. These postulates are the following:

1. *The Postulate of Stationary States.* Bohr assumed that the hydrogen atom can exist *without radiating* in any one of a discrete set of **stationary states** of fixed energy. This assumption of energy quantization flies in the face of classical theory but Bohr's attitude was, "Let's assume it anyway and see what happens." Note that this postulate says nothing at all about how we are to find the energies of these stationary states.

2. *The Frequency Postulate.* Bohr assumed that the hydrogen atom can emit or absorb radiation *only* when the atom changes from one of its stationary states to another. The energy of the emitted or absorbed photon is equal to the difference in energy between these two states. Thus if an atom changes from an initial state of energy E_i to a final state of (lower) energy E_f, the energy of the emitted photon is given by

$$hf_{if} = E_i - E_f$$

(Bohr frequency condition),   (43-20)

a relation known as the **Bohr frequency condition.** This postulate ties together neatly two new ideas (the photon hypothesis and energy quantization) with one familiar idea (the conservation of energy).

Bohr's next task was to select the stationary states by specifying their energies. Then, using Eq. 43-20, he could calculate the frequencies—and thus the wavelengths—of the spectral lines. How to find the energies? Bohr actually did this in a clever way, making use of the correspondence principle. We present his result here, without proof. In Section 43-10, however, we give a semiclassical proof—also due to Bohr—that leads to this result.

Bohr found that the energies of the stationary states of the hydrogen atom are given by

$$E = -\frac{me^4}{8\epsilon_0^2 h^2} \frac{1}{n^2}, \qquad n = 1, 2, 3, \ldots, \quad (43\text{-}21)$$

in which n is called a **quantum number.** The minus sign tells us that the hydrogen-atom states whose energies are given by this equation are bound states. That is, work must be done by an external agent to pull the atom apart. Although Bohr derived Eq. 43-21 in a semiclassical manner, exactly the same result follows from a rigorous derivation based on modern quantum theory.

Figure 43-13 is an energy level diagram for the hydrogen atom. The horizontal lines represent seven different energy states, and the vertical scale shows the energies of these states as calculated with Eq. 43-21; each level is labeled with its quantum number. The state of lowest energy, called the *ground state,* is found by putting $n = 1$ in Eq. 43-21; states of higher energy are called *excited states.* It is

easy to show that the ground-state energy $E_1 = -13.6$ eV, so Eq. 43-21 can be written as

$$E = -\frac{13.6 \text{ eV}}{n^2}, \qquad n = 1, 2, 3, \ldots. \quad (43\text{-}22)$$

The downward pointing arrows in Fig. 43-13 represent some transitions from one energy level to a lower level. These transitions can be grouped into several "series," each series having a particular level as its "home base." The Lyman series, for example, consists only of transitions to the ground state. Each of the series has a *series limit* that corresponds to a transition between the $n = \infty$ level and the home-base level for the series.

The energy emitted during a transition from one level to a lower one is, as we discussed, the difference in the energies of the two levels. We can find the wavelength of the emitted radiation by combining Eq. 43-21 with the Bohr frequency condition (Eq. 43-20), obtaining

$$hf = \frac{hc}{\lambda} = \frac{me^4}{8\epsilon_0^2 h^2} \left(\frac{1}{l^2} - \frac{1}{u^2} \right). \quad (43\text{-}23)$$

Here u and l are, respectively, the quantum numbers of the upper and the lower energy states in-

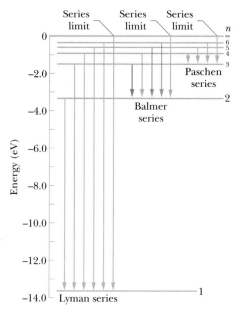

FIGURE 43-13 Some of the energy levels and transitions in the spectrum of atomic hydrogen.

volved in the transition whose wavelength is λ. We can recast Eq. 43-23 more compactly as

$$\frac{1}{\lambda} = R\left(\frac{1}{l^2} - \frac{1}{u^2}\right). \qquad (43\text{-}24)$$

The factor R, called the **Rydberg constant,** has the value

$$R = \frac{me^4}{8\epsilon_0^2 h^3 c} = 1.097 \times 10^7 \text{ m}^{-1}$$

$$= 0.01097 \text{ nm}^{-1}. \qquad (43\text{-}25)$$

In deriving Eq. 43-24, we considered the emission of light by a hydrogen atom, in which case the atom loses energy and undergoes a downward transition in the energy levels of Fig. 43-13. The equation is equally valid for absorption of light by a hydrogen atom, in which case the atom gains energy and undergoes an upward transition.

Equation 43-24 forms a bridge between the spectrum of Fig. 43-10 and the energy diagram of Fig. 43-13. As an example, the Balmer series (of emission and absorption) of the former can be defined by letting $l = 2$ in Eq. 43-24 and allowing u to take on the integer values 3, 4, 5, ...

We can now make sense of the difference in the spectra of the light sources captured in the photograph opening this chapter. The headlights on the car have incandescent filaments through which current flows; the electrons in the current cause the filament to heat until its temperature is high enough to produce light across the full visible spectrum. Because the atoms of the filament are not isolated, we do not find spectral lines in the emission of the filament; in fact, the emission is like that plotted in Fig. 43-8 for the cavity radiation experiment. The light of the stop light is similarly produced, but it must pass through a red plastic enclosure that absorbs all but the red end of the visible spectrum.

The mercury lamps on the house and on the street pole differ in their operation, because the mercury atoms that emit light are in a gaseous state and thus are approximately isolated from one another. When current flows through either lamp, the electrons in the current collide with mercury atoms, providing those atoms with energy so that they are then in excited states. The atoms quickly *de-excite* by emitting light and undergoing transitions to lower energy states and eventually ground state. In this

process, the atoms emit the spectral lines that are characteristic of mercury, as seen in the chapter-opening photograph.

Bohr quite properly pushed his theory for all it was worth, taking as his goal the development of a theoretical basis for the periodic table of the elements. However, beyond hydrogen and hydrogen-like atoms or ions such as He^+, Bohr had limited success. For the emission from a mercury lamp, for example, his theory could predict that the emission consists of discrete colors but could not predict *which* colors.

Bohr's "great and shining moment" was no doubt his prediction that element 72, then a blank space in the periodic table, should have chemical properties like those of zirconium. This led to the discovery of element 72 in zirconium ores and the naming of the new element *hafnium,* after an early name for Copenhagen, Bohr's home town. Word of this discovery came dramatically to Bohr in Stockholm, just hours before he was scheduled to receive the Nobel prize.

SAMPLE PROBLEM 43-7

a. What is the wavelength of the least energetic photon in the Balmer series?

SOLUTION We identify the Balmer series (see Fig. 43-13) by putting $l = 2$ in Eq. 43-24. The least energetic photon in that series (or any other series) is the photon that results from the smallest jump for the series on the energy diagram of Fig. 43-13. That smallest jump for the Balmer series occurs when $u = 3$ (the next larger quantum number). Equation 43-24 then gives us

$$\frac{1}{\lambda} = R\left(\frac{1}{l^2} - \frac{1}{u^2}\right)$$

$$= (0.01097 \text{ nm}^{-1})\left(\frac{1}{2^2} - \frac{1}{3^2}\right)$$

$$= 1.524 \times 10^{-3} \text{ nm}^{-1},$$

which gives us

$$\lambda = \frac{1}{1.524 \times 10^{-3} \text{ nm}^{-1}} = 656.3 \text{ nm}. \quad \text{(Answer)}$$

b. What is the wavelength of the series limit for the Balmer series?

SOLUTION Again we put $l = 2$ in Eq. 43-24. To find the series limit (see Fig. 43-13) we let $u \to \infty$. Equation 43-24 then becomes

$$\frac{1}{\lambda} = (0.01097 \text{ nm}^{-1})\left(\frac{1}{2^2}\right) = 2.743 \times 10^{-3} \text{ nm}^{-1},$$

which yields

$$\lambda = 364.6 \text{ nm.} \qquad \text{(Answer)}$$

43-10 BOHR'S DERIVATION (OPTIONAL)

Here we examine Bohr's derivation of Eq. 43-21, his formula for the energies of the stationary states of the hydrogen atom. As was indicated earlier, Bohr actually derived this formula in a clever way, using the correspondence principle and making no detailed assumptions about the nature of the quantum states of the hydrogen atom at low quantum numbers. Bohr also suggested an alternative derivation, which is the one that we examine here.

Let us apply Newton's second law ($F = ma$) to an electron revolving in a circular orbit of radius r, as in Fig. 43-11. We use Coulomb's law (Eq. 23-4) to find the force on the electron, obtaining

$$\frac{1}{4\pi\epsilon_0}\frac{(e)(e)}{r^2} = m\frac{v^2}{r}. \qquad (43\text{-}26)$$

We can now use Eq. 43-26 to find the kinetic energy of the electron. We get

$$K = \tfrac{1}{2}mv^2 = \frac{e^2}{8\pi\epsilon_0 r}. \qquad (43\text{-}27)$$

The electric potential energy of the electron–proton system is, from Eq. 26-35,

$$U = \frac{1}{4\pi\epsilon_0}\frac{(+e)(-e)}{r} = -\frac{e^2}{4\pi\epsilon_0 r}, \qquad (43\text{-}28)$$

and the total energy of the atom is

$$E = K + U = -\frac{e^2}{8\pi\epsilon_0 r}. \qquad (43\text{-}29)$$

This is as far as we can go with purely classical ideas. The total energy E of each stationary state depends on the orbit radius of that state and, unless we can find the orbit radii that correspond to the stationary states, we are stuck. In modern language, we need a quantization criterion.

Having come so far, you are entitled to suppose

that we are stuck at this point because the Planck constant h has not yet put in an appearance. If you think that, you are right. Bohr saw that the simplest way to introduce h was to quantize the angular momentum L of the atomic orbits, by assuming—quite arbitrarily—that L could have only values given by

$$L = n\frac{h}{2\pi}, \qquad n = 1, 2, 3, \ldots, \qquad (43\text{-}30)$$

in which n is a quantum number.* Note that, although they are quite different quantities, angular momentum and the Planck constant h have the same dimensions; this fact makes it less surprising that angular momentum should give us the simplest quantization rule.

From Eq. 12-25, we know that $L = mvr$ gives us the angular momentum of a particle of mass m moving in a circle of radius r at speed v. Using Eq. 43-26, we can find another expression for L for the electron in the hydrogen atom by writing

$$mv^2 r = \frac{e^2}{4\pi\epsilon_0}$$

and

$$\frac{(mvr)^2}{mr} = \frac{e^2}{4\pi\epsilon_0}.$$

Substituting L for mvr, we then have

$$L = mvr = \sqrt{\frac{me^2 r}{4\pi\epsilon_0}}. \qquad (43\text{-}31)$$

Combining Eqs. 43-30 and 43-31 gives us the radii of the allowed orbits as

$$r = n^2\frac{h^2\epsilon_0}{\pi me^2}, \qquad n = 1, 2, 3, \ldots. \qquad (43\text{-}32)$$

We can recast this more compactly as

$$r = n^2 r_B, \qquad n = 1, 2, 3, \ldots, \qquad (43\text{-}33)$$

in which r_B can be computed from the fundamental constants in Eq. 43-32 to be

$$r_B = 5.292 \times 10^{-11} \text{ m}$$

$$= 52.92 \text{ pm} = 0.05292 \text{ nm.} \qquad (43\text{-}34)$$

---

*The angular momentum of the electron in the hydrogen atom is indeed quantized, but the true quantization rule is not quite as simple as Eq. 43-30; see Section 45-5. Bear in mind that Bohr's approach is semiclassical and is an intermediate step in the development of a full quantum theory.

The quantity r_B, called the *Bohr radius,* is equal to the radius of the Bohr orbit for hydrogen in its ground state, defined by putting $n = 1$ in Eq. 43-32. Even though we no longer believe in such orbits, we still use the Bohr radius as a convenient measure of distance on the atomic scale. You should be impressed that, although Bohr put nothing into his theory that says how big atoms are, a length of about the right size comes out of it.

Finally, we can substitute the quantized orbit radius r from Eq. 43-32 into Eq. 43-29, obtaining as the total energies of the hydrogen energy states

$$E = -\frac{me^4}{8\epsilon_0^2 h^2}\frac{1}{n^2}, \qquad n = 1, 2, 3, \ldots ,$$

which is exactly Eq. 43-21, the formula that we set out to derive.

REVIEW & SUMMARY

Photons

In 1905, Einstein hypothesized that light is made up of concentrated bundles of energy, which we now call photons, and that each photon has energy E and momentum p with

$$E = hf \quad \text{and} \quad p = h/\lambda. \qquad (43\text{-}1, 43\text{-}4)$$

Here the Planck constant h has the value 6.63×10^{-34} J·s $= 4.14 \times 10^{-15}$ eV·s. This constant, although small, is not zero; this is the determining feature of modern quantum physics. Experiments that support this hypothesis are the following:

1. The **photoelectric effect,** in which electrons are ejected from a metal surface by incident light. Einstein's equation for this effect, based on his photon hypothesis, is

$$hf = \phi + K_m \qquad \text{(photoelectric equation).} \qquad (43\text{-}8)$$

Here hf is the energy of the photon absorbed by an electron in the metal surface. The **work function** ϕ is the energy needed to remove this electron from the metal; K_m is the maximum kinetic energy of the electron outside the surface.

2. The **Compton effect,** in which x-ray photons, scattered from free electrons, undergo a wavelength increase $\Delta\lambda$. This **Compton shift** (see Fig. 43-5) is given by

$$\Delta\lambda = \frac{h}{mc}(1 - \cos\phi). \qquad (43\text{-}14)$$

This equation follows from applying the laws of conservation of energy and momentum to a billiard-ball-like collision between a photon and a free electron, as in Fig. 43-6.

3. Measurement of the distribution with wavelength of the radiation emerging from heated cavities. These studies introduced the concept of **energy quantization** and brought the Planck constant h into the equations of physics for the first time.

Line Spectra

The absorption and emission of radiation at sharply defined wavelengths are characteristic of atoms, molecules, and nuclei. Representations of these wavelengths, like that of Fig. 43-10, are called *line spectra.* Classical physics cannot explain these phenomena.

Bohr's Quantum Postulates

Attempts to understand line spectra in quantum terms start with Bohr's **quantum postulates,** first introduced to explain the spectrum of hydrogen atoms: (1) an atom can exist *without radiating* in any one of a discrete set of **stationary states** of fixed energy, and (2) an atom can emit or absorb radiation only in a transition between these stationary states, the frequency of the radiation, and hence of the corresponding spectral line, being given by

$$hf_{if} = E_i - E_f \qquad \text{(Bohr frequency condition).} \qquad (43\text{-}20)$$

Here E_i and E_f are the energies of the initial and final states involved in the transition.

To find the energies of the stationary states in hydrogen atoms, Bohr assumed that the angular momentum L of the orbiting electron can have only the discrete values given by

$$L = n\frac{h}{2\pi}, \qquad n = 1, 2, 3, \ldots . \qquad (43\text{-}30)$$

The resulting energies of the allowed states are

$$E = -\left(\frac{me^4}{8\epsilon_0^2 h^2}\right)\frac{1}{n^2} = -\frac{13.6 \text{ eV}}{n^2}, \qquad n = 1, 2, 3, \ldots .$$
$$(43\text{-}21, 43\text{-}22)$$

Here n is called a quantum number. Combining Eqs. 43-20 and 43-21 leads to

$$\frac{1}{\lambda} = R\left(\frac{1}{l^2} - \frac{1}{u^2}\right) \qquad (43\text{-}24)$$

for the wavelengths of the lines of the hydrogen spectrum for a transition between an upper state with quantum number u to a lower state with quantum number l; $R = 0.01097$ nm^{-1} is the **Rydberg constant.**

QUESTIONS

1. How can a photon energy be given by $E = hf$ when the very presence of the frequency f in the formula implies that light is a wave?

2. Given that $E = hf$ for a photon, the Doppler shift in frequency of radiation from a receding light source would seem to indicate a reduced energy for the emitted photons. Is this in fact true? If so, what happened to the conservation of energy principle?

3. Photon A has twice the energy of photon B. What is the ratio of the momentum of A to that of B?

4. How does a photon differ from a material particle?

5. Show that the Planck constant has the dimensions of angular momentum. Does this necessarily mean that angular momentum is a quantized quantity?

6. For quantum effects to be "everyday" phenomena in our lives, what order of magnitude would the value of h need to have? (See G. Gamow, *Mr. Tompkins in Wonderland,* Cambridge University Press, Cambridge, 1957, for a delightful popularization of a world in which the physical constants c, G, and h make themselves obvious.)

7. An isolated metal plate yields photoelectrons when you first shine ultraviolet light on it, but later it doesn't give up any more. Explain.

8. In Fig. 43-2, why doesn't the photoelectric current rise to its maximum (saturation) value when the applied potential difference is just slightly more positive than V_0?

9. In the photoelectric effect, why does the existence of a cutoff frequency speak in favor of the photon theory and against the wave theory?

10. Why are photoelectric measurements so sensitive to the nature of the photoelectric surface?

11. Explain the statement that one's eyes could not detect faint starlight if light were not particle-like.

12. Consider the following procedures: (a) bombard a metal with electrons; (b) place a strong electric field near a metal; (c) illuminate a metal with light; (d) heat a metal to a high temperature. Which of the above procedures can result in the emission of electrons?

13. A certain metal plate is illuminated by light of a definite frequency. Whether or not photoelectrons are emitted as a result depends on which of the following features: (a) intensity of illumination, (b) length of time of exposure to the light, (c) thermal conductivity of the plate, (d) area of the plate, or (e) material of the plate?

14. Does Einstein's theory of photoelectricity, in which light is postulated to be a stream of photons, invalidate Young's double-slit interference experiment, in which light is postulated to be a wave?

15. What is the direction of a Compton-scattered electron with maximum kinetic energy, compared with the direction of the incident photon?

16. In Compton scattering, why would you expect $\Delta \lambda$ to be independent of the material of which the scatterer is composed?

17. Why don't we observe a Compton effect with visible light?

18. Light from distant stars is Compton-scattered many times by free electrons in outer space before reaching us. This shifts the light toward the red. How can this shift be distinguished from the Doppler red shift due to the motion of receding stars?

19. "Pockets" formed by the piled-together coals in a coal fire seem brighter than the coals themselves. Is the temperature in such pockets appreciably higher than the surface temperature of an exposed glowing coal? Explain this common observation.

20. If we look into a cavity whose walls are maintained at a constant temperature, no details of the interior are visible. Explain.

21. We claim that all objects radiate energy by virtue of their temperature, and yet we cannot see all objects in the dark. Why?

22. Only a relatively small number of Balmer lines can be observed from laboratory discharge lamps, whereas a large number are observed in the spectra of stars. Explain this in terms of the small density, high temperature, and large volume of gases in stellar atmospheres.

23. In Bohr's theory for the hydrogen atom, what is the implication of the fact that the potential energy is negative and is greater in magnitude than the kinetic energy? Is this a result of quantum physics or is it true classically as well?

24. Some lines in the hydrogen spectrum are brighter than others. Why?

25. Consider a hydrogenlike atom in which a (positively charged) positron orbits a (negatively charged) antiproton. In what ways, if any, would you expect the emission spectrum of this "antimatter atom" to differ from the spectrum of a normal hydrogen atom?

26. Radioastronomers observe lines in the hydrogen spectrum that originate in hydrogen atoms that are in states with $n = 350$ or so. Why can't hydrogen atoms in states with such high quantum numbers be produced and studied in the laboratory?

27. Can a hydrogen atom absorb a photon whose energy exceeds its binding energy (13.6 eV)?

28. List and discuss the assumptions made by Planck in connection with cavity radiation, by Einstein in connection with the photoelectric effect, by Compton in connection with the Compton effect, and by Bohr in connection with the structure of the hydrogen atom.

29. Describe several methods that can be used to experimentally determine the value of the Planck constant h.

30. According to classical mechanics, an electron moving in an orbit should be able to do so with any angular momentum whatever. According to Bohr's theory of the hydrogen atom, however, the angular momentum is quantized according to $L = nh/2\pi$. Reconcile these two statements, using the correspondence principle.

EXERCISES & PROBLEMS

SECTION 43-2 EINSTEIN MAKES A PROPOSAL

1E. Show that the energy E of a photon (in eV) is related to its wavelength λ (in nm) by

$$E = \frac{1240}{\lambda}.$$

This result can be useful in solving many problems.

2E. The orange-colored light from a highway sodium lamp has a wavelength of 589 nm. How much energy is possessed by an individual photon from such a lamp?

3E. Consider monochromatic light falling on photographic film. The incident photons will be recorded if they have enough energy to dissociate a AgBr molecule in the film. The minimum energy required to do this is about 0.6 eV. Find the highest wavelength of light that will be recorded. In what region of the spectrum does this wavelength fall?

4E. (a) A spectral emission line that is important in radioastronomy has a wavelength of 21 cm. What is its corresponding photon energy? (b) At one time the meter was defined as 1,650,763.73 wavelengths of the orange light emitted by a light source containing krypton-86 atoms. What is the corresponding photon energy of this radiation?

5E. A particular x-ray photon has a 35.0-pm wavelength. Calculate the photon's (a) energy, (b) frequency, and (c) momentum.

6P. Under ideal conditions the normal human eye will record a visual sensation at 550 nm if incident photons are absorbed at a rate as low as 100 photons per second. To what power does this correspond?

7P. What are (a) the frequency, (b) the wavelength, and (c) the momentum of a photon whose energy equals the rest energy of the electron?

8P. In the photon model of radiation, show that if two parallel beams of light of different wavelengths have the same intensity, then the rates per unit area at which photons pass through any cross section of the beams are in the same ratio as the wavelengths.

9P. An ultraviolet light bulb, emitting at 400 nm, and an infrared light bulb, emitting at 700 nm, are both rated at 400 W. (a) Which bulb radiates photons at the greater rate? (b) How many more photons does it generate per second than the other bulb?

10P. A satellite in Earth orbit maintains a panel of solar cells of 2.60-m² area at right angles to the direction of the sun's rays. Solar energy arrives at the rate of 1.39 kW/m². (a) At what rate does solar energy strike the panel? (b) At what rate do solar photons strike the panel? Assume that the solar radiation is monochromatic with a wavelength of 550 nm. (c) How long would it take for a "mole of photons" to strike the panel?

11P. A special kind of light bulb emits monochromatic light at a wavelength of 630 nm. It is rated at 60 W and is 93% efficient in converting electrical energy to light. How many photons will the bulb emit over its 730-h lifetime?

12P. The emerging beam from a 1.5-W argon laser ($\lambda = 515$ nm) has a diameter d of 3.0 mm. (a) At what rate per square meter do photons pass through any cross section of the beam? (b) The beam is focused by a lens system whose effective focal length f_L is 2.5 mm. The focused beam forms a circular diffraction pattern whose central disk has a radius R given by $1.22f_L\lambda/d$. It can be shown that 84% of the incident power lies within this central disk, the rest falling in the fainter, concentric diffraction rings that surround the central disk. At what rate per square meter do photons pass through the central disk of the diffraction pattern?

13P. Assume that a 100-W sodium-vapor lamp radiates its energy uniformly in all directions in the form of photons with a wavelength of 589 nm. (a) At what rate are photons emitted from the lamp? (b) At what distance from the lamp will the average flux of photons be 1.00 photon/ (cm²·s)? (c) At what distance from the lamp will the average density of photons be 1.00 photon/cm³? (d) What are the photon flux and the photon density 2.00 m from the lamp?

SECTION 43-3 THE PHOTOELECTRIC EFFECT

14E. You wish to pick a substance for a photocell that will operate via the photoelectric effect with visible light. Which of the following will do (work function in parentheses): tantalum (4.2 eV), tungsten (4.5 eV), aluminum (4.2 eV), barium (2.5 eV), lithium (2.3 eV)?

15E. A satellite or spacecraft in orbit about the Earth can become charged due, in part, to the loss of electrons caused by the photoelectric effect induced by sunlight on the vehicle's outer surface. Suppose that a satellite is coated with platinum, a metal with one of the largest work functions: $\phi = 5.32$ eV. Find the longest-wavelength photon that can eject a photoelectron from platinum. (Satellites must be designed to minimize such charging.)

16E. (a) The energy needed to remove an electron from metallic sodium is 2.28 eV. Does sodium show a photoelectric effect for red light, with $\lambda = 680$ nm? (b) What is the cutoff wavelength for photoelectric emission from sodium, and to what color does this wavelength correspond?

17E. Find the maximum kinetic energy of photoelectrons from a certain material if the work function is 2.3 eV and the frequency of the radiation is 3.0×10^{15} Hz.

18E. Incident photons strike a sodium surface having a work function of 2.2 eV, causing photoelectric emission. The stopping potential is 5.0 V. What is the wavelength of the incident photons?

19E. The work function of tungsten is 4.50 eV. Calculate the speed of the fastest of the photoelectrons emitted when photons of energy 5.80 eV are incident on a sheet of tungsten.

20E. Light of wavelength 200 nm falls on an aluminum surface. In aluminum 4.20 eV is required to remove an electron. What is the kinetic energy of (a) the fastest and (b) the slowest emitted photoelectrons? (c) What is the stopping potential? (d) What is the cutoff wavelength for aluminum?

21E. (a) If the work function for a metal is 1.8 eV, what is its stopping potential for light of wavelength 400 nm? (b) What is the maximum speed of the emitted photoelectrons at the metal's surface?

22E. The wavelength associated with the cutoff frequency for silver is 325 nm. Find the maximum kinetic energy of electrons ejected from a silver surface by ultraviolet light of wavelength 254 nm.

23P. The stopping potential for photoelectrons emitted from a surface illuminated by light of wavelength 491 nm is 0.710 V. When the incident wavelength is changed to a new value, the stopping potential is found to be 1.43 V. (a) What is this new wavelength? (b) What is the work function for the surface?

24P. In a photoelectric experiment in which a sodium surface is used, you find a stopping potential of 1.85 V for a wavelength of 300 nm, and a stopping potential of 0.820 V for a wavelength of 400 nm. From these data find (a) a value for the Planck constant, (b) the work function for sodium, and (c) the cutoff wavelength for sodium.

25P. In about 1916, R. A. Millikan (see Section 24-8) found the following stopping-voltage data for lithium in his photoelectric experiments:

Wavelength (nm)	433.9	404.7	365.0	312.5	253.5
Stopping potential (V)	0.55	0.73	1.09	1.67	2.57

Use these data to make a plot like Fig. 43-3 (which is for sodium) and find (a) the Planck constant and (b) the work function for lithium.

26P. Photosensitive surfaces are not necessarily very efficient. Suppose the fractional efficiency of a cesium surface (with work function 1.80 eV) is 1.0×10^{-16}; that is, one photoelectron is produced for every 10^{16} photons striking the surface. What would be the photocurrent from such a cesium surface if it were illuminated with 600-nm light from a 2.00-mW laser and all the photoelectrons produced took part in charge flow?

27P. X rays with a wavelength of 71 pm eject photoelectrons from a gold foil, the electrons originating from deep within the gold atoms. The ejected electrons move in circular paths of radius r in a region of uniform magnetic field **B**. Experiment shows that $Br = 1.88 \times 10^{-4}$ T·m. Find (a) the maximum kinetic energy of the photoelectrons and (b) the work done in removing the electrons from the gold atoms that make up the foil.

28P*. Show, by analyzing a collision between a photon and a free electron (using relativistic mechanics), that it is impossible for a photon to give all its energy to the free electron. In other words, the photoelectric effect cannot occur for completely free electrons; the electrons must be loosely bound in a solid or in an atom.

SECTION 43-4 THE COMPTON EFFECT

29E. Photons of wavelength 2.4 pm are incident on a target containing free electrons. (a) Find the wavelength of a photon that is scattered at 30° from the incident direction. (b) Do the same for a scattering angle of 120°.

30E. A 0.511-MeV gamma-ray photon scatters via the Compton effect from a free electron in an aluminum block. (a) What is the wavelength of the incident photon? (b) What is the wavelength of the scattered photon? (c) What is the energy of the scattered photon? Assume a scattering angle of 90.0°.

31E. An x-ray photon of wavelength 0.01 nm strikes an electron head on ($\phi = 180°$). Determine (a) the change in wavelength of the photon, (b) the change in energy of the photon, and (c) the kinetic energy imparted to the electron.

32E. The quantity h/mc in Eq. 43-14 is often called the *Compton wavelength* λ_C of the scattering particle and that equation is written as

$$\Delta\lambda = \lambda_C(1 - \cos\phi).$$

Calculate λ_C of (a) an electron and (b) a proton. What is the energy of a photon whose wavelength is equal to the Compton wavelength of (c) the electron and (d) the proton?

33E. Calculate the percent change in photon energy for a

Compton collision with ϕ in Fig. 43-4 equal to 90° for radiation in (a) the microwave range, with $\lambda = 3.0$ cm, (b) the visible range, with $\lambda = 500$ nm, (c) the x-ray range, with $\lambda = 25$ pm, and (d) the gamma-ray range, the energy of the gamma-ray photons being 1.0 MeV. What are your conclusions about the importance of the Compton effect in these various regions of the electromagnetic spectrum, judged solely by the criterion of energy loss in a single Compton encounter? (*Hint:* See Eq. 43-15.)

34E. What fractional increase in wavelength leads to a 75% loss of photon energy in a Compton collision with a free electron? (*Hint:* See Eq. 43-15.)

35E. Find the maximum wavelength shift for a Compton collision between a photon and a free *proton*.

36P. A 6.2-keV x-ray photon falling on a carbon block is scattered by a Compton collision and its frequency is shifted by 0.010%. (a) Through what angle is the photon scattered? (b) What kinetic energy is imparted to the electron involved in the collision?

37P. Show that $\Delta E/E$, the fractional loss of energy of a photon during a Compton collision, is given by

$$(hf'/mc^2)(1 - \cos \phi).$$

38P. Through what angle must a 200-keV photon be scattered by a free electron so that it loses 10% of its energy?

39P. Show that when a photon of energy E scatters from a free electron, the maximum kinetic energy of the electron is given by

$$K_{max} = \frac{E^2}{E + mc^2/2}.$$

40P. What is the maximum kinetic energy of electrons knocked out of a thin copper foil, via the Compton effect, by an incident beam of 17.5-keV x rays?

41P. Carry out the algebra needed to eliminate v and θ from Eqs. 43-10, 43-12, and 43-13 and thus to derive Eq. 43-14, the equation for the Compton shift.

SECTION 43-5 PLANCK AND HIS CONSTANT: HISTORICAL ASIDE

42E. The wavelength λ_{max} at which the spectral radiancy of a cavity radiator has its maximum value for a particular temperature T (see Fig. 43-8) is given by the Wien displacement law,

$$\lambda_{max} T = 2898 \ \mu m \cdot K = a \ constant.$$

The effective surface temperature of the sun is 5800 K. At what wavelength would you expect the sun to radiate most strongly? In what region of the spectrum is this? Why then does the sun appear yellow?

43E. Sensitive infrared detectors allow antiaircraft missiles to respond to the low-intensity radiation emitted by a target aircraft's airframe, and not just to the hot exhaust. This makes attack from any angle feasible. To what wavelength should a missile seeker be most sensitive if the tar-

get temperature is 290 K? Ignore atmospheric absorption. (See Exercise 42.)

44E. At what temperature is cavity radiation most visible to the human eye if the eye is most sensitive to yellow-green light of wavelength 550 nm? (See Exercise 42.)

45E. In 1983 the Infrared Astronomical Satellite (IRAS) detected a cloud of solid particles surrounding the star Vega, radiating maximally at a wavelength of 32 μm. What is the temperature of this cloud of particles? (See Exercise 42.)

46E. Low-temperature physicists would not consider a temperature of 2.00 mK to be particularly low. (a) At what wavelength would a cavity whose walls were at this temperature radiate most copiously? (See Exercise 42.) (b) To what region of the electromagnetic spectrum would this radiation belong? (c) What are some of the practical difficulties of operating a cavity radiator at such a low temperature?

47E. Calculate the wavelength of maximum spectral radiancy (see Exercise 42) and identify the region of the electromagnetic spectrum to which it belongs for each of the following: (a) the 3.0-K microwave background radiation, a remnant of the primordial fireball; (b) your body, assuming a skin temperature of 20°C; (c) a tungsten lamp filament at 1800 K; (d) an exploding thermonuclear device, at an assumed fireball temperature of 10^7 K; (e) the universe immediately after the Big Bang, at an assumed temperature of 10^{38} K.

48P. Show that the wavelength λ_{max} at which Planck's spectral radiation law (Eq. 43-18) has its maximum is given by

$$\lambda_{max} = (2898 \ \mu m \cdot K)/T.$$

(*Hint:* Set $dS/d\lambda = 0$; an equation will be encountered whose numerical solution is 4.965.)

49P. (a) By integrating Planck's radiation law (Eq. 43-18) over all wavelengths, show that the rate at which energy is radiated per square meter of a cavity surface is given by

$$P = \left(\frac{2\pi^5 k^4}{15h^3 c^2}\right) T^4 = \sigma T^4.$$

(*Hint:* Make a change in variables, letting $x = hc/\lambda kT$. A definite integral will be encountered that has the value

$$\int_0^\infty \frac{x^3 \ dx}{e^x - 1} = \frac{\pi^4}{15}.$$

(b) Verify that the numerical value of the constant σ is 5.67×10^{-8} W/(m²·K⁴).

50P. Calculate the rate at which thermal energy is radiated from a fireplace, assuming an effective radiating surface of 0.50 m² and an effective temperature of 500°C. (See Problem 49.)

51P. (a) Show that a human body of surface area 1.8 m² and temperature 31°C radiates radiation at the rate of 872 W. (b) Why, then, do people not glow in the dark? (See Problem 49.)

52P. A cavity at absolute temperature T_1 radiates energy at a power of 12.0 mW. At what power does the same cavity radiate at temperature $2T_1$? (See Problem 49.)

53P. A *thermograph* is a medical instrument used to measure radiation from the skin. Its usefulness stems from, for example, the fact that normal skin radiates at a temperature of 34°C while the skin over a tumor radiates at a slightly higher temperature. Derive an expression for the fractional difference in the radiance between adjacent areas of the skin that are at slightly different temperatures. Evaluate the expression for a temperature difference of 1°C. (See Problem 49.)

54P. An oven with inside temperature $T_o = 227$°C is in a room having a temperature $T_r = 27$°C. There is a small opening of area 5.0 cm² in one side of the oven. At what net rate is energy transferred from the oven to the room? (*Hint:* Consider both the oven and the room as cavities. See Problem 49.)

SECTION 43-9 NIELS BOHR AND THE HYDROGEN ATOM

55E. An atom absorbs a photon of frequency 6.2×10^{14} Hz. By what amount does the energy of the atom increase?

56E. An atom absorbs a photon having a wavelength of 375 nm and immediately emits another photon having a wavelength of 580 nm. How much net energy was absorbed by the atom in this process? Ease the computation by using the result of Exercise 1.

57E. A line in the x-ray spectrum of gold has a wavelength of 18.5 pm. The emitted x-ray photons correspond to a transition of the gold atom between two stationary states, the upper one of which has the energy -13.7 keV. What is the energy of the lower stationary state?

58E. (a) By direct substitution of the numerical values of the fundamental constants, verify that the energy of the ground state of the hydrogen atom is -13.6 eV. See Eqs. 43-21 and 43-22. (b) Similarly, show that the value of the Rydberg constant R is 0.01097 nm^{-1}, as asserted in Eq. 43-25.

59E. Answer the questions of Sample Problem 43-7, but for the Lyman series.

60E. What are (a) the energy, (b) the momentum, and (c) the wavelength of the photon that is emitted when a hydrogen atom undergoes a transition from the state with $n = 3$ to that with $n = 1$?

61E. Using Bohr's formula (Eq. 43-24) calculate the three longest wavelengths of the Balmer series.

62E. Find the ratio of the shortest wavelength of the Balmer series to the shortest wavelength of the Lyman series.

63E. A hydrogen atom is excited from the state with $n = 1$ to that with $n = 4$. (a) Calculate the energy that must be absorbed by the atom. (b) Calculate and display on an energy-level diagram the different photon energies that may be emitted if the atom returns to the $n = 1$ state.

64P. How much work must be done by an external agent to pull apart a hydrogen atom if the electron initially is (a) in the ground state and (b) in the state with $n = 2$?

65P. (a) What are the wavelength intervals over which the Lyman, Balmer, and Paschen series extend? (Each interval extends from the longest wavelength to the series limit.) (b) What are the corresponding frequency intervals?

66P. Light of wavelength 486.1 nm is emitted by a hydrogen atom. (a) What transition of the atom is responsible for this radiation? (b) To what series does this radiation belong?

67P. Show, on an energy-level diagram for hydrogen, the quantum numbers corresponding to a transition in which the wavelength of the emitted photon is 121.6 nm.

68P. A hydrogen atom in a state having a *binding energy* (the energy required to remove an electron) of 0.85 eV makes a transition to a state with an *excitation energy* (the difference in energy between a state and the ground state) of 10.2 eV. (a) Find the energy of the emitted photon. (b) Show this transition on an energy-level diagram for hydrogen, labeling the levels with the appropriate quantum numbers.

69P. Calculate the speed at which an initially stationary hydrogen atom recoils (owing to photon emission) if the electron makes a transition from the $n = 4$ state directly to the ground state. (*Hint:* Apply the principle of conservation of linear momentum.)

70P. A neutron, with kinetic energy of 6.0 eV, collides with a resting hydrogen atom in its ground state. Show that this collision must be elastic (that is, energy must be conserved). (*Hint:* Show that the atom cannot be raised to a higher excitation state as a result of the collision.)

71P. From the energy-level diagram for hydrogen, explain the observation that the frequency of the second Lyman-series line is the sum of the frequencies of the first Lyman-series line and the first Balmer-series line. This is an example of the empirically discovered *Ritz combination principle*. Use the diagram to find some other valid combinations.

72P. (a) Show that the smallest lower quantum number of two adjacent energy levels in hydrogen between which a transition produces radio-frequency waves is $n = (2R\lambda)^{1/3}$, where λ is the wavelength of the radio wave. Note that for radio-wave emission, n must be large. (b) If the 21.0-cm radio emission from interstellar hydrogen were due to such a transition (it isn't), what would be the value of n?

SECTION 43-10 BOHR'S DERIVATION

73E. Verify the numerical value of r_B, given in Eq. 43-34, by direct computation of its equivalent expression given in Eq. 43-32.

74E. What is the quantum number for a hydrogen atom that has an orbital radius of 0.847 nm?

75E. (a) If the angular momentum of the Earth due to its motion around the sun were quantized according to

Bohr's relation $L = nh/2\pi$, what would the quantum number be? (b) Could such quantization be detected if it existed?

76P. In the ground state of the hydrogen atom, according to Bohr's theory, what are (a) the quantum number, (b) the electron's orbit radius, (c) its angular momentum, (d) its linear momentum, (e) its angular velocity, (f) its linear speed, (g) the force on the electron, (h) the acceleration of the electron, (i) the electron's kinetic energy, (j) the potential energy, and (k) the total energy?

77P. How do the quantities (b) to (k) in Problem 76 vary with the quantum number?

78P. A diatomic gas molecule consists of two atoms of mass m separated by a fixed distance d rotating about an

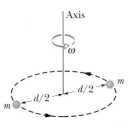

FIGURE 43-14 Problem 78.

axis as indicated in Fig. 43-14. Assuming that its angular momentum is quantized as in the Bohr atom, determine (a) the possible angular velocities and (b) the possible quantized rotational energies.

ADDITIONAL PROBLEMS

79. An initially free electron with a kinetic energy of 3.0 eV falls into orbit around a proton, forming a hydrogen atom that is in ground state. What is the frequency of the light that is emitted by this process?

80. Consider an atom with two closely spaced excited states A and B. If it jumps to ground state from A or from B, it emits a wavelength of 500 nm or 510 nm, respectively. What is the energy difference between states A and B?

81. When an electron is emitted by an initially neutral metal plate via the photoelectric effect, it leaves behind a net positive charge that is, at any given instant, effectively as far behind the surface of the plate as the electron is in front of it. The electric field due to that positive charge does work on the escaping electron. Assume that the electron escapes to an infinite distance and that the metal is sodium, which has a work function of 1.8 eV. For what least initial separation between the electron and the positive charge does the work done on the electron equal that work function?

82. Experiments show that the universe is filled with electromagnetic radiation which was released near the beginning of the universe and which is in accordance with the Big Bang model of that beginning. The spectral radiancy of the universe presently has a maximum that corresponds to a temperature of 3 K. Using the Wien displacement law of Exercise 42, find the photon energy at the maximum of the spectral radiancy.

83. A muon of charge $-e$ and mass $m = 207m_e$ (where m_e is the mass of an electron) orbits the nucleus of a singly ionized helium atom (He$^+$). Assuming that the Bohr model of the hydrogen atom can be applied to this muon–helium system, verify that the energy levels of the system are given by

$$E = -\frac{11.3 \text{ keV}}{n^2}.$$

84. An electron of mass m and velocity $\mathbf{v}$ undergoes a head-on collision with a gamma ray of energy hf_0, scattering the gamma ray back along the gamma ray's path. Verify that the energy of the scattered gamma ray, as measured in the laboratory system, is

$$E = hf_0 \bigg/ \left(1 + \frac{2hf_0}{mc^2}\sqrt{\frac{1 + v/c}{1 - v/c}}\right).$$

QUANTUM PHYSICS—II

44

Tracks of tiny vapor bubbles in this bubble-chamber image reveal where electrons (tracks color-coded green) and positrons (red) moved. A gamma-ray photon (which left no track when it entered at the top) kicked an electron out of one of the hydrogen atoms that filled the chamber and converted to an electron-positron pair. Another photon underwent another pair production farther down. These tracks (curved because of a magnetic field) clearly show that electrons and positrons are particles that move along narrow paths. Yet, those particles can also be interpreted in terms of waves. How can a particle be a wave?

44-1 LOUIS VICTOR DE BROGLIE MAKES A SUGGESTION

Physicists have rarely gone wrong in banking on the underlying symmetry of nature. For example, when you learn that a changing magnetic field produces an electric field it is a good bet to guess—and it turns out to be true—that a changing electric field will produce a magnetic field. For another example: The electron has an *antiparticle*, that is, a particle of the same mass but of opposite charge. Is it not reasonable to expect that the proton should also have an antiparticle? A 5-GeV proton accelerator was built at the University of California at Berkeley to search for the antiproton. It was found.

In 1924 Louis de Broglie, a physicist and a member of a distinguished French family, puzzled over the fact that light seemed to have a dual wave–particle aspect while matter—at that time—seemed to be entirely particle-like. This did not seem to jibe with the fact that both light and matter are forms of energy, that each can be transformed into the other, and that they are both governed by the spacetime symmetries of the theory of relativity. He then began to think that matter should also have a dual character and that particles such as electrons should have wavelike properties.

If we want to describe a moving particle as a wave, our first task is to answer the question: What is its wavelength? De Broglie made the bold suggestion that the relation

$$\lambda p = h \qquad (44\text{-}1)$$

applies both to light and to matter. If we solve Eq. 44-1 for p we have

$$p = \frac{h}{\lambda} \qquad \text{(momentum of a photon)}, \qquad (44\text{-}2)$$

which we obtained as Eq. 43-4 and used in Section 43-4 to assign a momentum to a photon of known wavelength.

If we solve Eq. 44-1 for λ we have

$$\lambda = \frac{h}{p} \qquad \text{(wavelength of a particle)}, \qquad (44\text{-}3)$$

which (said de Broglie) we can use to assign a wavelength to a particle of known momentum. A wavelength calculated from Eq. 44-3 is called a **de Broglie wavelength.** Look again at Eq. 44-1 and note the central role played by the Planck constant h in connecting the wave and particle aspects of light and matter.

SAMPLE PROBLEM 44-1

What is the de Broglie wavelength of an electron whose kinetic energy is 120 eV?

SOLUTION We can find the de Broglie wavelength from Eq. 44-3 if we know the momentum of the electron. For such relatively slow electrons ($K = 120$ eV is small), we can relate to the kinetic energy by combining $K = \frac{1}{2}mv^2$ and $p = mv$, obtaining

$$p = \sqrt{2mK}$$

$$= \sqrt{(2)(9.11 \times 10^{-31} \text{ kg})(120 \text{ eV})(1.60 \times 10^{-19} \text{ J/eV})}$$

$$= 5.91 \times 10^{-24} \text{ kg} \cdot \text{m/s}.$$

From Eq. 44-3 then

$$\lambda = \frac{h}{p} = \frac{6.63 \times 10^{-34} \text{ J} \cdot \text{s}}{5.91 \times 10^{-24} \text{ kg} \cdot \text{m/s}}$$

$$= 1.12 \times 10^{-10} \text{ m} = 112 \text{ pm}. \qquad \text{(Answer)}$$

This is about the size of a typical atom. As you will see, we can take advantage of this fact to verify the wave nature of such slow electrons experimentally in the laboratory.

SAMPLE PROBLEM 44-2

What is the de Broglie wavelength of a 150-g baseball traveling at 35.0 m/s?

SOLUTION From Eq. 44-3 we have

$$\lambda = \frac{h}{p} = \frac{h}{mv} = \frac{6.63 \times 10^{-34} \text{ J} \cdot \text{s}}{(150 \times 10^{-3} \text{ kg})(35.0 \text{ m/s})}$$

$$= 1.26 \times 10^{-34} \text{ m}. \qquad \text{(Answer)}$$

Because there is no hope of measuring such a small length, we do not speak of the wave nature of objects as large as baseballs.

PROBLEM SOLVING

TACTIC 1: ENERGY ADJECTIVES

When an energy is used as an adjective for a particle of matter, it refers to the kinetic energy of the particle, not the total energy or the rest energy. In Sample Problem 44-1, for example, we consider a 120-eV electron.

When an energy is used as an adjective for a photon, it can refer only to the photon energy $E = hf$, the only energy a photon can have, since its rest energy is zero. For example, we might refer to a 25-MeV gamma ray, a 25-keV x ray, or a 2.5-eV visible photon.

44-2 TESTING DE BROGLIE'S HYPOTHESIS

If you want to prove that you are dealing with a wave, a convincing thing to do is to measure the wavelength in the laboratory. That is what Thomas Young did in 1801 to establish the wave nature of visible light; it is what Max von Laue did in 1912 to establish the wave nature of x rays.

To measure a wavelength, you need two or more diffracting centers (pinholes, slits, or atoms) separated by a distance that is about the same size as the wavelength you are trying to measure. Sample Problem 44-2 shows at once that it is hopeless to try to measure the wavelength of a pitched baseball. You would need a pair of "slits" spaced about 10^{-34} m apart! That is why our daily experiences with large moving objects give no clue to the wave nature of matter. Sample Problem 44-1, however, suggests that we *should* be able to measure the wavelength of a moving electron, using the atoms of a crystal as a three-dimensional diffraction grating.

The Davisson–Germer Experiment

Figure 44-1 shows the apparatus used by C. J. Davisson and L. H. Germer of what is now the AT&T Bell Laboratories to measure the de Broglie wavelengths of slow electrons. In 1937 Davisson shared the Nobel prize in physics for this work, one of seven such prizes awarded (as of 1992) to scientists associated with this remarkable laboratory.

In the apparatus of Fig. 44-1 electrons from heated filament F are accelerated by an adjustable potential difference V. The resultant beam, made up of electrons whose kinetic energy is eV, then travels to a crystal C which was of nickel. The beam, "reflected" from the crystal surface, enters detector D and is recorded as a current I.

The experimenters set V to a particular arbitrary value and read the detector current I for various angular settings ϕ of the detector. They then set V to other values and repeated the angular sweep of the

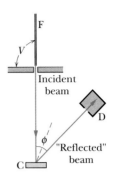

FIGURE 44-1   The apparatus of Davisson and Germer, used to demonstrate the wave nature of electrons. The electrons, emitted from heated filament F, are accelerated by an adjustable potential difference V. After reflection from crystal C they are recorded by detector D, which can be set to various angular positions ϕ.

detector each time. Figure 44-2 is a polar plot of $I(\phi)$ for $V = 54$ V; a strong diffracted beam at $\phi = 50°$ is evident. If the accelerating potential is either increased slightly or decreased slightly, the intensity of the diffracted beam drops.

Figure 44-3 is a simplified representation of the nickel crystal C of Fig. 44-1. The diffracted beam is formed by Bragg reflection of the electron matter wave from a particular family of atomic planes within the crystal. However, except in the case of normal incidence, the electron beam bends as it enters the crystal surface and also as it leaves; thus its angle of reflection within the crystal is not the same as angle ϕ of Fig. 44-3, which is measured outside the crystal. It can be shown that when this surface bending is taken into account, as it was to obtain the 50° value,

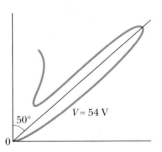

FIGURE 44-2   One set of results for the Davisson–Germer experiment of Fig. 44-1. Here the potential difference V is 54 V. The current I recorded by detecor D is plotted radially from the origin against the angle ϕ in a polar plot. A sharp diffraction maximum (of the measured current I) occurs for $\phi = 50°$. If V is changed from its set value of 54 V, the intensity of the diffraction maximum decreases.

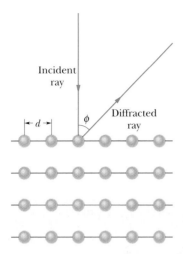

FIGURE 44-3 A schematic view of the atoms that make up crystal C in Fig. 44-1. The crystal behaves like a diffraction grating, the lines of atoms on its surface being separated by a distance d. For the crystal used by Davisson and Germer, $d = 215$ pm.

the crystal behaves like a two-dimensional diffraction grating; its grating lines are the parallel lines of atoms lying on the crystal surface, and its grating spacing is the interval marked d in Fig. 44-3. The principal maxima for such a grating must satisfy Eq. 41-18,

$$d \sin \phi = m\lambda, \qquad \text{for } m = 0, 1, 2, 3, \ldots . \quad (44\text{-}4)$$

For their crystal Davisson and Germer knew that $d = 215$ pm. For $m = 1$, which corresponds to a first-order diffraction peak, Eq. 44-4 leads to

$$\lambda = \frac{d \sin \phi}{m} = \frac{(215 \text{ pm})(\sin 50°)}{1} = 165 \text{ pm}.$$

The expected de Broglie wavelength for an electron of energy 54 eV, calculated as in Sample Problem 44-1, is 167 pm, in good agreement with the measured value. De Broglie's prediction was confirmed.

G. P. Thomson's Experiment

In 1927 George P. Thomson, working at the University of Aberdeen in Scotland, independently confirmed de Broglie's equation, using a somewhat different method. As Fig. 44-4a shows, he directed a monoenergetic beam of either x rays or electrons through a thin metal target foil. The target was specifically *not* a single large crystal (as in the Davisson–

Germer experiment) but was made up of a large number of tiny, randomly oriented crystals. With this arrangement there will always, by chance, be a certain number of crystals oriented at the proper angle to produce a diffracted beam.

If a photographic plate is placed perpendicular to the incident beam, as in Fig. 44-4a, it will show a central spot surrounded by diffraction rings. Figure 44-4c shows this pattern for a target of powdered aluminum and a beam of electrons of energy 15 eV. Figure 44-4b shows the pattern when the electron beam is replaced with an x-ray beam of the same wavelength. A simple glance at these two diffraction patterns leaves no doubt that both originated in the same way. Measurement and analysis of the patterns confirm de Broglie's hypothesis in every detail.

Thomson shared the 1937 Nobel prize with Davisson for his electron diffraction experiments. George P. Thomson was the son of J. J. Thomson, who won the Nobel prize in 1906 for his discovery of the electron and for his measurement of its charge-to-mass ratio. It has been written that

one may feel inclined to say that Thomson, the father, was awarded the Nobel prize for having shown that the electron is a particle, and Thomson, the son, for having shown that the electron is a wave.

Matter Waves: Some Applications

Today the wave nature of matter is taken for granted, and diffraction studies involving beams of electrons or neutrons are used routinely to study the atomic structure of solids and liquids. Figure 44-5 shows a commercial electron diffraction apparatus, of a kind found in many chemical, physical, and metallurgical analytical laboratories.

Matter waves are a valuable supplement to x rays in studying the atomic structure of solids. Electrons, for example, are less penetrating than x rays and so are particularly useful for studying surface features. Again, x rays interact largely with the electrons in a target, and for that reason it is not easy to use them to locate light atoms—particularly hydrogen—which have few electrons. Neutrons, on the other hand, interact largely with the nucleus of a target atom and can be used when x rays cannot. Figure 44-6 shows the structure of solid benzene as deduced from neutron diffraction studies. The six carbon atoms that form the familiar benzene ring are clearly there to see, as are the six hydrogen atoms coupled to them.

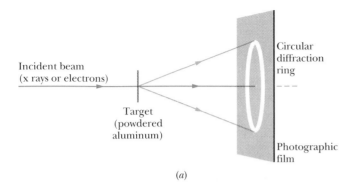

(a)

FIGURE 44-4 (a) The experimental arrangement used by Thomson to demonstrate a diffraction pattern characteristic of the atomic arrangements in a target of powdered aluminum. (b) The diffraction pattern if the incident beam in (a) is an x-ray beam. (c) The diffraction pattern if the incident beam is an electron beam. Note that the main features of the patterns in (b) and (c) are the same. The electron energy was chosen so that the de Broglie wavelength of the electrons in (c) matched the wavelength of the incident x rays in (b).

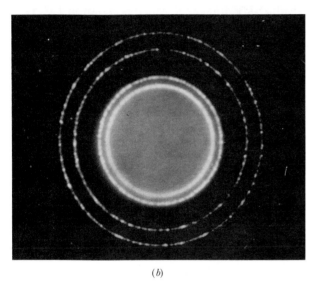

(b)

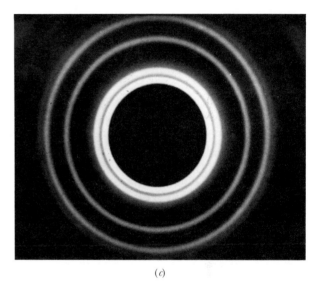

(c)

FIGURE 44-5 The central unit of a commercial apparatus for analyzing the atomic structure of solids via low-energy electron diffraction (LEED). Pumps for creating the necessary vacuum, the power supply, and the data recording instrumentation are not shown. If electrons did not have a wavelike aspect, this equipment would not work.

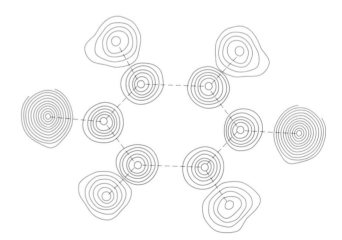

FIGURE 44-6 The atomic structure of solid benzene as deduced by neutron diffraction methods. The familiar benzene ring of six carbon atoms is clearly present, as are the hydrogen atoms coupled to them. The lines are "contour lines" that identify the electron density at various positions surrounding each nucleus.

44-3 THE WAVE FUNCTION

When Thomas Young measured the wavelength of light in 1801 he had no idea of the nature of the beam of sunlight that fell on the two pinholes in his interference apparatus. It was more than half a century later that Maxwell postulated that light is a traveling configuration of electric and magnetic fields.

We are in exactly the same situation at this stage of our introduction to matter waves. We can measure the wavelength associated with an electron or a neutron but—to put it loosely—we do not know what is waving. That is, we do not know what quantity in a matter wave corresponds to the electric field in an electromagnetic wave, to the transverse displacement in a wave traveling along a stretched string, or to the local pressure variation in a sound wave traveling through an air-filled pipe.

For the time being we shall use the term **wave function** for the quantity whose variation with position and time represents the wave aspect of a moving particle and assign it the symbol* ψ. Later, we shall interpret the wave function in a physical way by developing an analogy with light:

> Matter wave *is to* particle
>
> *as*
>
> Light wave *is to* photon.   (44-5)

First, let us develop a useful theorem that applies to waves of all kinds. When we discussed waves on strings, we saw that you can send a *traveling* wave of *any* wavelength down a stretched string of *infinite* length. However, if you deal with a taut string of *finite* length, only *standing* waves can be set up and these occur only at a *discrete set* of wavelengths. We summarize this general experience with waves by saying:

> Localizing the extent of a wave in space has the result that only a discrete set of wavelengths— and, correspondingly, a discrete set of frequencies—can occur. That is, *localization leads to quantization*.

---
*The lowercase symbol ψ refers only to the space-varying portion of the wave function. That is the only portion that concerns us in this chapter.

This theorem holds not only for waves in strings but for waves of all kinds, including electromagnetic waves and—as we shall see—matter waves.

Figure 44-7 shows a few of the standing wave patterns that can exist when a taut string is restricted to a finite length L, with its ends secured in rigid clamps. As you learned in Section 17-13, each such pattern has a well-defined wavelength, given by

$$\lambda = \frac{2L}{n}, \qquad \text{for } n = 1, 2, 3, \ldots , \quad (44\text{-}6)$$

in which the integer n, which is shown in Fig. 44-7, defines the oscillation mode. We shall come to call such integers *quantum numbers*. The frequencies corresponding to these wavelengths are also quantized and are given by

$$f = \frac{v}{\lambda} = \frac{v}{2L}\, n, \qquad \text{for } n = 1, 2, 3, \ldots , \quad (44\text{-}7)$$

in which v is the wave speed.

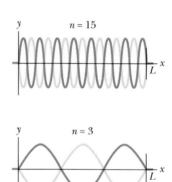
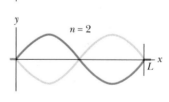
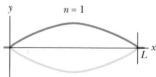

FIGURE 44-7 Four standing wave patterns for a taut string of length L, clamped rigidly at each end. The label n appears in Eqs. 44-6 and 44-7. The pattern for $n = 1$ represents the largest possible wavelength (and the lowest possible frequency) at which the string can oscillate. We can view these patterns as *stationary states* of the vibrating string, occurring at frequencies that are *quantized* according to Eq. 44-7, n being a *quantum number*.

44-4 LIGHT WAVES AND PHOTONS

We can set up standing electromagnetic waves, exactly like those shown in Fig. 44-7 for a stretched string, by trapping some radiation between two parallel, perfectly reflecting mirrors. In the visible or near-visible region, standing waves set up in the cavity of a gas laser serve nicely as an example. (We could also set up such standing waves in the microwave region of the spectrum, using parallel copper sheets as mirrors.)

For convenience in what follows, we shall deal only with the mode of oscillation that has the longest wavelength and hence the lowest frequency, corresponding to the wave with $n = 1$ in Fig. 44-7. Figure 44-8a (a copy of that wave) shows a plot of the wave amplitude E_{max} of our standing electromagnetic wave as a function of position for this mode. We see that exactly half a wave fits between the mirrors, which are located at O and L, so that the wavelength λ is $2L$.

Figure 44-8b shows a plot of E^2_{max} for this same oscillation mode. In view of Eq. 27-23 ($u = \frac{1}{2}\epsilon_0 E^2$), we can also interpret Fig. 44-8b as a plot of the *energy density* in the standing electromagnetic wave. We may think of the energy density at any point as being due to photons that are located at that point, each photon carrying the same energy hf. We can then conclude that the square of the wave amplitude at any point in a standing electromagnetic wave is pro-

portional to the density of photons at that point. You could test this conclusion by exploring the region between the mirrors with a photon probe. You would find a maximum density of photons halfway between the mirrors (halfway between O and L in Fig. 44-8b) and a density approaching zero just in front of each mirror.

If the total energy in the standing wave pattern is so low that it corresponds to the energy of a single photon, you would be led to conclude:

> The probability of detecting a photon at any location is proportional to the square of the amplitude that the electromagnetic wave has at that location.

Note that our knowledge of the photon position is inherently statistical. That is, we cannot say exactly where a photon is at a given moment; we can speak only of the relative probability that a photon will be in a certain region of space. As you will see, this statistical limitation is fundamental for both light and matter, for both photons and particles.

44-5 MATTER WAVES AND ELECTRONS

In considering the relation between matter waves and particles, we use the electron as a prototype and are guided by our analogy (Eq. 44-5) between light and matter waves.

How might we set up a standing matter wave using an electron? Recalling the localization–quantization theorem developed in Section 44-3, we are led to try to confine the electron, with electric forces, to a certain region of space—in some sort of *electron trap*. The matter waves associated with the electron should then occur as a set of standing matter waves, each at a specified frequency (like the waves in Fig. 44-7).

Atoms are just such electron traps. In fact, most of the electrons in the atoms that make up our planet and the living things that inhabit it have been trapped since before the solar system was formed.

For our purpose, however, we imagine a simple one-dimensional electron trap, in which a single electron is confined by electrical forces to move back and forth along an x axis, between two rigid "walls" separated by a distance L. If such a trap could actually be constructed, the potential energy function $U(x)$ of the electron in the trap would be that

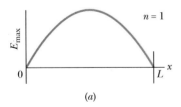

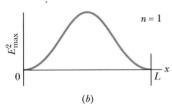

FIGURE 44-8 Light is trapped between two parallel mirrors separated by a distance L, forming a standing wave pattern. (a) A graph of the amplitude of oscillation versus position for the lowest-frequency oscillation mode, corresponding to $n = 1$ in Fig. 44-7. (b) A graph of the square of this amplitude, which is proportional at any point to the density of photons at that point.

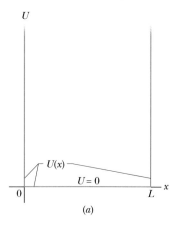

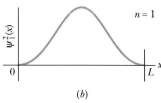

FIGURE 44-9 (a) An infinite well of width L. (b) The function $\psi^2(x)$ for an electron trapped in this well; the electron is in its ground state, corresponding to $n = 1$ in Figs. 44-7 and 44-8.

graphed in Fig. 44-9a: the potential energy would be zero within the trap* but would rise rapidly to an infinitely great value at $x = 0$ and $x = L$. (We discussed a similar sort of potential energy trap for a particle in Section 8-5; see also Fig. 8-12, which shows such traps.) The trap of Fig. 44-9a is more formally known as an infinitely deep *potential well*, or an *infinite well*, which is what we shall call it from now on.

For a trapped electron in its $n = 1$ state we expect a graph of its wave function ψ to look like Fig. 44-8a, and the square of that function to look like Fig. 44-9b—which is identical to Fig. 44-8b except that the former applies to matter waves instead of to light waves. Reasoning by analogy with light waves and photons, we conclude that

> The probability of finding the electron at any given location is proportional to the square of the amplitude that the matter wave has at that location.

*Recall that our choice of a configuration to which we assign zero potential is arbitrary; only potential differences count.

In particular, the probability of finding the electron in the interval that lies between x and $x + dx$ is proportional to the quantity $\psi^2(x)\ dx$. For our purposes, the square of the wave function—which we call the **probability density**—is more important than the wave function itself because the square tells us where the electron is likely to be. The probability that the electron will be *somewhere* in the infinite well of Fig. 44-9a is unity, which represents a certainty. Thus we have

$$\int_0^L \psi^2(x)\ dx = 1. \qquad (44\text{-}8)$$

The integral in Eq. 44-8 is simply the area under the curve of Fig. 44-9b. Because this area has the numerical value unity, it is said to be *normalized*.

The Energies of the Allowed States

Figure 44-10 shows the probability densities for four of the allowed standing matter wave patterns, corresponding to the four oscillation modes of Fig. 44-7. Let us find the energies of these allowed states.

Figure 44-9a shows us that the potential energy of the trapped electron is constant within the infi-

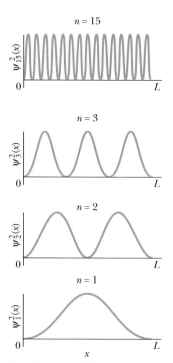

FIGURE 44-10 The probability density for four states of an electron trapped in the infinite well of Fig. 44-9a. The quantum numbers are indicated.

nite well and has the value zero. Thus the total electron energy is equal to its kinetic energy, and we have

$$E = K = \frac{p^2}{2m}. \tag{44-9}$$

We find the momentum of the trapped electron from Eq. 44-2 ($p = h/\lambda$). The de Broglie wavelength of an electron in a particular state is related to the quantum number n of that state by Eq. 44-6 ($\lambda = 2L/n$). Thus we have

$$p = \frac{h}{\lambda} = \frac{hn}{2L}.$$

The total energy is then obtained from Eq. 44-9 as

$$E_n = n^2 \frac{h^2}{8mL^2}, \quad \text{for } n = 1, 2, 3, \ldots . \tag{44-10}$$

We see that the state with $n = 1$, whose probability density is sketched in Fig. 44-9b, is the state with the lowest total energy; we call it the *ground state*.

The Zero-Point Energy

We note that, contrary to classical expectation, *the electron cannot be at rest in the well.* This is because its lowest energy is its ground-state energy, corresponding to $n = 1$ in Eq. 44-10, which gives us

$$E_1 = \frac{h^2}{8mL^2} \quad \text{(zero-point energy)}. \tag{44-11}$$

This energy is called the **zero-point energy.** What this result means is that the energy of the electron in the well cannot be zero and thus the electron cannot be stationary, not even at the absolute zero of temperature.

Equation 44-11 tells us that we can make the zero-point energy as small as we like by making the well wider, that is, by increasing L. In the limit of $L \to \infty$, which corresponds to a free particle, the zero-point energy approaches zero. We also see from Eq. 44-11 that, if we lived in a world (not ours!) in which the Planck constant were zero, there would be no such thing as a zero-point energy and the electron could indeed be at rest in its well. This involvement of the Planck constant shows us that the phenomenon of a zero-point energy—which turns out to be completely general—is strictly a quantum phenomenon.

SAMPLE PROBLEM 44-3

An electron is confined in an infinite well whose width L is 120 pm, about the diameter of an atom. What are the energies of the four states whose probability densities are displayed in Fig. 44-10?

SOLUTION From Eq. 44-10, with $n = 1$, we have

$$E_n = n^2 \frac{h^2}{8mL^2}$$

$$= (1)^2 \frac{(6.63 \times 10^{-34} \text{ J·s})^2}{(8)(9.11 \times 10^{-31} \text{ kg})(120 \times 10^{-12} \text{ m})^2}$$

$$= 4.19 \times 10^{-18} \text{ J} = 26.2 \text{ eV}. \quad \text{(Answer)}$$

The energy of the state with $n = 2$ is $2^2 \times 26.2$ eV or 105 eV. Similarly, the energies of the states with $n = 3$ and $n = 15$ are, respectively, 236 eV and 5900 eV.

SAMPLE PROBLEM 44-4

A 1.5-μg speck of dust moves back and forth between two rigid walls separated by 0.10 mm. It moves so slowly that it takes 120 s to cross this gap. Let us view this motion as that of a particle trapped in an infinite well. What quantum number describes the motion?

SOLUTION Solving Eq. 44-10 for n yields

$$n = \sqrt{\frac{8mEL^2}{h^2}}.$$

The energy of the particle is entirely kinetic. Noting that the speed of the particle is

$$v = \frac{0.10 \times 10^{-3} \text{ m}}{120 \text{ s}} = 8.33 \times 10^{-7} \text{ m/s},$$

we find the energy to be

$$E \,(= K) = \tfrac{1}{2}mv^2 = (\tfrac{1}{2})(1.5 \times 10^{-9} \text{ kg})$$
$$\times (8.33 \times 10^{-7} \text{ m/s})^2$$

$$= 5.2 \times 10^{-22} \text{ J}.$$

The quantum number n is then

$$n = \sqrt{\frac{8mEL^2}{h^2}}$$

$$= \sqrt{\frac{(8)(1.5 \times 10^{-9} \text{ kg})(5.2 \times 10^{-22} \text{ J})(0.10 \times 10^{-3} \text{ m})^2}{(6.63 \times 10^{-34} \text{ J·s})^2}}$$

$$= 3.8 \times 10^{14}. \quad \text{(Answer)}$$

This is a very large number indeed. It is impossible to distinguish experimentally between $n = 4 \times 10^{14}$ and $(4 \times 10^{14}) + 1$. Even this tiny speck of dust is a gross macroscopic object when compared to an electron. Quantum physics and classical physics give the same answers in this problem. We are in a region governed by the correspondence principle.

44-6 THE HYDROGEN ATOM

Let us now extend our analysis of an electron trapped in an infinite well to the more realistic case of an electron trapped in an atom. We choose the simplest atom, hydrogen.

The hydrogen atom consists of a single electron, bound to its nucleus (a single proton) by the attractive Coulomb force. The potential energy function $U(r)$ for this system is

$$U(r) = -\frac{1}{4\pi\epsilon_0} \frac{e^2}{r}, \qquad (44\text{-}12)$$

in which e is the magnitude of the charges of the electron and the proton, and r is the distance between them. Figure 44-11 is a graph of Eq. 44-12 for $r \leq 400$ pm.

This hydrogen atom trap, unlike the (one-dimensional) infinite well of Fig. 44-9a, is three-dimensional. It has spherical symmetry, so that the potential energy depends on only one variable—the separation r between the electron and the (relatively massive) central proton. Note also that the potential

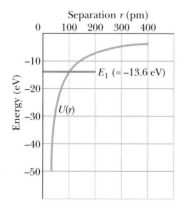

FIGURE 44-11 The potential well that governs the motion of the electron in the hydrogen atom. Compare it with the infinite well of Fig. 44-9a. The horizontal green line represents the state of lowest energy (the *ground state*), corresponding to $n = 1$ in Eq. 44-13.

energy given by Eq. 44-12 is negative for all values of r. This comes about because we have (arbitrarily) chosen our zero of potential to correspond to $r = \infty$. In Fig. 44-9a, on the other hand, we (arbitrarily) chose to assign $U(x) = 0$ to the region inside the well.

Quantum mechanics yields, for the energies of the allowed states of the hydrogen atom,

$$E_n = -\left(\frac{me^4}{8\epsilon_0^2 h^2}\right)\frac{1}{n^2}, \qquad \text{for } n = 1, 2, 3, \ldots$$
$$(44\text{-}13)$$

The hydrogen atom, like the electron in an infinite well, also exhibits a zero-point energy, corresponding to $n = 1$ in Eq. 44-13; all other values of n yield higher (that is, less negative) energies.

The probability density for the ground state of the hydrogen atom, given by quantum mechanics, is

$$\psi^2(r) = \frac{1}{\pi r_B^3} e^{-2r/r_B}, \qquad (44\text{-}14)$$

in which r_B is the **Bohr radius,** a convenient measure of distance on the atomic scale. Its value, as given in Eq. 43-34, is

$$r_B = 5.292 \times 10^{-11} \text{ m} = 52.92 \text{ pm}. \quad (44\text{-}15)$$

The physical meaning of Eq. 44-14 is that $\psi^2(r)\, dV$ is proportional to the probability that the electron will be found in any specified infinitesimal volume element dV. Suppose we want to evaluate $\psi^2(r)\, dV$. Because the probability density $\psi^2(r)$ depends only on r, it makes sense to choose as a volume element dV the volume between two concentric spherical shells whose radii are r and $r + dr$. That is, we define a volume element dV as

$$dV = (4\pi r^2)(dr). \qquad (44\text{-}16)$$

We now define a **radial probability density** $P(r)$ such that $P(r)\, dr$ gives the probability that we will find the electron in the volume element defined by Eq. 44-16. Thus, from Eqs. 44-14 and 44-16,

$$P(r)\, dr = \psi^2(r)\, dV = \frac{4}{r_B^3} r^2 e^{-2r/r_B}\, dr \quad (44\text{-}17)$$

or

$$P(r) = \frac{4}{r_B^3} r^2 e^{-2r/r_B}. \qquad (44\text{-}18)$$

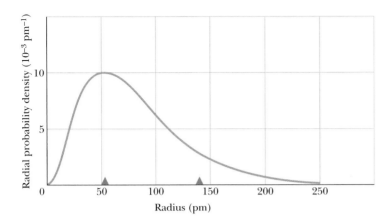

FIGURE 44-12 The radial probability density for the hydrogen atom in its ground state. Note that the electron is more likely to be found close to the Bohr radius ($r = 52.9$ pm, marked with a small triangle) than to any other position. The radius of the 90% sphere, within which the electron will be found 90% of the time, is also marked, at 2.67 Bohr radii or 141 pm.

Figure 44-12 shows a plot of Eq. 44-18. We can show that

$$\int_0^\infty P(r)\ dr = 1, \qquad (44\text{-}19)$$

so that the area under the curve of Fig. 44-12 is unity; this ensures that the hydrogen atom's electron must lie *somewhere* (see discussion at Eq. 44-8).

In the semiclassical theory of Bohr, the electron in its ground state simply revolved in a circular orbit of radius r_B. In wave mechanics, however, we discard this mechanical picture. We think instead of the hydrogen atom as a tiny nucleus surrounded by a *probability cloud* whose value $P(r)$ at any point is given by Eq. 44-18. We ask not "Is the electron near this point?" but "What are the odds that the electron is near this point?" This probabilistic information is all that we can ever learn about the electron; as it turns out, it is also all that we ever need to know.

As Fig. 44-12 shows, the radial probability density—interestingly enough—has its maximum value at the classical Bohr radius. The odds are that the electron will be farther away from the nucleus than this value 68% of the time and that it will be closer the remaining 32% of the time.

It is not easy for a beginner to look at subatomic particles in this probabilistic and statistical way. The difficulty is our natural impulse to regard an electron as something like a tiny marble or a tiny jelly bean, being at a certain place at a certain time and following a well-defined path. Electrons and other subatomic particles simply do not behave in this way. In the sections that follow we will do what we can to help dispel this pervasive *jelly bean fallacy*, as some call it.

SAMPLE PROBLEM 44-5

Show that the radial probability density has a maximum at $r = r_B$.

SOLUTION The radial probability density is given by Eq. 44-18,

$$P(r) = \frac{4}{r_B^3}\ r^2\ e^{-2r/r_B}.$$

If we differentiate this with respect to r we find, using the rule for products,

$$\frac{dP}{dr} = \frac{4}{r_B^3}\ r^2(-2/r_B)\ e^{-2r/r_B} + \frac{4}{r_B^3}\ (2r)\ e^{-2r/r_B}$$

$$= \frac{8}{r_B^4}\ r(r_B - r)\ e^{-2r/r_B}.$$

At the maximum of this function we must have $dP/dr = 0$; inspection of the function shows that this does indeed occur at $r = r_B$, which is what we sought to prove. Note that we also have $dP/dr = 0$ at $r = 0$ and at $r \to \infty$. However, these conditions correspond to minima, as can be seen in Fig. 44-12.

SAMPLE PROBLEM 44-6

The probability $p(r)$ that the electron in the ground state of the hydrogen atom will be found not just at r but inside a spherical shell of radius r is given by

$$p(r) = 1 - e^{-2x}(1 + 2x + 2x^2),$$

in which x, a dimensionless quantity, is equal to r/r_B. Find r for $p(r) = 0.90$.

SOLUTION We seek the radius of a sphere for which $p(r)$ is 0.90; see Fig. 44-12. From the foregoing expression for $p(r)$ we have

$$0.90 = 1 - e^{-2x}(1 + 2x + 2x^2)$$

or

$$10e^{-2x}(1 + 2x + 2x^2) = 1,$$

and we must find the value of x that satisfies this equality. It is not possible to solve explicitly for x, but a little trial and error with a pocket calculator (write a small program for it) quickly yields $x = 2.67$. This means that the radius of a sphere that the electron will be inside of 90% of the time is $2.67 r_B$.

44-7 BARRIER TUNNELING

Figure 44-13a sets the stage for an interesting quantum surprise. It shows an **energy barrier** (often called a **potential barrier,** where *potential* refers to potential energy) of height U and thickness L. An electron of total energy E approaches the barrier from the left. Classically, because $E < U$, the electron would be reflected from the barrier and would move back in the direction from which it came. In

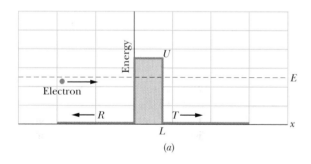

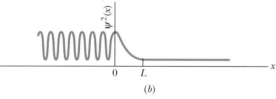

(b)

FIGURE 44-13 (*a*) An energy diagram showing an energy barrier of height U and thickness L, and the total energy E of an electron that approaches the barrier from the left. The electron has a probability R of being reflected from the barrier and a probability T of being transmitted through it via tunneling. (*b*) The probability density for the matter wave of the electron in (*a*). The pattern to the left of $x = 0$ is the interference pattern due to the incident and reflected matter waves.

quantum mechanics, however, there is a finite chance that the electron will appear on the other side of the barrier and continue its motion to the right.

It is as if you were to slide a jelly bean over a frictionless surface toward a frictionless hill for which the jelly bean lacked the energy to pass and —to your surprise—it materialized on the other side of the hill. Don't expect this to happen for jelly beans. However, electrons are not jelly beans and such **barrier tunneling,** as it is called, certainly *does* happen for electrons.

We can assign a reflection coefficient R and a transmission coefficient T to the incident electron in Fig. 44-13a; the sum of these two quantities is necessarily unity. Thus, if $T = 0.02$, of every 1000 electrons fired at the barrier, 20 (on average) will tunnel through it and 980 will be reflected.

Let us represent the electron as a matter wave. Figure 44-13b shows the appropriate probability density curve. To the left of the barrier, there is an incident matter wave moving to the right and a (somewhat less intense) reflected matter wave moving to the left. These two waves interfere, producing the interference pattern that we see in the $x < 0$ region of Fig. 44-13b. Within the barrier the probability density decreases exponentially. On the far side of the barrier ($x > L$) we have only a matter wave traveling to the right, with the reduced but constant amplitude shown.

The **transmission coefficient** T can be shown to be given approximately (for small values of T) by

$$T = e^{-2kL}, \qquad (44\text{-}20)$$

in which

$$k = \sqrt{\frac{8\pi^2 m(U - E)}{h^2}}. \qquad (44\text{-}21)$$

The value of T is very sensitive to the energy of the incident particle and to the height U and width L of the barrier.

Barrier tunneling is a puzzle only if you cling to the jelly bean fallacy. The proper way to look at barrier tunneling is to think of it as a matter-wave problem, the wave being related to the electron only in a probabilistic sense. That is, if the probability density on the far side of the barrier is not zero, there is a specific probability that you will find the electron there if you look for it.

For a simple example of barrier tunneling, consider a bare copper wire that has been cut and the two resulting ends then twisted together. The wire

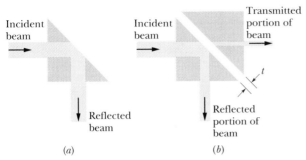

FIGURE 44-14 (*a*) A light wave is totally reflected from a glass–air interface. (*b*) If a second glass prism is held close to the first, the light wave can tunnel through the barrier associated with the air film between them. For this to happen, the thickness *t* of the air film must be no more than a few wavelengths of the incident light.

will conduct electricity readily, in spite of the fact that the wires may become coated with a thin layer of copper oxide, an insulator. The electrons simply tunnel through this thin insulating barrier.

Among other examples, we may list the tunnel diode, in which the flow of electrons (by tunneling) through a device can be rapidly turned on or off by controlling the height of the barrier. This can be done very quickly (within 5 ps) so that the device is suitable for applications where high-speed response is critical. The 1973 Nobel prize was shared by three "tunnelers," Leo Esaki (for tunneling in semiconductors), Ivar Giaever (for tunneling in superconductors), and Brian Josephson (for the Josephson junction, a quantum switching device based on tunneling). The 1986 Nobel prize was also awarded (to Gerd Binnig and Heinrich Rohrer) to recognize a device based on tunneling, the *scanning tunneling microscope*. In later chapters you will see the importance of tunneling in understanding certain kinds of radioactive decay, nuclear fission, and nuclear fusion.

Barrier tunneling occurs not only for matter waves but for waves of all kinds, including water waves and light waves. Figure 44-14*a* shows a light wave falling on a glass–air interface at an angle of incidence such that total internal reflection occurs. When we treated this subject in Section 39-3, we assumed that the incident light did not penetrate into the air space beyond the interface. However, that treatment was based on geometrical optics, which, as we know, is an approximation, being a limiting case of the more general wave optics. In much the same way, Newtonian mechanics (with its raylike trajectories) is a limiting case of the more general wave mechanics.

When a scanning tunneling microscope scans a surface to make an image, a wire tip is brought close enough to the surface for electrons to tunnel through the potential energy barrier between the tip and the surface. The tip–surface separation then determines the flow of electrons through the barrier. As the tip is moved over the surface, it is also moved up and down so as to maintain a constant flow of electrons. It thereby maps the contours of the surface. The scan shown here reveals individual xenon atoms arranged on a nickel crystal surface. The 5 nm letters, which spell I.B.M., were fashioned by using the electric force from the wire tip to drag the xenon atoms from their originally random locations on the crystal.

If we analyze total internal reflection from the wave point of view, we learn that light *does* penetrate beyond the interface, for a distance of a few wavelengths. Speaking very loosely, we can say that such a penetration is necessary because the incident wave must "feel out" the situation locally before it can "know for sure" that there *is* an interface.

In Fig. 44-14*b*, we place the face of a second glass prism parallel to the interface, the gap between them being no more than a few wavelengths. The incident wave can then "tunnel" through this narrow "barrier" (the gap) and generate a transmitted wave *T*. Specialists in optics call this phenomenon *frustrated total internal reflection* (FTIR).

SAMPLE PROBLEM 44-7

An electron whose total energy *E* is 5.1 eV is approaching a barrier whose height *U* is 6.8 eV and whose thickness *L* is 750 pm; see Fig. 44-15.

a. What is the de Broglie wavelength of the incident electron?

SOLUTION Before the electron reaches the barrier, its total energy *E* is entirely kinetic, the potential energy to

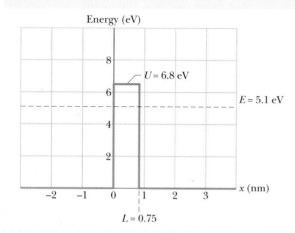

FIGURE 44-15 Sample Problem 44-7. An energy diagram showing an energy barrier of height $U = 6.8$ eV and thickness $L = 750$ pm, and the total energy $E = 5.1$ eV of an electron that approaches the barrier from the left.

the left of the barrier being zero. Proceeding as in Sample Problem 44-1 we find

$$\lambda = 540 \text{ pm.} \qquad \text{(Answer)}$$

Thus the barrier is about 750 pm/540 pm or about 1.4 de Broglie wavelengths thick.

b. What transmission coefficient follows from Eqs. 44-20 and 44-21?

SOLUTION From Eq. 44-21, we have

$$k = \sqrt{\frac{8\pi^2 m(U - E)}{h^2}}.$$

The numerator in the radical is

$$8\pi^2(9.11 \times 10^{-31} \text{ kg})(6.8 \text{ eV} - 5.1 \text{ eV})$$
$$\times (1.60 \times 10^{-19} \text{ J/eV})$$
$$= 1.956 \times 10^{-47} \text{ J} \cdot \text{kg}.$$

So, k is given by

$$k = \sqrt{\frac{1.956 \times 10^{-47} \text{ J} \cdot \text{kg}}{(6.63 \times 10^{-34} \text{ J} \cdot \text{s})^2}}$$
$$= 6.67 \times 10^9 \text{ m}^{-1}.$$

The quantity $2kL$ is then

$$2kL = (2)(6.67 \times 10^9 \text{ m}^{-1})(750 \times 10^{-12} \text{ m}) = 10.0$$

and the transmission coefficient, from Eq. 44-20, is

$$T = e^{-2kL} = e^{-10.0} = 45 \times 10^{-6}. \qquad \text{(Answer)}$$

Thus, of every million electrons striking the barrier, about 45 will tunnel through it.

c. What would be the transmission coefficient if the incident particle were a proton?

SOLUTION Carrying out the calculation once more but with the proton mass (1.67×10^{-27} kg) substituted for the electron mass yields $T \approx 10^{-186}$. The transmission coefficient is very substantially reduced indeed for this more massive particle. Imagine how small it would be for a jelly bean!

44-8 HEISENBERG'S UNCERTAINTY PRINCIPLE

The "jelly bean fallacy" way of thinking is a natural extension of our experiences with objects like baseballs that we can see and touch. As you have seen, however, this model simply doesn't work at the subatomic level. If an electron were like a tiny jelly bean, we should—in principle—be able to measure both its position and its momentum at any instant, with unlimited precision. But

IT CAN'T BE DONE.

We are not balked by the practical difficulties of measurement because we assume ideal measuring instruments. Nor is it as if the electron *has* an infinitely precise position and momentum but that—for some reason—nature will not let us find it out. What we are dealing with is a fundamental limitation on the concept of "particle."

Heisenberg's uncertainty principle provides a quantitative measure of this limitation. Suppose that you try to measure both the position and the momentum of an electron constrained to move along the x axis. Let Δx be the uncertainty in your measurement of its position and Δp_x your uncertainty in the measurement of its momentum. Heisenberg's principle states that

$$\Delta x \cdot \Delta p_x \approx h \qquad \text{(uncertainty principle).*} \qquad (44\text{-}22)$$

---

*The symbol $\approx$ is sometimes replaced by $\geqslant$, to recognize the fact that, in practice, you can never actually do as well as the quantum limit. Also, some formulations of the principle put $h/2\pi$ or $h/4\pi$ in place of h. These small differences need not concern us here.

That is, if you design an experiment to pin down the position of an electron as closely as possible (by making Δx smaller), you will find that you are not able to measure its momentum very well (Δp_x will get bigger). If you tinker with the experiment to improve the precision of your momentum measurement, the precision of your position measurement will deteriorate. *There is nothing that you can do about it.* The product of the two uncertainties must remain fixed, and this fixed product is nothing other than the Planck constant. Because momentum and position are vectors, a relation like Eq. 44-22 holds for the y and z coordinates as well.

The uncertainty principle seems strange only if you cling to the jelly bean fallacy and think of the electron as a tiny dot. Richard Feynman, in a footnote to a published series of lectures, puts the situation with characteristic clarity and forcefulness:*

> **I would like to put the uncertainty principle in its historical place: When the revolutionary ideas of quantum physics were first coming out, people still tried to understand them in terms of old-fashioned ideas [that is, the jelly bean fallacy] . . . . But at a certain point the old-fashioned ideas would begin to fail, so a warning was developed that said, in effect, "Your old-fashioned ideas are no damn good . . . . If you get rid of all these old-fashioned ideas and instead use the ideas that I'm explaining in these lectures . . . there is no need for an uncertainty principle!"**

What Feynman is saying, in effect, is: "Think in terms of matter waves. Throw out the notion of the electron as a tiny dot. When you want to think of electrons, do so statistically, being guided by the probability density of the matter wave."

Heisenberg's Principle: Another Formulation

Another way to formulate this principle is in terms of energy and time, both scalars. The relation is

$$\Delta E \cdot \Delta t \approx h \qquad \text{(uncertainty principle).} \qquad (44\text{-}23)$$

Thus if you try to measure the energy of a particle, allowing yourself a time interval Δt to do so, your energy measurement will be uncertain by an amount

*Richard P. Feynman, *QED — The Strange Theory of Light and Matter* (Princeton University Press, 1985), p. 55.

ΔE given by $h/\Delta t$. To improve the precision of your energy measurement, you must allow more time.

Another way to look at Eq. 44-23 is this: you can violate the law of conservation of energy by "borrowing" an energy amount ΔE *provided* that you "pay back" the borrowed energy within a time Δt given by $h/\Delta E$.

Let's apply this idea to the barrier tunneling problem of Section 44-7. If the electron in Fig. 44-13a only had an additional energy $U - E$ it could climb over the barrier in accordance with classical rules. The uncertainty principle tells us that the electron can "borrow" this amount of energy if it "pays it back" in the time it would take for the electron to travel a distance equal to the barrier thickness. Thus the electron finds itself on the other side of the barrier, the energy books are balanced again, and nobody is the wiser!

SAMPLE PROBLEM 44-8

An electron of kinetic energy 12 eV can be shown to have a speed of 2.05×10^6 m/s. Assume that you can measure this speed with a precision of 1.50%. With what uncertainty can you simultaneously measure the position of the electron?

SOLUTION The electron's momentum is

$$p = mv = (9.11 \times 10^{-31}\ \text{kg})(2.05 \times 10^6\ \text{m/s})$$

$$= 1.87 \times 10^{-24}\ \text{kg} \cdot \text{m/s}.$$

The uncertainty Δp in momentum is 1.50% of this, or 2.80×10^{-26} kg·m/s. The uncertainty in position is then, from Eq. 44-22,

$$\Delta x \approx \frac{h}{\Delta p} = \frac{6.63 \times 10^{-34}\ \text{J} \cdot \text{s}}{2.80 \times 10^{-26}\ \text{kg} \cdot \text{m/s}}$$

$$= 2.37 \times 10^{-8}\ \text{m} = 23.7\ \text{nm}, \qquad \text{(Answer)}$$

which is about 200 atomic diameters. Given your measurement of the electron's momentum, there is simply no way to pin down its position to any greater precision than this.

SAMPLE PROBLEM 44-9

A golf ball has a mass of 45 g and a speed, which you can measure to a precision of 1.5%, of 35 m/s. What limits does the uncertainty principle place on your ability to measure the position of the golf ball?

SOLUTION This example is like Sample Problem 44-8, except that the golf ball is much more massive and much slower than the electron of that example. The same calculation yields, in this case,

$$\Delta x \approx 3 \times 10^{-32} \text{ m.} \qquad \text{(Answer)}$$

This is about 10^{17} times smaller than the diameter of a typical atomic nucleus. Where large objects are concerned, the uncertainty principle sets no meaningful limit to the precision of measurement. All this is in accord with the correspondence principle, which tells us that, in situations in which classical physics is known to give correct answers (that is, golf balls), the predictions of quantum physics must merge with those of classical physics.

44-9 THE UNCERTAINTY PRINCIPLE: TWO CASE STUDIES

Here we explore the uncertainty principle by trying our best to beat it. We shall not succeed.

A Particle in a Box

Let us try to pin down the position of an electron by trapping it in a box from which it cannot escape and then shrinking the walls of the box, which we assume we can do without limit. We work in one dimension so that our "box" becomes the familiar infinite well of Fig. 44-9a, its walls separated by a distance L. For the uncertainty in our position measurement we then have

$$\Delta x \approx L \qquad \text{(uncertainty in position).} \quad (44\text{-}24)$$

As Eq. 44-11 shows, if we decrease L we increase the energy (that is, the zero-point energy) of the trapped electron. You might say:

"It is true that the energy gets bigger (and so does the momentum), but at least I know *exactly* what this larger energy is; it is given by Eq. 44-11 and there is no uncertainty about it."

The flaw in your argument is that you are not taking fully into account the fact that momentum is a vector. You may know the *magnitude* of the momentum exactly but you do not know its *direction*. That is, the electron may be bouncing back and forth between

its confining walls but, at any instant, you do not know whether it is moving from left to right or from right to left.

In the wave model, the standing matter wave that represents the trapped electron is made up of two traveling waves traveling in opposite directions, each carrying momentum. The magnitude of the uncertainty in momentum is then

$$\Delta p \approx (+p) - (-p) = 2p.$$

The momentum of the electron is given by Eq. 44-2 ($p = h/\lambda$), and $\lambda = 2L$, so

$$\Delta p = 2p = \frac{2h}{\lambda} = \frac{2h}{2L}$$

from which

$$\Delta p \approx \frac{h}{L} \qquad \text{(uncertainty in momentum).} \quad (44\text{-}25)$$

If we multiply Eqs. 44-24 and 44-25, we have

$$\Delta x \cdot \Delta p \approx (L)(h/L),$$

or

$$\Delta x \cdot \Delta p \approx h,$$

which is exactly the uncertainty principle. We have failed in our attempt to pin down the position and momentum of our trapped electron. We knew we would fail, but let's try again anyway, with a different attack.

A Particle Passing Through a Slit

Let an electron, represented by an incident matter wave, pass through a slit of width Δy in screen A of Fig. 44-16. We are going to try to pin down the *vertical* position and momentum components of the electron at the instant it passes through the slit.

Because the electron got through the slit, we know its vertical position at the instant it did so with an uncertainty Δy. By reducing the slit width, we can pin down the vertical position of the electron as closely as we like.

However, matter waves—like all other waves—flare out by diffraction when they pass through a slit. Furthermore, the narrower the slit, the more they flare out. From the particle point of view, this "flaring out" means that the electron acquires a vertical component of momentum as it passes through the slit. Some electrons acquire only a little vertical mo-

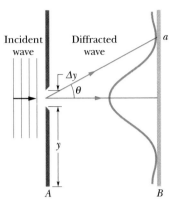

FIGURE 44-16 An arrangement for trying to beat the uncertainty principle. (It doesn't work.) An incident beam of electrons is diffracted at the slit in screen A, forming a typical diffraction pattern on screen B. The narrower the slit, the wider the pattern.

mentum and others acquire a lot, so that there is an uncertainty.

There is one particular value of the vertical momentum component that will carry the electron to the first minimum of the diffraction pattern, point a on screen B of Fig. 44-16. Let us take this value as a measure of the uncertainty Δp in our knowledge of the vertical momentum component of the electron.

From Eq. 41-1, with the slit width now being Δy, we know that the first minimum of the diffraction pattern occurs at an angle θ given by

$$\sin \theta = \frac{\lambda}{\Delta y}.$$

If θ is small enough, we can replace $\sin \theta$ with θ. Also, from Eq. 44-3, we have $\lambda = h/p$. Our equation then becomes

$$\theta \approx \frac{h}{p \, \Delta y}, \qquad (44\text{-}26)$$

in which p is the horizontal momentum component. To reach the first minimum, θ must be such that

$$\theta = \frac{\Delta p}{p}. \qquad (44\text{-}27)$$

If we equate Eqs. 44-26 and 44-27, we find

$$\Delta y \cdot \Delta p \approx h,$$

which is once more the uncertainty principle. Foiled again!

44-10 WAVES AND PARTICLES

How can an electron (or a photon) be wavelike under some circumstances and particlelike under others? Our mental images of "wave" and "particle" are drawn from our familiarity with large-scale objects such as ocean waves and tennis balls. In a way it is fortunate that we are able to extend these concepts (separately!) into the subatomic domain and apply them to entities such as the electron, which we can neither see nor touch. But you must understand that no single concrete mental image, combining the features of *both* wave and particle, is possible in the quantum world. As Paul Davies, physicist and science writer, has written: "It is impossible to visualize a wave-particle, so don't try."

Niels Bohr, who not only played a major role in the development of quantum mechanics but also served as its major philosopher and interpreter, has shown a way to feel comfortable with the wave–particle duality problem. It is embodied in his *principle of complementarity*, which states:

> The wave and particle aspects of a quantum entity are both necessary for a complete description. However, both aspects cannot be revealed simultaneously in a single experiment. Which aspect is revealed is determined by the nature of the experiment being done.

Consider a beam of light, perhaps from a laser, that passes across a laboratory table. What is the nature of the light beam? Is it a wave or a stream of particles? You cannot answer this question unless you interact with the beam in some way.

If you put a diffraction grating in the path of the beam, you reveal it as a wave. If you interpose a photoelectric apparatus such as that of Fig. 43-1, you will need to regard the beam as a stream of particles (photons) to interpret your measurements in a satisfactory way. Try as you will, there is no single experiment that you can carry out with the beam that will require (or allow) you to interpret it as a wave *and* as a particle *at the same time*. You may not like having to switch back and forth between wave and particle descriptions, depending on the experiment you are doing, but there is at least no confusion caused by overlap of the two models.

Complementarity: A Case Study

Let us see how complementarity works by trying to set up an experiment that will force nature to reveal both the wave and the particle aspects of electrons at the same time. We didn't succeed in beating the uncertainty principle and we won't succeed here either, but we'll try!

In Fig. 44-17 a beam of electrons falls on a double-slit arrangement in screen *A* and sets up a pattern of interference fringes on screen *B*. This is proof enough of the wave nature of the incident electrons.

Suppose now that we replace screen *B* with a small electron detector, designed to generate and record a "click" every time an electron hits it. We find that such clicks do indeed occur. If we move the detector up and down in Fig. 44-17 we can, by plotting the click rate against the detector position, trace out the pattern of interference fringes. Have we not succeeded in demonstrating both wave and particle? We see the fringes (wave) and we hear the clicks (particle).

We have not. A mere "click" is not enough evidence that we are dealing with a particle. The concept of "particle" involves the concept of "trajectory" and a mental image of a dot following a prescribed path. As a minimum, we want to be able to know which of the two slits in screen *A* the electron passed through on its way to generating a click in the detector. Can we find out?

We can, in principle, by putting a very thin detector in front of each slit, designed so that, if an electron passes through the slit, the detector will generate an electronic signal. We can then try to correlate each click, or "screen arrival signal," with a "slit passage signal," thus identifying the path of the electron involved.

If we succeed in modifying the apparatus to do this, we find a surprising thing. *The interference fringes have disappeared!* In passing through the slit detectors the electrons were affected in ways that destroyed the interference pattern. Although we have now shown the particle nature of the electron, the evidence for its wave nature has vanished.

The converse to our thought experiment is also true. If we start with an experiment that shows that electrons are particles and if we tinker with it to bring out the wave aspect, we will always find that the evidence for particles has vanished. Also, our experiment would work in precisely the same way if we substituted a light beam for the incident electron beam in Fig. 44-17.

A Quantum Puzzle Solved

At the beginning of this chapter you were asked how tracks such as those shown in the image at the start of this chapter, made up of tiny bubbles and so clearly suggesting the wake of a fast charged particle, can be associated with waves.

As the beginnings of an answer we look again at the thought experiment of Fig. 44-17, in which the pattern of fringes on screen *B* is neatly accounted for by the alternating constructive and destructive interference of matter wavelets radiating from each of the two slits in screen *A*. We can think of these as "guiding waves," their connection with the particles being that the square of their associated wave function at any point measures the probability that a particle will be found at that point. Thus, on screen *B*, electrons will pile up at those places where this probability is large, and they will be found in lesser abundance at those places where it is small. Figure 44-18 shows how the fringes build up with time.

These considerations apply even if the incident beam is deliberately made so weak that, by calculation, there should be—on average—*only one electron in the apparatus at any instant.* You might think that, because the single electron that chances to be in the apparatus must go through one slit or the other, the fringes must vanish; after all—you might reason— the electron cannot interfere with itself and there is nothing else for it to interfere with. However, exper-

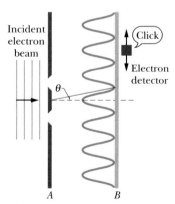

FIGURE 44-17 An arrangement for trying to prove that an incident electron beam is simultaneously both wavelike and particle-like. (It doesn't work.) You can modify the apparatus to show one aspect or the other, but not both at the same time.

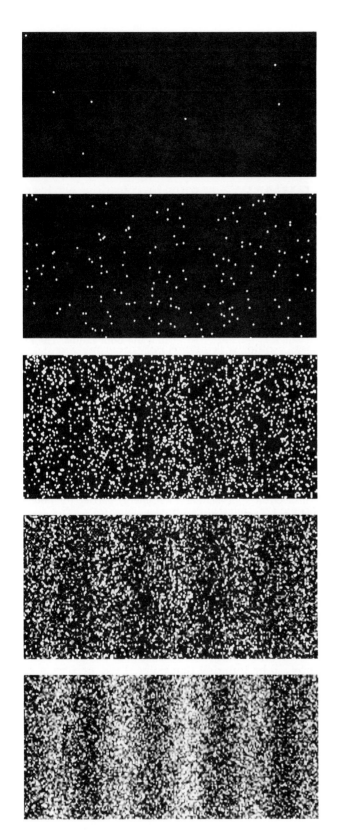

FIGURE 44-18 The buildup of an interference pattern by electrons in an actual two-slit interference experiment, from top to bottom: 7 electrons, 100 electrons, 3000 electrons, 20,000 electrons, and 70,000 electrons.

iment shows that the fringes will *still* be formed, built up slowly as electron after electron falls on screen *B*. Even under these conditions the associated matter wave always passes through *both* slits and *this* fact is what determines where the electrons are likely to fall on screen *B*. The single electron *does* interfere with itself. But don't try to visualize how it does so!

With this background we are ready to answer the question about the wave nature of the particles and their tracks in the bubble-chamber image opening this chapter. To simplify the situation, let us turn off the magnetic field in the bubble chamber so that the tracks left by the particles are straight.

When an electron travels through the chamber, it collides with some of the hydrogen atoms in the liquid hydrogen that fills the bubble chamber, ionizing those atoms. The liquid hydrogen is slightly above its vaporization (boiling) point but is still liquid. However, the sudden presence of ions at scattered points along the electron's route causes rapid vaporization of the liquid at those points, which produces vapor bubbles marking the route. In detecting these bubbles, we detect the electron's passage.

How does the electron traverse the space between two successive detection points (two successive bubbles) where it is not detected? We are tempted to answer that it travels as a particle along the straight line connecting those points. Indeed, because the series of bubbles forms a line, this answer seems compelling.

However, the quantum-mechanical answer is that matter waves of the electron travel along all possible paths connecting two successive detection points. This is suggested by a few such paths in Fig. 44-19, where an electron is detected by causing a bubble first at point *I* and then at point *F*. Along the straight line connecting *I* and *F*, the matter waves undergo constructive interference. At any point off the straight line, the matter waves undergo destructive interference.

This means that if you were to somehow place an obstacle along the straight line between *I* and *F*, the obstacle would be at a point of constructive interference of the waves and could intercept the elec-

FIGURE 44-19 An electron moves from *I* to *F*. From the wave point of view, all possible paths are explored, the resultant track being the superposition of all these paths.

tron. But if you place the obstacle off that line, it would be at a point of destructive interference and could not intercept the electron. Thus, you need not consider the track of bubbles as being left by an electron that continuously travels as a particle. In-

stead, you can think of the track as being a series of detection points where the matter waves undergo constructive interference. In other words, you can think of an electron as being a wave instead of being a particle.

REVIEW & SUMMARY

The Wave Nature of Matter
Beams of electrons and other forms of matter exhibit wave properties, including interference and diffraction, with a **de Broglie wavelength** given by

$$\lambda = h/p \qquad \text{(wavelength of a particle).} \qquad (44\text{-}3)$$

These wave properties are most easily shown by the diffraction, similar to x-ray diffraction, that occurs during reflection from atomic planes in crystals.

The Wave Function
Matter waves are described by a **wave function** ψ. Such waves describe particle motion in much the same way that electromagnetic waves describe photons. In particular, the probability of finding an electron at any location, called the **probability density,** is proportional to the square of the wave function at that location.

A Particle Trapped Between Rigid Walls
A simple one-dimensional introduction to matter waves is the study of the motion of a particle trapped between rigid walls. We can study the relevant wave function because of its close mathematical relationship to two classical problems: the standing-wave oscillations of a short constrained string and the electromagnetic oscillations inside a cavity with perfectly reflecting walls. In this one-dimensional case, $\psi^2(x)\,dx$ is proportional to the probability of finding the particle in the interval between x and $x + dx$. Furthermore,

$$\int_0^L \psi^2(x)\,dx = 1 \qquad (44\text{-}8)$$

is the probability of finding the particle anywhere between the walls, located at $x = 0$ and $x = L$. Figure 44-9b shows the probability density for the lowest-energy wave function. The trapped particle's energy is limited to quantized values given by

$$E_n = n^2 \frac{h^2}{8mL^2}, \qquad \text{for } n = 1, 2, 3, \ldots \quad (44\text{-}10)$$

The lowest energy, $E_1 = h^2/8mL^2$, is the **zero-point energy,** the energy retained by the particle even at 0 K.

The Hydrogen Atom
The energies of the allowed states of an electron in a hydrogen atom are

$$E_n = -\left(\frac{me^4}{8\epsilon_0^2 h^2}\right)\frac{1}{n^2}, \qquad \text{for } n = 1, 2, 3, \ldots . \quad (44\text{-}13)$$

The **probability density** for the ground state (the lowest-energy state) is

$$\psi^2 = \frac{1}{\pi r_B^3} e^{-2r/r_B}, \qquad (44\text{-}14)$$

in which $r_B = 5.292 \times 10^{-11}$ m is the **Bohr radius.** The **radial probability density** $P(r)$ for this atom is

$$P(r) = \frac{4}{r_B^3} r^2 e^{-2r/r_B}. \qquad (44\text{-}18)$$

$P(r)\,dr$ is the probability of finding the electron between two concentric shells of radii r and $r + dr$.

Barrier Tunneling
An electron approaching a flat **energy barrier** (or **potential barrier**) of height U and thickness L has a finite probability T (the **transmission coefficient**) of penetrating the barrier even if the electron's kinetic energy E is less than the height of the barrier. The probability is

$$T = e^{-2kL}, \qquad (44\text{-}20)$$

in which

$$k = \sqrt{\frac{8\pi^2 m(U - E)}{h^2}}. \qquad (44\text{-}21)$$

Heisenberg's Uncertainty Principle
The uncertainty principle suggests that the very concept of "particle" (as in "particle between rigid walls") is inherently fuzzy. In particular, we cannot, even in principle, simultaneously measure both a particle's position **r** and momentum **p** with arbitrary precision. In fact, the uncertainties associated with each component of **r** and **p** must obey a relationship of the form

$$\Delta x \cdot \Delta p_x \approx h \qquad \text{(uncertainty principle).} \quad (44\text{-}22)$$

The principle also applies to energy and time measurements in the form

$$\Delta E \cdot \Delta t \approx h \qquad \text{(uncertainty principle).} \quad (44\text{-}23)$$

QUESTIONS

1. How can the wavelength of an electron be given by $\lambda = h/p$? Doesn't the very presence of the momentum p in this formula imply that the electron is a particle?

2. In a repetition of Thomson's experiment for measuring e/m for the electron (see Section 30-3), electrons are collimated (made to form a beam) by passage through a slit. Why is the beamlike character of the emerging electrons not destroyed by diffraction of the electron wave at this slit?

3. Considering the wave behavior of electrons, we should expect to be able to construct an "electron microscope" using short-wavelength electrons to provide high resolution. This, indeed, has been done. (a) How might an electron beam be focused? (b) What advantages might an electron microscope have over a light microscope? (c) Why not make a proton microscope? Why not a neutron microscope?

4. If the following particles all have the same energy, which has the shortest wavelength: electron, alpha particle, neutron, proton?

5. What expression can be used for the momentum of either a photon or a particle?

6. Discuss the analogy between (a) wave optics and geometrical optics and (b) wave mechanics and classical mechanics.

7. Does a photon have a de Broglie wavelength? Explain.

8. Discuss similarities and differences between a matter wave and an electromagnetic wave.

9. Can the de Broglie wavelength associated with a particle be smaller than the size of the particle? Larger? Is there necessarily any relation between these two quantities of a particle?

10. If, in the de Broglie formula $\lambda = h/mv$, we let $m \rightarrow \infty$, do we get the classical result for particles of matter?

11. How could Davisson and Germer be sure that the "54-eV" peak of Fig. 44-2 was a first-order diffraction peak, that is, that $m = 1$ in Eq. 44-4?

12. The allowed energies for a particle confined between rigid walls are given by Eq. 44-10. First, convince yourself that as n increases these energy levels become farther apart. Then explain how this can possibly be. The correspondence principle would seem to require that they move closer together as n increases, approaching a continuum.

13. How can the predictions of quantum mechanics be so exact if the only information we have about the positions of the electrons in atoms is statistical?

14. In the $n = 1$ state, for a particle confined between rigid walls, what is the probability that the particle will be found in a small-length element just at the surface of either wall?

15. Given Fig. 44-10, what do you imagine the curve for $\psi^2(x)$ for $n = 100$ looks like? Convince yourself that these curves approach classical expectations as $n \rightarrow \infty$.

16. We have seen that barrier tunneling works for matter waves and for electromagnetic waves. Do you think that it also works for water waves? For sound waves?

17. Comment on the statement, "A particle can't be detected while tunneling through a barrier, so it doesn't make sense to say that such a thing actually happens."

18. A proton and a deuteron, each having 3 MeV of energy, attempt to penetrate a rectangular potential barrier of height 10 MeV. Which particle has the higher probability of succeeding? Explain in qualitative terms.

19. A laser projects a beam of light across a laboratory table. If you put a diffraction grating in the path of the beam and observe the spectrum, you declare the beam to be a wave. If instead you put a clean metal surface in the path of the beam and observe the ejected photoelectrons, you declare this same beam to be a stream of particles (photons). Which description of the beam is correct if you don't put anything in its path?

20. State and discuss (a) the correspondence principle, (b) the uncertainty principle, and (c) the complementarity principle.

21. In Fig. 44-17, why would you expect the electrons from each slit to arrive at the screen over a range of positions? Shouldn't they all arrive at the same place? How does your answer relate to the complementarity principle?

22. Several groups of experimenters are trying to detect gravity waves, perhaps coming from our galactic center, by measuring small distortions in a massive object through which the hypothesized waves pass. They seek to measure displacements as small as 10^{-21} m. (The radius of a proton is about 10^{-15} m, a million times larger!) Does the uncertainty principle put any restriction on the precision with which this measurement can be carried out?

23. Figure 44-10 shows that for $n = 3$ the probability density $\psi^2(x)$ for a particle confined between rigid walls is zero at two points between the walls. How can the particle ever move across these positions? (*Hint:* Consider the implications of the uncertainty principle.)

24. Why does the concept of Bohr orbits violate the uncertainty principle? (*Hint:* see Problem 47.)

25. (a) Give examples of how the process of measurement disturbs the system being measured. (b) Can the disturbances be taken into account ahead of time by suitable calculations?

EXERCISES & PROBLEMS

SECTION 44-1 LOUIS VICTOR DE BROGLIE MAKES A SUGGESTION

1E. A bullet of mass 40 g travels at 1000 m/s. (a) What wavelength can we associate with it? (b) Why does the wave nature of the bullet not reveal itself through diffraction effects?

2E. Using the classical relation between momentum and kinetic energy, show that the de Broglie wavelength of an electron can be written (a) as

$$\lambda = \frac{1.226 \text{ nm}}{\sqrt{K}},$$

in which K is the kinetic energy in electron-volts, or (b) as

$$\lambda = \sqrt{\frac{1.50}{V}},$$

where λ is in nanometers, and V is the accelerating potential in volts.

3E. In an ordinary color television set, electrons are accelerated through a potential difference of 25.0 kV. What is the de Broglie wavelength of such electrons? (*Hint:* Ignore relativistic effects.)

4E. Calculate the wavelength of (a) a 1-keV electron, (b) a 1-keV photon, and (c) a 1-keV neutron.

5E. An electron and a photon each have a wavelength of 0.20 nm. Calculate their (a) momenta and (b) energies.

6E. The wavelength of the yellow spectral emission line of sodium is 590 nm. At what kinetic energy would an electron have the same de Broglie wavelength?

7E. Thermal neutrons have an average kinetic energy of $\frac{3}{2}kT$, where T may be taken to be 300 K. Such neutrons are in thermal equilibrium with their normal surroundings. (a) What is the average energy of a thermal neutron? (b) What is the corresponding de Broglie wavelength?

8E. If the de Broglie wavelength of a proton is 0.100 pm, (a) what is the speed of the proton and (b) through what electric potential would the proton have to be accelerated to acquire this speed?

9P. Consider a balloon filled with (monatomic) helium gas at room temperature and pressure. (a) Calculate the average de Broglie wavelength of the helium atoms and the average distance between atoms under these conditions. The average kinetic energy of an atom is equal to $\frac{3}{2}kT$. (b) Can the molecules be treated as particles under these conditions?

10P. (a) A photon has an energy of 1.00 eV, and an electron has a kinetic energy of that same amount. What are their wavelengths? (b) Repeat for an energy of 1.00 GeV.

11P. (a) If a photon and an electron both have a wavelength of 1.00 nm, what is the energy of the photon and the kinetic energy of the electron? (b) Repeat for a wavelength of 1.00 fm.

12P. Singly charged sodium ions are accelerated through a potential difference of 300 V. (a) What is the momentum acquired by the ions? (b) Calculate their de Broglie wavelength.

13P. The 20-GeV electron accelerator at Stanford provides an electron beam of small wavelength, suitable for probing the fine details of nuclear structure via scattering. What is the wavelength of the electrons, and how does it compare with the radius of an average nucleus (about 5.0 fm)? (At this energy it is sufficient to use the extreme relativistic relationship between momentum and energy, namely, $p = E/c$. This is the same relationship used for light and is justified when the kinetic energy of a particle is much greater than its rest energy, as in this case.)

14P. The existence of the atomic nucleus was discovered in 1911 by Ernest Rutherford, who properly interpreted some experiments in which a beam of alpha particles was scattered from a metal foil of atoms such as gold. (a) If the alpha particles had a kinetic energy of 7.5 MeV, what was their de Broglie wavelength? (b) Should the wave nature of the incident alpha particles have been taken into account in interpreting these experiments? The mass of an alpha particle is 4.00 u, and its distance of closest approach to the nuclear center in these experiments was about 30 fm. (The wave nature of matter was not postulated until more than a decade after these crucial experiments were first performed.)

15P. A nonrelativistic particle is moving three times as fast as an electron. The ratio of the de Broglie wavelength of the particle to that of the electron is 1.813×10^{-4}. By calculating its mass, identify the particle.

16P. The highest achievable resolving power of a microscope is limited only by the wavelength used; that is, the smallest detail that can be separated has dimensions about equal to the wavelength. Suppose one wishes to "see" inside an atom. Assuming the atom to have a diameter of 100 pm, this means that we wish to resolve detail of separation of, say, 10 pm. (a) If an electron microscope is used, what minimum energy of electrons is needed? (b) If a light microscope is used, what minimum energy of photons is needed? (c) Which microscope seems more practical for this purpose? Why?

17P. What accelerating voltage would be required for electrons in an electron microscope to obtain the same ultimate resolving power as that which could be obtained from a gamma-ray microscope using 100-keV gamma rays? (*Hint:* See Problem 16.)

18P. (a) Calculate, according to the Bohr model, the speed of the electron in the ground state of the hydrogen atom. (b) Calculate the corresponding de Broglie wavelength. (c) Comparing the answers to (a) and (b), find a relation between the de Broglie wavelength λ and the radius r of the ground-state Bohr orbit.

SECTION 44-2 TESTING DE BROGLIE'S HYPOTHESIS

19E. A potassium chloride (KCl) crystal is cut so that the layers of atomic planes parallel to its surface have an interplanar spacing of 0.314 nm. A beam of 380-eV electrons is incident normally on the crystal surface. Calculate the angles ϕ at which the detector must be positioned to record strongly diffracted beams of all orders present.

20P. In the experiment of Davisson and Germer, (a) at what angles would the second- and third-order diffracted beams corresponding to a strong maximum in Fig. 44-2 occur, provided they are present? (b) At what angle would the first-order diffracted beam occur if the accelerating potential were changed from 54 to 60 V?

SECTION 44-5 MATTER WAVES AND ELECTRONS

21E. (a) A proton or (b) an electron is trapped in a one-dimensional box of 100-pm length. What is the minimum energy each of these particles can have?

22E. What must be the width of an infinite well such that the energy of an electron trapped therein in the $n = 3$ state has an energy of 4.7 eV?

23E. (a) Calculate the smallest allowed energy of an electron were it trapped inside an atomic nucleus (diameter about 1.4×10^{-14} m). (b) Compare this with the several MeV of energy binding protons and neutrons inside the nucleus; on this basis should we expect to find electrons inside nuclei?

24E. The ground-state energy of an electron in an infinite well is 2.6 eV. What will the ground-state energy be if the width of the well is doubled?

25E. An electron, trapped in an infinite well of width 0.250 nm, is in the ground ($n = 1$) state. How much energy must it absorb to jump up to the third excited ($n = 4$) state?

26P. (a) What is the separation in energy between the lowest two energy levels for a container 20 cm on a side containing argon atoms? Assume, for simplicity, that the argon atoms are trapped in a one-dimensional well 20 cm wide. The molar mass of argon is 39.9 g/mol. (b) How does this energy separation compare with the thermal energy of the argon atoms at 300 K? (c) At what temperature does the thermal energy equal the energy separation?

27P. Consider a conduction electron in a cubical crystal of a conducting material. Such an electron is free to move throughout the volume of the crystal but cannot escape to the outside. It is trapped in a three-dimensional infinite well. The electron can move in three dimensions, so that its total energy is given by

$$E = \frac{h^2}{8L^2 m}\,(n_1^2 + n_2^2 + n_3^2),$$

in which n_1, n_2, n_3 each take on the values 1, 2, . . . (compare with Eq. 44-10). Calculate the energies of the lowest five distinct states for a conduction electron moving in a cubical crystal of edge length $L = 0.250\ \mu m$.

28P. The wave function of a particle confined to an infinite well and in the lowest energy state is $\psi = A \sin(\pi x/L)$. Use the "normalization condition" expressed by Eq. 44-8 to show that $A = \sqrt{2/L}$.

29P. A particle is confined between rigid walls separated by a distance L. The particle is in the lowest energy state; the wave function for this state is given in Problem 28. Use this wave function to calculate the probability that the particle will be found between the points (a) $x = 0$ and $x = L/3$, (b) $x = L/3$ and $x = 2L/3$, and (c) $x = 2L/3$ and $x = L$.

SECTION 44-6 THE HYDROGEN ATOM

30E. In Fig. 44-12, verify the plotted values of $P(r)$ at (a) $r = 0$, (b) $r = r_B$, and (c) $r = 2r_B$.

31E. In the ground state of the hydrogen atom, what is the probability that the electron will be found within a sphere whose radius is that of the first Bohr orbit? See Sample Problem 44-6.

32E. For the ground state of the hydrogen atom, evaluate the probability density $\psi^2(r)$ and the radial probability density $P(r)$ for the positions (a) $r = 0$ and (b) $r = r_B$. Explain what these quantities mean.

33E. Use the result of Sample Problem 44-6 to calculate the probability that the electron in a hydrogen atom, in the ground state, will be found between the spheres with radii $r = r_B$ and $r = 2r_B$.

34P. For an electron in the ground state of the hydrogen atom, (a) verify Eq. 44-19 and (b) calculate the radius of a sphere for which the probability that the electron will be found inside the sphere equals the probability that the electron will be found outside the sphere. (*Hint:* See Sample Problem 44-6.)

35P. For the ground state of the hydrogen atom show that the probability $p(r)$ that the electron lies within a sphere of radius r is given by

$$p(r) = 1 - e^{-2x}(1 + 2x + 2x^2),$$

in which $x = r/r_B$, a dimensionless ratio.

36P. In atoms there is a finite, though very small, probability that, at some instant, an orbital electron will actually be found inside the nucleus. In fact, some unstable nuclei use this occasional appearance of the electron to decay by *electron capture*. Assuming that the proton itself is a sphere of radius 1.1×10^{-15} m and that the wave function of the hydrogen atom's electron holds all the way to the proton's center, use the ground-state wave function to calculate the probability that the hydrogen atom's electron is inside its nucleus. (*Hint:* When $x \ll 1$, $e^{-x} \approx 1$.)

SECTION 44-7 BARRIER TUNNELING

37E. A proton and a deuteron (which has the same charge as a proton but twice the mass) are incident on an energy barrier of thickness 10 fm and height 10 MeV. Each particle has a kinetic energy of 3.0 MeV. Find the transmission probabilities for them.

38P. Consider an energy barrier such as that of Fig. 44-13a, but whose height U is 6.0 eV and whose thickness L is 0.70 nm. Calculate the energy of an incident electron such that its transmission probability is 0.001.

39P. Suppose that a beam of 5.0-eV protons is incident on an energy barrier of height 6.0 eV and thickness 0.70 nm, at a rate equivalent to a current of 1.0 kA. How long would you have to wait—on the average—for one proton to be transmitted?

40P. Consider the barrier-tunneling situation in Sample Problem 44-7. What fractional change in the transmission coefficient occurs for a 1.00% increase in (a) the barrier height, (b) the barrier thickness, and (c) the incident energy of the electron?

SECTION 44-8 HEISENBERG'S UNCERTAINTY PRINCIPLE

41E. A microscope using photons is employed to locate an electron in an atom to within a distance of 10 pm. What is the minimum uncertainty in a measurement of the momentum of the electron located in this way?

42E. The uncertainty in the position of an electron is given as 50 pm, which is about the radius of the first Bohr orbit in hydrogen. What is the uncertainty in a measurement of the momentum of the electron?

43E. Imagine playing baseball in a universe where Planck's constant was 0.60 J·s. What would be the uncertainty in the position of a 0.50-kg baseball that is moving at 20 m/s with an uncertainty of 1.0 m/s? Why would it be hard to catch such a ball?

44E. Consider an electron trapped in an infinite well whose width is 100 pm. If it is in a state with $n = 15$, what are (a) its energy, (b) the uncertainty in its momentum, and (c) the uncertainty in its position?

45E. The lifetime of an electron in the state with $n = 2$ in hydrogen is about 10^{-8} s. What is the uncertainty in the energy of the $n = 2$ state? Compare this with the energy of this state.

46P. Show that if the uncertainty in the location of a particle is equal to its de Broglie wavelength, then the uncertainty in its velocity is equal to its velocity.

47P. Suppose that we wish to test the possibility that electrons in atoms move in orbits by "viewing" them with photons with sufficiently short wavelengths, say 10.0 pm or less. (a) What would be the energy of such photons? (b) How much energy would such a photon transfer to a free electron in head-on Compton scattering? (c) What does this tell you about the possibility of confirming orbital motion by "viewing" an atomic electron at two or more points along its path?

ADDITIONAL PROBLEMS

48. The wave function for a particle confined to a one-dimensional box of length L is $\psi = A$, where A is a constant. Find A.

49. The "average" radial distance of the electron in a hydrogen atom can be defined as

$$r_{av} = \int_0^\infty r\, P(r)\, dr,$$

where $P(r)$ is the radial probability density of Eq. 44-18 for hydrogen in the ground state. Find r_{av} in terms of the Bohr radius r_B.

50. An electron of kinetic energy 60 eV passes from a zero-potential region into a region in which the electric potential is 100 V, speeding up as it crosses the (assumed) sharp boundary between the two regions. The electron's incident path is at 50° to a normal to the boundary. Assuming that Snell's law (Eq. 39-2) applies to the de Broglie wave of the electron, find (a) the ratio of the index of refraction of the 100-V region to that of the zero-potential region, and (b) the angle of refraction of the electron.

51. When the energy E of an incident particle exceeds the barrier height U of a square energy barrier of thickness L, the transmission coefficient is given by

$$\frac{1}{T} = 1 + \frac{1}{4}\frac{U^2}{E(E-U)}\sin^2(k_2 L),$$

where

$$\frac{\hbar^2 k_2^2}{2m} = E - V.$$

Classically, if $E > U$ the particle should continue in its direction of motion and hence have a transmission coefficient $T = 1$. (a) If $U > 0$, what is the minimum energy of a particle that has a transmission coefficient $T = 1$? (b) What is the next highest energy?

WINNING THE NOBEL PRIZE

Ivar Giaever
Rensselaer Polytechnic Institute

What follows is an excerpt from Dr. Giaever's Nobel prize acceptance speech, describing the research that led to his exciting discoveries in the understanding of electronic tunneling.

In my laboratory notebook is the entry, dated May 2, 1960: "Friday, April 22, I performed the following experiment aimed at measuring the forbidden gap in a superconductor." [See Section 28-9.] This was an extraordinary event, not only because I rarely write in my notebook, but also because of the success of that experiment. I shall try to recollect some of the events and thoughts that led to my notebook entry, though it is difficult to describe what now appears as fortuitous.

An Oslo paper headline read approximately as follows: "Master in billiards and bridge, almost flunked physics—gets Nobel Prize." The paper refers to my student days and I have to admit that the reporting is reasonably accurate; therefore I shall not attempt a "cover up," but confess that I almost flunked mathematics as well. In those days I wasn't very interested in mechanical engineering or school in general, but I managed to graduate in 1952. My wife and I emigrated to Canada, where I was employed by the Canadian General Electric Co. A three-year course in engineering and applied mathematics was offered to me. I realized that this time school was for real, and since it probably would be my last chance I really studied hard.

When I was 28 years old I moved to Schenectady, New York, where I discovered that it was possible to make a good living as a physicist. After working on various assignments in applied mathematics, I realized that the mathematics was much more advanced than the physical systems we applied it to. Thus, I decided to

learn about physics and, even though I was still an engineer, I was given the opportunity to try it at the General Electric Research Laboratory.

The assignment I had was to work with thin films. To me "films" meant photography. However, I was fortunate to be working with John Fisher, who obviously had other things in mind. Fisher started out as a mechanical engineer, but turned his attention to theoretical physics. He believed that useful electronic devices could be made using thin film technology; soon I was working with metal films separated by thin film insulating layers and trying to do tunneling experiments.

The concept that a particle can go through a barrier seemed odd to me, since I was struggling with quantum mechanics at Rensselaer Polytechnic Institute (RPI). For an engineer, it sounds rather strange that if you throw a tennis ball against a wall enough times it will eventually go through without damaging either the wall or itself. The trick is to use very tiny tennis balls, and lots of them. If we can place two metals very close together without making a short, the electrons in the metals can be considered as the balls; the wall is represented by the space between the metals (Fig. 1).

Neither Fisher nor I had much background in experimental physics—none to be exact. We made several false starts. To measure a tunneling current, the two metals must be spaced no more than about 10 nm apart. To avoid vibration problems, we decided not to use air or vacuum between the metals. After all, we both had training in mechanical engineering! We tried to keep the two metals apart by using a variety of thin insulator films. Invariably, these films had pinholes and the mercury counter electrode we used would short the films. Thus we spent time measuring very interesting but always non-

Ivar Giaever was born in Norway and educated as a mechanical engineer. He emigrated to Canada and then to the United States, where he worked for General Electric for 30 years. He is best known for his work on electron tunneling, for which he shared the 1973 Nobel prize in physics with Leo Esaki and Brian D. Josephson. While working for General Electric he obtained his Ph.D. in physics from Rensselaer Polytechnic Institute, where he is now an Institute Professor of Science. For the last 15 years, most of his research has been in biophysics.

reproducible current–voltage characteristics, referred to as miracles since each occurred only once. After a few months we hit the correct idea: to use evaporated metal films and to separate them by a naturally grown oxide layer.

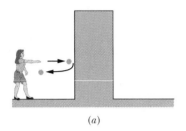

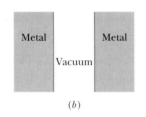

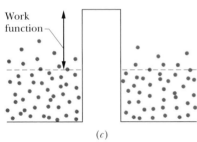

FIGURE 1 (a) If a person throws a ball against a wall, the ball bounces back. The laws of physics allow the ball to penetrate, or "tunnel," through the wall, but the chance of this happening is infinitesimally small because the ball is a macroscopic object. (b) Two metals separated by a vacuum will approximate the above situation. The electrons in the metals are the "balls"; the vacuum represents the wall. (c) A pictorial energy diagram of the two metals. The electrons do not have enough energy to escape into the vacuum. The two metals can, however, exchange electrons by tunneling. If the metals are spaced close together the probability for tunneling is large because the electron is a microscopic particle.

To carry out our ideas we needed an evaporator, so I purchased my first piece of experimental equipment. While waiting for the evaporator to arrive I worried a lot—I was afraid about getting stuck in experimental physics tied to this expensive machine. My plans were to switch into theory as soon as I had acquired enough knowledge. The premonition was correct: I did get stuck with the evaporator (Fig. 2), not because it was expensive, but because it fascinated me. To prepare a tunnel junction, we evaporated an aluminum strip onto a glass slide. This film was removed from the vacuum system and heated to oxidize the surface rapidly. Cross strips of aluminum were deposited over the first film, making several junctions at once. The steps in the sample preparation are illustrated in Fig. 3. This procedure solved two problems: there were no pinholes in the oxide, because it was self-healing; and we got rid of mechanical problems with the mercury oxide counter electrode.

By April 1959, we had performed several successful tunneling experiments. The current–voltage charac-

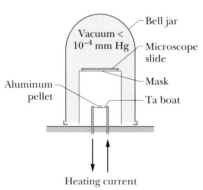

FIGURE 2 A vacuum system for depositing metal films. If aluminum is heated resistively in a tantalum boat, the aluminum first melts, then boils and evaporates. The aluminum vapor will solidify on any cold substrate placed in the vapor stream. The most common substrates are ordinary microscope glass slides. Patterns can be formed on the slides by suitably shielding them with a metal mask.

teristics of our samples were reasonably reproducible and conformed to theory (Fig. 4). Several checks were made, such as varying the area and the oxide thickness of the junction, as

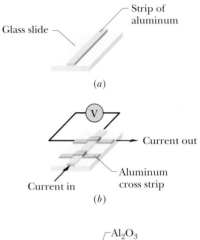

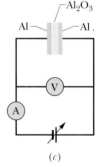

FIGURE 3 (a) A microscope glass slide with a vapor-deposited aluminum strip down the middle. As soon as the aluminum film is exposed to air, a protective insulating oxide forms on the surface. The thickness of the oxide depends on such factors as time, temperature, and humidity. (b) After a suitable oxide has formed, cross strips of aluminum are evaporated over the first film, sandwiching the oxide between the two metal films. Current is passed along one aluminum film up through the oxide and out through the other film, while the voltage drop is monitored across the oxide. (c) A schematic circuit diagram. We are measuring the current–voltage characteristics of the capacitor-like arrangement formed by the two aluminum films and the oxide. When the oxide thickness is less than 50 Å or so, an appreciable direct current will flow through the oxide.

well as changing the temperature. However, there were many physicists at the laboratory who questioned my experiment. How did I know I didn't have metallic shorts? Ionic current? Semiconduction rather than tunneling? Of course, I didn't know, and even though theory and experiments agreed, I had doubts about validity.

I continued to try out my ideas on John Fisher, who was now looking into the problems of fundamental particles with his characteristic optimism and enthusiasm. In addition, I received guidance from Charles Bean and Walter Harrison, both physicists with the uncanny ability of making things clear as long as a piece of chalk and a blackboard were available.

While taking courses at RPI, one day in a solid-state physics course taught by Professor Huntington, we got to superconductivity. What really caught my attention was the mention of the ''energy gap'' in superconductors, central to the Bardeen–Cooper–Schrieffer (BCS) theory. If the theory was valid and my tunneling experi-

ments were any good, it was obvious that by combining the two, some pretty interesting things should happen. Back at the laboratory, I tried this simple idea out on my friends; it didn't look as good to them. The energy gap was really a many-body effect and couldn't be interpreted literally as I had done. Even though there was skepticism, everyone urged me to go ahead. Then I realized that I did not know what the size of the gap was in units I understood—electron-volts. This was easily solved by my usual method: first asking Bean and then Harrison. When they agreed on a few millielectron-volts, I was happy, because it was an easily measured voltage range.

I had never done an experiment requiring low temperatures and liquid helium—that seemed like complicated business. One great advantage of being associated with General Electric's research laboratory is that the people there are knowledgeable in almost any field. Better still, they are willing to lend a hand. In my case, all I had to do was go to the end of the hall, where Warren DeSorbo was already experimenting with superconductors.

It took a day or two to set up the helium Dewars I borrowed (Fig. 5). People unfamiliar with low-temperature work believe that this field is pretty esoteric, but all it requires is access to liquid helium, which was readily available at the laboratory. I made my samples using aluminum–aluminum oxide and put lead strips on top. Both lead and aluminum are superconductors. Lead is superconducting at 7.2 K; thus all you need to make it superconducting is liquid helium, which boils at 4.2 K. Aluminum becomes superconducting only below 1.2 K, and to reach this temperature a more complicated experimental setup is required.

The first two experiments I tried were failures, because I used oxide layers which were too thick. I didn't get enough current through the thick oxide to measure it reliably with the instruments I had—a standard voltmeter and a standard ammeter. In

the third attempt, rather than deliberately oxidizing the first aluminum strip, I exposed it to air for a few minutes, and put it back into the evaporator to deposit the cross strips of lead. This way the oxide was no more than about 3 nm thick, and I could readily measure the current–voltage characteristic.

To me the greatest moment in an experiment is always just before I learn whether a particular idea is good or bad. Thus even failures are exciting, and most of my ideas have been wrong. But this time it worked! The current–voltage characteristic changed markedly when the lead changed from the normal to the superconducting state. That was exciting! I immediately repeated the ex-

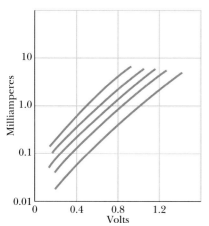

FIGURE 4 Current–voltage characteristics of five different tunnel junctions all with the same thickness, but with five different areas. The current is proportional to the area of the junction. This was one of the first clues that we were dealing with tunneling rather than with short circuits. In the early experiments we used a relatively thick oxide; thus very little current would flow at low voltages.

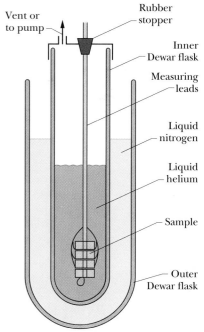

FIGURE 5 A standard experimental arrangement used for low-temperature experiments. It consists of two Dewars: the outer one contains liquid nitrogen, the inner one, liquid helium. Helium boils at 4.2 K at atmospheric pressure. The temperature can be lowered to about 1 K by reducing the pressure. The sample simply hangs in the liquid helium, supported by the measuring leads.

periment using a different sample—same results! Another sample—still the same results—everything looked good! But how to make certain?

It was well known that superconductivity is destroyed by a magnetic field, but my simple setup of Dewars made that experiment impossible. This time I had to see Israel Jacobs, who studied magnetism at low temperatures. Again I was lucky enough to go right into an experimental rig where both the temperature and the magnetic field could be controlled and I could quickly do all the proper experiments. Everything held together and the whole group was very excited. In particular, I remember Bean enthusiastically spreading the news through the halls, and also patiently explaining to me the significance of the experiment.

I wasn't the first person to measure the energy gap in a superconductor, and I soon became aware of the work done by M. Tinkham and his students using infrared transmission. I was worried that the size of the gap that I measured didn't quite agree with those previous measurements. Bean set me straight by saying that from then on people would have

to agree with me; my experiment would set the standard. I felt pleased and like a physicist for the first time! That was a very exciting time in my life; we had great ideas for improvements and we wanted to extend the experiment to other materials: normal metals, magnetic materials, and semiconductors. There were many informal discussions over coffee about what to try next, and one of these sessions is shown in Fig. 6. To be honest, the picture was staged; we weren't normally so dressed up, and rarely did I find myself in charge at the blackboard!

The superconducting experiment was charmed and always worked. It looked like the tunneling probability was directly proportional to the density of states in a superconductor. Now if this were true, it did not take much imagination to realize that tunneling between two superconductors should display a negative resistance characteristic—an increase in voltage results in a decrease in current. A negative resistance characteristic meant I had to get more complex equipment since nobody around me had facilities to pump on the helium sufficiently to make aluminum be-

come superconducting. This time I had to reactivate a low-temperature setup in an adjacent building. As soon as the aluminum went superconducting, a negative resistance appeared, and the notion that the tunneling probability was directly proportional to the density of states was experimentally correct. Now things looked very good, because all sorts of electronic devices operative at low temperatures could be made using this effect.

I hope that this personal account may provide some insight into the nature of scientific discovery. My own beliefs are that the road to a scientific discovery is seldom direct, and that it does not necessarily require a great expertise. In fact, I am convinced that often newcomers to a field have a great advantage because they are ignorant of the complicated reasons why a particular experiment shouldn't be attempted. It is essential to get advice and help from experts when you need it. For me the most important ingredients were being at the right place at the right time and finding so many friends who unselfishly supported me.

FIGURE 6 Informal discussion over a cup of coffee. From left: Ivar Giaever, Walter Harrison, Charles Bean, and John Fisher. (See Essay 10.)

ALL ABOUT ATOMS

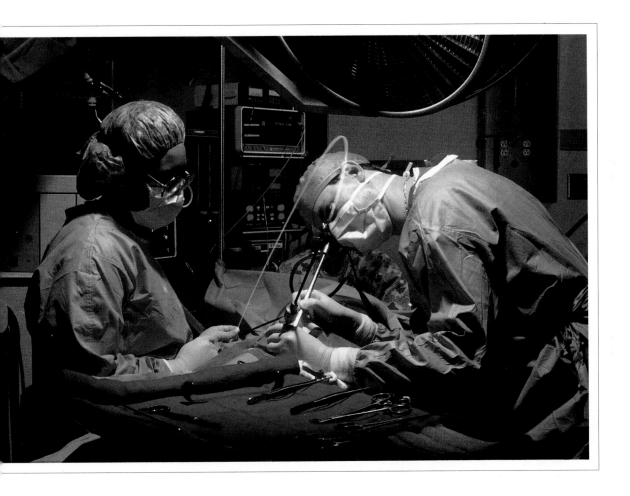

Soon after lasers were invented in the 1960s, they became novel sources of light in uncountable research laboratories. But today, lasers are ubiquitous, being found in such diverse applications as voice and data transmission, surveying, welding, and grocery-store price scanning. The photograph shows surgery being performed with laser light transmitted via optical fibers. Light from a laser and light from any other source are both due to emissions by atoms. What, then, is so different about the light from a laser?

45-1 ATOMS AND THE WORLD AROUND US

What would you think if your physics or chemistry instructor told you that she or he did not believe in atoms? In the early years of this century, quite a few prominent scientists held just that view. Today, however, no well-informed person doubts that the material world around us is made up of atoms.

Why do we believe in these tiny objects that—it is often alleged—we cannot see? For one thing, with modern techniques we now *can* see individual atoms. Figure 45-1, taken with a high-resolution electron microscope, and Fig. 2-13, taken with a scanning transmission microscope, leave little doubt. Even more to the point than these convincing pictures is the steady piling up of mountains of experimental information about atoms, all of it totally understandable in terms of modern quantum theory.

45-2 SOME PROPERTIES OF ATOMS

Here we describe some of the properties of atoms that any theory of atomic structure must be able to explain.

FIGURE 45-1 A photo of a very thin crystalline sample containing copper, chlorine, and nitrogen atoms, taken with a high-resolution electron microscope. The copper atoms show up well at the centers of the "rosettes" formed by 16 chlorine atoms. Nitrogen atoms occupy intermediate positions.

Atoms Are Put Together Systematically

The existence of the *periodic table of the elements* (see Appendix E), with its remarkable repetitive sequences of chemical and physical properties, is evidence enough that the atoms of the elements are constructed systematically. Figure 45-2 shows one simple example of a systematic (repetitive) property. It is a plot of the *ionization energy* of the elements (that is, the work required to remove a single electron from a neutral atom) as a function of the position of the element in the periodic table.

The periodic table contains six complete* horizontal periods of elements, each period starting with a highly reactive alkali metal (lithium, sodium, potassium, and so on) and ending with a chemically inert noble gas (neon, argon, krypton, and so on). The numbers of elements in these periods are:

$$2, 8, 8, 18, 18, \text{ and } 32.$$

As you will see, quantum physics predicts these numbers and leads us to a general understanding of the periodic table and thus of much of physics and nearly all of chemistry. Because the life processes that sustain us as thinking beings are (bio)chemical, the influence of quantum physics in our lives runs deeply.

Atoms Emit and Absorb Light

Another central feature of atoms is that they emit and absorb light at sharply defined frequencies. As discussed in Section 43-6, atoms can exist only in certain discrete quantum states, each state with its characteristic energy. An atom emits light when it transfers from one of these states to another state, of lower energy. The frequency f of the emitted light is given by the *Bohr frequency condition*, Eq. 43-20,

$$hf = E_i - E_f. \qquad (45\text{-}1)$$

Here E_i and E_f are the energies of the higher (initial) and lower (final) energy states, respectively, and h is the Planck constant.

Thus the problem of finding the frequencies of the light emitted (or absorbed) by an atom reduces to the problem of finding the energy levels for that atom. The laws of quantum physics allow us—in principle at least—to calculate these energies.

*The last horizontal period, starting with element 87 (francium), is incomplete.

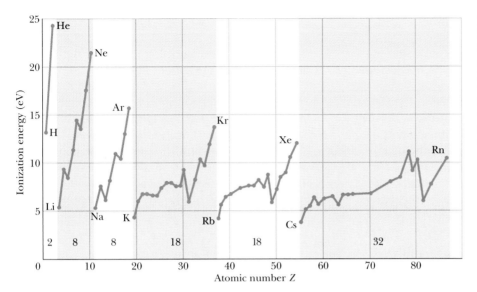

FIGURE 45-2 A plot of the ionization energies of the elements as a function of atomic number, showing the periodic repetition of properties through the six complete horizontal periods of the periodic table. The numbers of elements in these periods are indicated.

Atoms Have Angular Momentum and Magnetism

Electrons in atoms behave classically like tiny current loops and have both an *orbital angular momentum* and an *orbital magnetic moment* associated with this motion. In Section 34-2 we pointed out that an electron also has an *intrinsic* angular momentum, called its *spin angular momentum*. The electron behaves classically like a spinning negative charge, thus giving rise to an intrinsic *spin magnetic moment*. Because the electron is negatively charged, the orbital and spin magnetic moments are directed opposite their corresponding angular momenta, as discussed in Sections 34-2 and 34-3.

The spin and orbital angular momenta of the individual electrons in an atom combine to produce a net angular momentum for the atom as a whole. Associated with this net angular momentum is a net magnetic moment. For some atoms (neon, for example), the effects of the various electrons cancel each other so that the net angular momentum and the associated net magnetic moment are zero. For many other atoms, however, the cancellation is not complete, and these atoms exhibit a net angular momentum and a net magnetic moment.

The magnetism of atoms—at least in the special case of ferromagnetism—is familiar to all. The *angular momentum* of atoms, however, is not so familiar. It occurred to Einstein that, if the atomic magnets in an iron bar were aligned, their associated angular momenta should also be aligned (in the opposite sense) and should exhibit large-scale external ef-

fects. In 1915, Einstein and W. J. de Haas carried out an ingenious experiment based on this idea.

In an ordinary iron bar such as that of Fig. 45-3a, the atoms (and thus their atomic magnets) are randomly oriented, so their magnetic effects cancel

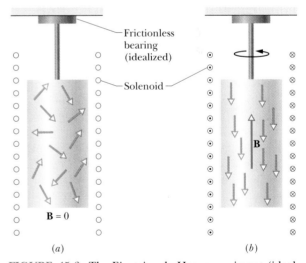

FIGURE 45-3 The Einstein–de Haas experiment (idealized). (*a*) Initially, the magnetic field is zero and the atomic angular momentum vectors in the iron bar are randomly oriented, as the figure shows. The direction of the atomic magnetic moment vectors (not shown) is opposite the direction of the atomic angular momentum vectors. (*b*) When an axial magnetic field is applied, the alignment of the magnetic moment vectors causes the atomic angular momentum vectors to line up as shown. Because the bar is isolated from external torques, angular momentum is conserved and the bar as a whole must rotate as shown.

at all external points. Suppose, however, that these atomic magnets are suddenly aligned by switching on a current in the solenoid shown in that figure. That alignment causes a corresponding alignment of the angular momenta of the individual atoms. Because angular momentum must be conserved, the bar as a whole must rotate so that the angular momentum of that rotation cancels the atomic angular momentum. With this clever experiment, Einstein and de Haas demonstrated quantitatively the intimate connection between atomic angular momentum and atomic magnetism.

45-3 SCHRÖDINGER'S EQUATION AND THE HYDROGEN ATOM

How do we use quantum theory to calculate numerical values for the atomic properties outlined in the previous section? In particular, how do we calculate the energies, the angular momenta, and the magnetic moments of the quantum states of an atom?

Let us start with hydrogen. Physicists love this atom because it is so simple, consisting of a single electron bonded by the electrostatic force to a single central proton. From the early days of Bohr theory to modern quantum electrodynamics, this atom has served as a laboratory for testing, in exquisite detail, the depth of our understanding of the nature of matter.

As you will see, four quantum numbers are required to describe completely a quantum state of the hydrogen atom. *Four quantum numbers are also required to identify a quantum state of a single electron in a multi-electron atom.* Thus you can carry over much of what you learn about the hydrogen atom to atoms with more than one electron.

How to proceed? For a problem in classical mechanics, we use Newton's laws. For a problem in electromagnetism, we use Maxwell's equations. For a problem in wave mechanics, we use *Schrödinger's wave equation,* a relation first advanced by Austrian physicist Erwin Schrödinger in 1926.

Instead of writing down and analyzing the Schrödinger equation, we will simply discuss how it is used. Imagine, as in Fig. 45-4, a computer programmed to solve this equation. For its INPUT we insert the potential energy function that defines the problem at hand. For the problem of defining the quantum states of the hydrogen atom, that function is the familiar Coulomb potential energy, given by Eq. 26-35:

$$U = -\frac{1}{4\pi\epsilon_0}\frac{e^2}{r}. \tag{45-2}$$

Here e is the magnitude of the charge of the electron and of the proton and r is the distance between these two particles.

When we run the program (that is, when we solve the equation), the computer generates a PRINTOUT of the wave functions that define the quantized hydrogen atom states. Printed alongside each wave function are the corresponding energy, angular momentum, and magnetic moment for the atom when it is in that state. Let us discuss these quantities in more detail.

45-4 THE ENERGIES OF THE HYDROGEN-ATOM STATES

The Schrödinger equation has an infinite number of solutions, but most of them do not make any sense physically. In solving the equation, we program our computer to deliberately discard all solutions *except* those for which the wave function approaches zero as r in Eq. 45-2 approaches infinity. This is equivalent to recognizing that, beyond a certain distance,

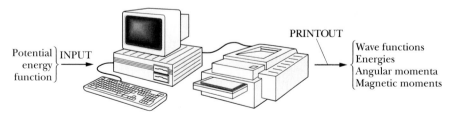

FIGURE 45-4 A computer programmed to solve Schrödinger's equation. The INPUT is the potential energy function that defines the problem, along with suitable boundary conditions suggested by the physics of the situation. The PRINTOUT is the wave functions, the energies, the angular momenta, and the magnetic moments of the quantized states.

as you move away from the central proton you are less and less likely to find the electron. The existence of quantized states having well-defined energies is a direct consequence of imposing this sensible requirement.

This is another example of the idea that *localization leads to quantization*, which was first stated as a theorem in Section 44-3 but which you saw at work as early in this text as Section 17-13. There we noted that a wave of *any* frequency can be propagated along a stretched string of *infinite* length, but only *discrete* frequencies of standing waves can be set up in a string of *finite* length. That is, localizing the wave quantizes the frequency. In the case of the hydrogen atom, localizing the wave function quantizes the energy.

The energies of the hydrogen-atom states are given by Eq. 44-13,

$$E_n = -\frac{me^4}{8\epsilon_0^2 h^2}\frac{1}{n^2}$$

$$= -\frac{13.6 \text{ eV}}{n^2}, \qquad n = 1, 2, 3, \dots, \quad (45\text{-}3)$$

in which the integer n is called the **principal quantum number;** it is the first of the four quantum numbers that we need to identify fully the allowed quantum states of the hydrogen atom.

45-5 ORBITAL ANGULAR MOMENTUM AND MAGNETIC MOMENT

Each state of the hydrogen atom has an associated orbital angular momentum **L**. We discuss first the magnitude and then the direction of **L**.

The Magnitude of L

In solving the Schrödinger equation we learn that the magnitude of the orbital angular momentum of the hydrogen-atom states is quantized. Its allowed values are

$$L = \sqrt{\ell(\ell + 1)}\ \hbar, \qquad (45\text{-}4)$$

in which $\hbar$ (pronounced *h-bar*) is an abbreviation for $h/2\pi$ and ℓ is the **orbital quantum number;** it is the second of the four quantum numbers that we seek. The allowed values of ℓ depend on the value of the principal quantum number n and are

$$\ell = 0, 1, 2, \dots, n - 1. \qquad (45\text{-}5)$$

That is, ℓ may take on only nonnegative integer values less than n. Thus for $n = 1$, only $\ell = 0$ is permitted. And for $n = 2$, only $\ell = 0$ and $\ell = 1$ are permitted.

The Direction of L

States with the same values of n and ℓ may have different wave functions because their angular momentum vectors **L** have different directions. For an isolated hydrogen atom, there is no obvious direction in space with respect to which the orientation of its angular momentum vector can be measured. To supply one, it is convenient to imagine that the atom is immersed in a weak but uniform magnetic field whose direction we may take as a z axis.

According to the rules of wave mechanics, the angular momentum vector **L** cannot make *any* angle with the z axis, but can make only those angles that yield a component along this axis given by

$$L_z = m_\ell \hbar. \qquad (45\text{-}6)$$

Here m_ℓ, the **magnetic quantum number,** is restricted to the values

$$m_\ell = 0, \pm 1, \pm 2, \dots, \pm \ell. \qquad (45\text{-}7)$$

The magnetic quantum number is the third of the four quantum numbers that we are seeking.

Figure 45-5 shows the allowed values of L_z and the associated directions of **L** for $\ell = 1$, 2, and 10. Note that, for a given value of ℓ, there are $2\ell + 1$ different values of m_ℓ. For $\ell = 10$, we begin to merge with the classical limit, in which the correspondence principle requires that *any* orientation of the angular momentum vector be allowed. The restriction imposed by quantum theory on the direction of the angular momentum vector is called *space quantization;* an early experimental demonstration of the existence of space quantization is described in Section 45-8.

A Useful Vector Model

Figure 45-6 suggests a classical vector model that helps us to visualize the space quantization of **L**. It shows the angular momentum vector precessing about the z direction, like a top precessing about a vertical axis in the Earth's gravitational field. Its pro-

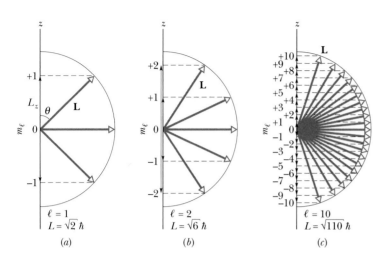

FIGURE 45-5 The allowed values of L_z for $\ell = 1$, 2, and 10. The numbers on the z axis are the values of the magnetic quantum number m_ℓ. The figures are drawn to different scales.

jection L_z on the z axis remains constant as the motion proceeds.

Heisenberg's uncertainty principle in its angular form (compare Eq. 44-22) is

$$\Delta L_z \cdot \Delta \phi \approx h \qquad (z \text{ component}), \qquad (45\text{-}8)$$

in which ϕ is the angle of rotation about the z axis in Fig. 45-6. Once we have specified the magnetic quantum number, L_z is *precisely* known; that is, $\Delta L_z = 0$. Equation 45-8 then requires that $\Delta \phi$ be infinitely great, which means that we have no information at all about the angular position about the z axis of the precessing angular momentum vector **L**. We know the magnitude of **L** and its projection L_z on the z axis, and *nothing else*.

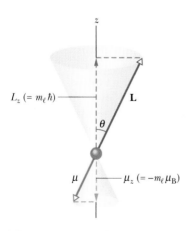

FIGURE 45-6 A vector model designed to represent the space quantization of the angular momentum and magnetic moment vectors. Note that the magnitudes L and μ and their projections L_z and μ_z remain constant as the vectors precess about the z axis.

Orbital Magnetic Moment

As Fig. 45-6 suggests, the orbital magnetic moment of an electron is related to its orbital angular momentum. Because the orbital angular momentum vector is restricted to a discrete set of components along the z axis, the orbital magnetic moment vector is similarly restricted. The allowed projections on the z axis of the orbital magnetic moment vector (compare Eq. 45-6) are

$$\mu_{\ell,z} = -m_\ell \mu_B, \qquad (45\text{-}9)$$

in which μ_B is the Bohr magneton, defined in Eq. 34-1 as

$$\mu_B = eh/4\pi m$$
$$= 9.274 \times 10^{-24} \, \text{J/T}$$
$$= 5.788 \times 10^{-5} \, \text{eV/T}. \qquad (45\text{-}10)$$

It is a convenient measure of atomic magnetism, just as $\hbar$ is a convenient measure of atomic angular momentum and r_B (the *Bohr radius*) is a convenient measure of atomic distance. The minus sign in Eq. 45-9 shows that (as we expect for an orbiting negative charge) the angular momentum vector and the magnetic moment vector are oppositely directed.

SAMPLE PROBLEM 45-1

What is the minimum angle θ in Fig. 45-6 between the angular momentum vector and the z axis? Calculate the result for $\ell = 1$, 10^2, 10^3, 10^4, and 10^9.

SOLUTION From Fig. 45-5 we see that the minimum angle occurs when $m_\ell = \ell$ in Eq. 45-6 and is

$$\theta_{min} = \cos^{-1} \frac{L_{z,max}}{L} = \cos^{-1} \frac{\ell\hbar}{\sqrt{\ell(\ell+1)}\hbar}$$

$$= \cos^{-1} \left(1 + \frac{1}{\ell}\right)^{-1/2}.$$

Substituting for ℓ in this equation leads to:

ℓ	θ_{min}
1	45°
10^2	5.7°
10^3	1.8°
10^4	0.57°
10^9	0.0018°

(Answer)

For a macroscopic object like a spinning top, ℓ would be enormously larger than 10^9 and θ_{min} would be so close to zero that there would be no hope of measuring it. The correspondence principle really works!

SAMPLE PROBLEM 45-2

a. For $n = 4$, what is the largest allowed value of ℓ?

SOLUTION From Eq. 45-5 it is

$$\ell_{max} = n - 1 \quad \text{or} \quad \ell_{max} = 3. \quad \text{(Answer)}$$

b. What is the magnitude of the corresponding angular momentum for $\ell = 3$?

SOLUTION From Eq. 45-4 it is

$$L = \sqrt{\ell(\ell+1)}\hbar = \sqrt{(3)(3+1)}\hbar = 2\sqrt{3}\hbar. \quad \text{(Answer)}$$

c. How many different projections on the z axis may this angular momentum vector have?

SOLUTION From Eq. 45-7, we see that the number is

$$(2\ell + 1) = (2 \times 3 + 1) = 7. \quad \text{(Answer)}$$

d. What is the magnitude of the largest projected component?

SOLUTION This follows from Eq. 45-6, with m_ℓ given its largest value, which is ℓ. Thus

$$L_{z,max} = \ell\hbar = 3\hbar. \quad \text{(Answer)}$$

e. What is the smallest angle θ that the angular momentum vector can make with the z axis?

SOLUTION From the equation for θ_{min} from Sample Problem 45-1 and (b) and (d) above we have

$$\theta_{min} = \cos^{-1} \frac{L_{z,max}}{L} = \cos^{-1} \frac{3\hbar}{2\sqrt{3}\hbar}$$

$$= \cos^{-1} \sqrt{3}/2 = 30°. \quad \text{(Answer)}$$

45-6 SPIN ANGULAR MOMENTUM AND MAGNETIC MOMENT

Whether or not it is trapped in an atom, an electron has an intrinsic angular momentum of its own. This **spin angular momentum,** as it is called, is also space quantized and can have components in the z direction given by

$$S_z = m_s \hbar, \quad (45\text{-}11)$$

in which the *spin quantum number* m_s can have only the values $+\frac{1}{2}$ and $-\frac{1}{2}$. (The electron is said to have

TABLE 45-1
THE HYDROGEN-ATOM QUANTUM NUMBERS

NAME	SYMBOL	ALLOWED VALUES	ASSOCIATED WITH	NUMBER OF VALUES
Principal	n	1, 2, 3, . . .	Energy	∞
Orbital	ℓ	0, 1, 2, . . . , $(n-1)$	Orbital angular momentum	n
Magnetic	m_ℓ	0, ±1, ±2, . . . , ±ℓ	Orbital angular momentum	$2\ell + 1$
Spin	m_s	±$\frac{1}{2}$	Spin angular momentum	2

a *spin* of $\frac{1}{2}$ in units of $\hbar$.) We have used the symbol "S" for angular momentum associated with spin, to distinguish it from "L," the angular momentum associated with orbital motion. The spin quantum number is the fourth and last of the four quantum numbers needed to describe the states of the hydrogen atom. Table 45-1 summarizes them.

A host of experimental data require us to assume that the corresponding spin magnetic moment component $\mu_{s,z}$ can have only the values given by

$$\mu_{s,z} = -2m_s\mu_B, \qquad (45\text{-}12)$$

in which μ_B is the Bohr magneton. The factor 2 in Eq. 45-12 tells us that:

Spin angular momentum is twice as effective as orbital angular momentum in generating magnetism.

(Compare Eq. 45-12 with Eq. 45-9.) This experimental result is fully supported by relativistic quantum theory.

45-7 THE HYDROGEN-ATOM WAVE FUNCTIONS

To complete our discussion of the hydrogen atom, let us examine the wave functions for a few of its states. We start with the ground state, for which the quantum numbers* are $n = 1$, $\ell = 0$, and $m_\ell = 0$. The wave function and probability density for this state, as we saw in Section 44-6, depend only on r. It is reasonable that such a spherically symmetrical "billiard-ball" state should have zero angular momentum, because in this state all directions through the center of the atom are completely equivalent.

The *radial probability density* for the ground state is, from Eq. 44-18,

$$P(r) = \left(\frac{4r^2}{r_B^3}\right) e^{-2r/r_B}. \qquad (45\text{-}13)$$

Figure 45-7a is a plot of this function. Recall that the radial probability density is defined so that $P(r)\, dr$ gives the probability that the electron will be found between shells whose radii are r and $r + dr$. Figure 45-7a shows that $P(r)$ has its maximum value for $r = r_B = 52.9$ pm.

Consider next the state with $n = 2$, $\ell = 0$, and $m_\ell = 0$. Like all states with $\ell = 0$, this state is a

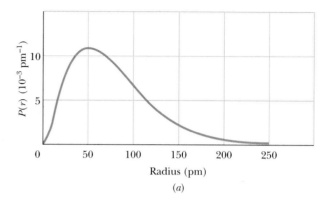

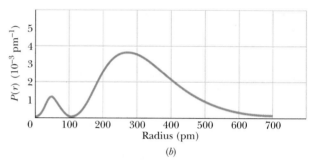

FIGURE 45-7 (*a*) The radial probability density for the ground state of the hydrogen atom, for which $n = 1$, $\ell = 0$, and $m_\ell = 0$. (*b*) The radial probability density for the state of the hydrogen atom with $n = 2$, $\ell = 0$, and $m_\ell = 0$.

spherically symmetric, or "billiard-ball," state, its radial probability density being given by

$$P(r) = \left(\frac{r^2}{8r_B^3}\right) \left(2 - \frac{r}{r_B}\right)^2 e^{-r/r_B}. \qquad (45\text{-}14)$$

Figure 45-7b shows a plot of this function. Inspection of Eq. 45-14 shows that $P(r) = 0$ for $r = 2r_B$.

For $n = 2$, states with $\ell = 1$ are also permitted. There are three such states, defined by the following sets of quantum numbers:

n	ℓ	m_ℓ
2	1	$+1$
2	1	0
2	1	-1

---

*In this section, we omit consideration of electron spin. Its quantum number, m_s, has no effect on the wave functions and simply doubles the number of states defined by the quantum numbers n, ℓ, and m_ℓ.

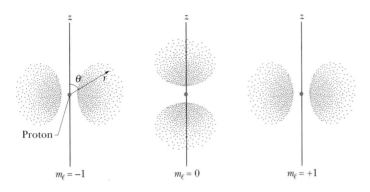

FIGURE 45-8 "Dot plots" for the three states of the hydrogen atom with $n = 2$ and $\ell = 1$. The values of m_ℓ correspond to the three allowed orientations in space of the angular momentum vector corresponding to $\ell = 1$. The patterns are symmetrical about the z axis. The density of dots at any point is proportional to the probability density at that point. (Although the probability densities in the three patterns are functions of both r and θ, the sum of all three patterns is spherically symmetrical, being a function of r alone.)

The three values of m_ℓ represent the three allowed orientations of the orbital angular momentum vector corresponding to $\ell = 1$; these orientations are shown in Fig. 45-5a.

The probability densities for these three states are *not* spherically symmetric. That is, as Fig. 45-8 shows, the probability density at any point depends not only on the length r of the radius vector to that point but also on the angle θ between the radius vector and the z axis. The density of the dots at any point in the "dot plots" of Fig. 45-8 is proportional to the probability density at that point; all three plots have rotational symmetry about the z axis.

45-8 THE STERN–GERLACH EXPERIMENT

In 1922, several years before the development of wave mechanics, space quantization was verified experimentally by Otto Stern and Walter Gerlach. Figure 45-9 shows their apparatus.

Silver is vaporized in an electrically heated "oven," and silver atoms spray into the external vacuum of the apparatus from a small hole in the oven wall. The atoms (which are electrically neutral but which have a magnetic moment) are formed into a narrow beam as they pass through a collimating slit. The beam then passes between the poles of an electromagnet, finally depositing itself on a glass detector plate.

A Dipole in a Nonuniform Field

The pole faces of the magnet in Fig. 45-9 are shaped to make the magnetic field as *nonuniform* as possible. We digress to ask what force acts on a magnetic dipole placed in such a field. Figure 45-10a shows a dipole of magnetic moment $\boldsymbol{\mu}$, making an angle θ with a *uniform* magnetic field. We can imagine the dipole to have north and south poles, the magnetic dipole moment vector $\boldsymbol{\mu}$ pointing (by convention) from the south to the north pole. We see that, for a uniform field, there is no net force on the dipole. The upward and downward forces on the poles are of the same magnitude and they cancel, no matter what the orientation of the dipole.

Figures 45-10b and 45-10c show the situation in a nonuniform field. Here the upward and downward forces do *not* have the same magnitude because the two poles are immersed in fields of different strengths. In this case there *is* a net force, both its magnitude and direction depending on the orientation of the dipole, that is, on the value of θ. In Fig. 45-10b this net force is up, and in Fig. 45-10c it is down. This tells us that the silver atoms in Fig. 45-9 will be deflected as they pass through the magnet, the direction and magnitude of the deflection depending on the orientation of their magnetic moment.

Now let us calculate the deflecting force. The magnetic potential energy of a dipole in a magnetic field **B** is given by Eq. 30-33 as

$$U = -\boldsymbol{\mu} \cdot \mathbf{B} = -\mu B \cos \theta, \qquad (45\text{-}15)$$

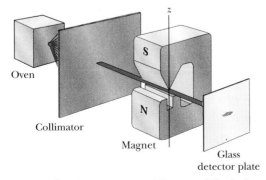

FIGURE 45-9 The apparatus of Stern and Gerlach, used to demonstrate space quantization.

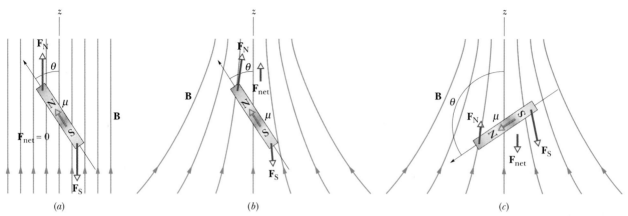

FIGURE 45-10   A magnetic dipole, represented as a small bar magnet with two poles, in (*a*) a uniform magnetic field and (*b,c*) a nonuniform field. The net force acting on the magnet is zero in (*a*), points up in (*b*), and points down in (*c*).

in which θ (see Fig. 45-10) is the angle between the directions of μ and **B**. From Eq. 8-17, the net force F_z on the atom is $-(dU/dz)$, so, from Eq. 45-15,

$$F_z = -(dU/dz) = \mu(dB/dz)\cos\theta. \quad (45\text{-}16)$$

In Fig. 45-10*b,c*, B increases as z increases, so dB/dz is positive. Thus the sign of the deflecting force F_z in Eq. 45-16 is determined by the angle θ. If $\theta < 90°$ (as in Fig. 45-10*b*), the atom will be deflected up; if $\theta > 90°$ (as in Fig. 45-10*c*) the deflection will be down.

The Experimental Results

When the electromagnet in Fig. 45-9 is turned off there are no deflections of the atoms and the beam forms a narrow line on the detecting plate. When the electromagnet is turned on, however, strong deflecting forces come into play. Then there are two possibilities, depending on whether space quantization exists or not. (Don't forget that the object of this experiment is to find out!) If there is no space quantization, the atomic magnetic dipoles will make a continuous distribution of angles θ with the direction of the magnetic field, and the beam will simply broaden.

On the other hand, if space quantization *does* exist, there will be only a discrete set of values for θ. This means that there will be only a discrete set of values for the deflecting force F_z in Eq. 45-16 and the beam will split into discrete components.

Figure 45-11 shows what happens. The beam does *not* broaden but splits cleanly into two sub-

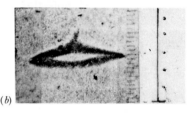

FIGURE 45-11   The results of the Stern–Gerlach experiment, showing the silver deposit on the glass detector plate of Fig. 45-9, with the magnetic field (*a*) turned off and (*b*) turned on. The beam has been split into two subbeams by the action of the field. The vertical bar at the right in (*b*) represents 1 mm.

beams.* Space quantization exists! Stern and Gerlach ended the published report of their work with the words: "We view these results as direct experimental verification of space quantization in a magnetic field." Physicists everywhere agree.

---

*The spin and orbital angular momenta of the electrons of a silver atom cancel out *except* for the spin angular momentum of its single valence electron. This spin can have only two orientations, described by $m_s = +\frac{1}{2}$ and $m_s = -\frac{1}{2}$: hence, two subbeams, and not some other number.

SAMPLE PROBLEM 45-3

In the magnet in a Stern–Gerlach experiment, the magnetic field gradient dB/dz through which the beam passed was 1.4 T/mm, and the length w of the beam path through the magnet was 3.5 cm. The temperature of the oven in which the silver was evaporated was adjusted so that the most probable speed v for the atoms in the beam was 750 m/s. Find the separation d between the two deflected subbeams as they emerge from the magnet. The mass M of a silver atom is 1.8×10^{-25} kg and the projection of its magnetic moment on the z axis is 1.0 Bohr magneton, or 9.27×10^{-24} J/T.

SOLUTION The acceleration of a silver atom as it passes through the electromagnet is (from Eq. 45-16)

$$a = \frac{F_z}{M} = \frac{(\mu \cos \theta)(dB/dz)}{M}.$$

The vertical deflection of either subbeam as it clears the magnet is

$$\tfrac{1}{2}d = \tfrac{1}{2}at^2 = \tfrac{1}{2}\frac{(\mu \cos \theta)(dB/dz)}{M}\left(\frac{w}{v}\right)^2,$$

so

$$d = \frac{(\mu \cos \theta)(dB/dz)w^2}{Mv^2}$$

$$= (9.27 \times 10^{-24}\,\text{J/T})(1.4 \times 10^3\,\text{T/m})$$

$$\times \frac{(3.5 \times 10^{-2}\,\text{m})^2}{(1.8 \times 10^{-25}\,\text{kg})(750\,\text{m/s})^2}$$

$$= 1.6 \times 10^{-4}\,\text{m} = 0.16\,\text{mm.} \qquad \text{(Answer)}$$

This is the order of magnitude of the separation displayed in Fig. 45-11; note the scale in that figure.

45-9 SCIENCE, TECHNOLOGY, AND SPIN: AN ASIDE

Most discoveries in pure science turn out to have applications in technology. On November 8, 1895, for example, William Röntgen, working in his physics laboratory at the University of Wurzburg in Germany, discovered x rays. Less than three months later a skater, having fallen on the ice of the Connecticut River, had his broken arm x-rayed at Dartmouth College in the first medical application of x rays in this country.

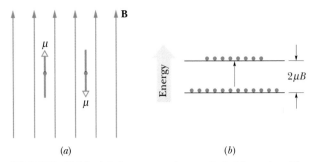

FIGURE 45-12 (*a*) A proton, whose spin is $\tfrac{1}{2}$ in units of $\hbar$, can occupy either of two quantized positions in an external magnetic field. If Eq. 45-17 is satisfied, the protons in the sample can flip from one orientation to the other. (*b*) Normally there are more protons in the lower energy state than in the higher energy state.

The spin of the electron was postulated in 1925 by two Dutch graduate students (George Uhlenbeck and Samuel Goudsmit), two years away from their doctorates at the University of Leiden. It took longer for spin than for x rays to make the journey from pure science to technology but it has now done so, in nuclear magnetic resonance (NMR) studies. Consider a proton, perhaps residing in a drop of water. The proton, like the electron, also has a spin of $\tfrac{1}{2}$ in units of $\hbar$. Its magnetic dipole moment $\boldsymbol{\mu}$ can occupy either of two quantized orientations with respect to a magnetic field **B**, as Fig. 45-12*a* shows. These two positions differ in energy by $2\mu B$, which is the work required to turn a magnetic dipole end for end in a magnetic field.

If a drop of water containing our proton is subjected to an electromagnetic field that is alternating with frequency f, transitions between the two orientations of the magnetic dipole moment—called *spin flips*—may be induced. For this to happen the equation

$$hf = 2\mu B \qquad (45\text{-}17)$$

must be satisfied. That is, the energy hf of the photons associated with the alternating electromagnetic field must be just equal to the energy difference between the two spin orientations.

The spin flip may go either way (that is, from up to down or from down to up in Fig. 45-12*a*) with equal probability. However, if the water drop is in thermal equilibrium, there will be more proton spins in the lower energy state than in the higher energy state, as Fig. 45-12*b* suggests. This means that there will be a net *absorption* of energy from the electromagnetic field.

The usefulness of magnetic resonance studies, which are carried out under conditions of high resolution, lies in the fact that the factor B in Eq. 45-17 is *not* the imposed external magnetic field B_{ext} but rather it is that field as uniquely modified by the small local internal magnetic field B_{local} due to the electrons and nuclei in the molecule of which our proton is a part. Thus we can rewrite Eq. 45-17 as

$$hf = 2\mu(B_{local} + B_{ext}). \qquad (45\text{-}18)$$

In NMR studies, it is customary to leave the frequency f of the electromagnetic oscillations fixed and to vary B_{ext} until Eq. 45-18 is satisfied and an absorption peak is recorded.

Figure 45-13 shows a **nuclear magnetic resonance spectrum,** as it is called, for ethanol, whose formula we may write as CH_3—CH_2—OH. The various resonance peaks all represent spin flips of protons. They occur at different values of B_{ext}, however, because the local environments of the protons within the ethanol molecule are different. The spec-

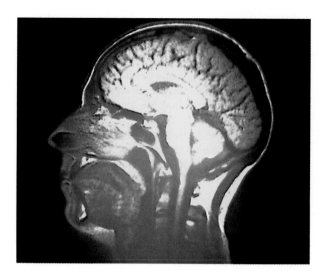

FIGURE 45-14   A cross section of a human head taken with magnetic resonance imaging reveals anatomical detail that would not show in an x-ray image, not even with a modern computed tomography scanner (CAT scan).

trum of Fig. 45-13 is a unique signature for ethanol. The nuclear magnetic resonance method is of great value as an analytical tool in organic chemistry.

Spin technology has also been applied to medical diagnostic imaging. The protons in the various tissues of the human body find themselves in different local magnetic environments. When the body, or part of it, is immersed in a strong external magnetic field, these environmental differences can be detected by spin-flip techniques and translated by computer processing into a x-ray-like image. Figure 45-14, for example, shows a cross section of a human head taken by this method, which is called magnetic resonance imaging (MRI). The method supplements x-ray imaging in many important ways.

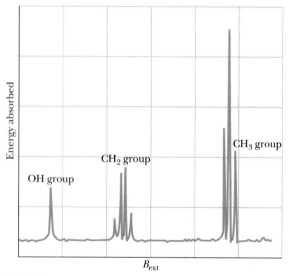

FIGURE 45-13   A nuclear magnetic resonance spectrum for ethanol. All the absorption lines are due to the spin flips of protons. The three groups of lines correspond, as indicated, to protons in the OH group, the CH_2 group, and the CH_3 group within the molecule. The entire horizontal axis encompasses considerably less than 10^{-4} T.

SAMPLE PROBLEM 45-4

A drop of water is suspended in a 1.80-T magnetic field and an alternating electromagnetic field is applied, its frequency chosen so as to produce spin flips for the protons in the sample. The magnetic moment of the proton is 1.41×10^{-26} J/T. Ignore local magnetic fields within the sample. What are the frequency and wavelength of the alternating field?

SOLUTION From Eq. 45-17 we have

$$f = \frac{2\mu B}{h}$$

$$= \frac{(2)(1.41 \times 10^{-26}\,\text{J/T})(1.80\,\text{T})}{6.63 \times 10^{-34}\,\text{J}\cdot\text{s}}$$

$$= 7.66 \times 10^7\,\text{Hz} = 76.6\,\text{MHz}. \qquad \text{(Answer)}$$

This frequency is in the VHF television band. The corresponding wavelength is

$$\lambda = \frac{c}{f} = \frac{3.00 \times 10^8\,\text{m/s}}{7.66 \times 10^7\,\text{Hz}} = 3.92\,\text{m}. \qquad \text{(Answer)}$$

45-10 MULTI-ELECTRON ATOMS AND THE PERIODIC TABLE

We turn now from the hydrogen atom to atoms with more than one electron, and we state without proof that:

> The four quantum numbers listed in Table 45-1 that identify the states of the hydrogen atom also serve to identify the states of individual electrons in atoms with more than one electron.

Just because states are described by the same quantum numbers does not mean that they have the same energies and wave functions; they do not. In multi-electron atoms, the potential energy associated with any given electron is determined not only by the atomic nucleus but also by the other electrons in the atom. When this is taken properly into account in solving the Schrödinger equation, it turns out that now the energy of a state depends not only on the principal quantum number n—as in Eq. 45-3 for the hydrogen atom states—but also on the orbital quantum number ℓ.

When we assign electrons to states in a multi-electron atom, we must be guided by the **Pauli exclusion principle,** which asserts:

> Only a single electron can be assigned to a given quantum state.

If this important principle did not hold, all the electrons in an atom would move to the state of lowest energy and the world would be a far different place.

The Orbitals

The electron states of a multi-electron atom can be organized into groups called *orbitals*, each characterized by a given value of n and of ℓ. Equation 45-7 tells us that, for a given value of ℓ, there are $2\ell + 1$ possible states, each with a different value of the magnetic quantum number m_ℓ. Each of these states has two choices of the spin quantum number m_s, so that the total number of states in an orbital with a given value of ℓ is $2(2\ell + 1)$. For $\ell = 0$, this number is two and for $\ell = 1$, it is six. Note that the number of states in an orbital depends only on ℓ but the energy of those states depends on both n and ℓ.

Neon

This atom has 10 electrons. Two of them occupy the two states of the orbital with $n = 1$ and $\ell = 0$ (the 1,0 orbital), thus *filling* the orbital completely. Two of the remaining eight electrons fill the orbital with $n = 2$ and $\ell = 0$ (the 2,0 orbital). The remaining six electrons fill the 2,1 orbital. Thus the neon atom in its lowest energy state has its electrons arranged in three filled orbitals.

In a filled orbital, the vectors for both the orbital angular momentum and the spin angular momentum point in all possible directions and thus cancel. So, for neon with its three filled orbitals, those vectors cancel for the atom as a whole. Similarly, neon has zero net magnetic dipole moment. With only filled orbitals, neon is chemically inert.

Sodium

Next after neon in the periodic table comes sodium, with 11 electrons. Ten of them form a neonlike core, leaving the remaining single electron in the 3,0 orbital. To a first approximation we can think of the sodium nucleus (whose charge is $+11e$) as being partially *screened* by the neonlike core (with a charge $-10e$) surrounding it, leaving a reduced net central charge to govern the motion of the outer electron. The entire angular momentum and magnetic dipole moment of the sodium atom are due to this single, relatively loosely bound electron. (Such loosely bound electrons are called valence electrons.) Because the electron is in a state with $\ell = 0$, the atom's angular momentum and magnetic dipole moment must be due entirely to the intrinsic spin (and none to the orbital motion) of this single electron.

The existence of a single electron in an outer, unfilled orbital means that sodium is chemically active. Only 5 eV is required to remove sodium's valence electron, much less than the 22 eV needed to remove an electron from chemically inert neon.

The Periodic Table

In adding electrons to a bare nucleus to form an atom, the orbitals are always filled in the order of increasing energy. For heavier atoms, however, this filling order is not always the logical sequence suggested by the quantum numbers. In krypton, for example, the 4,0 orbital lies lower in energy than the 3,2 orbital and thus the electrons in the 4,0 orbital lie deeper within the atom than do the 10 electrons in the 3,2 orbital. It is a major triumph of wave mechanics that, taking the filling order properly into account, we can account for the entire periodic table of the elements shown in Appendix E.

SAMPLE PROBLEM 45-5

Account for the number of elements in the six complete horizontal periods of the periodic table in terms of the populations of orbitals.

SOLUTION   As Appendix E shows, the numbers of elements in the six horizontal rows of the periodic table are 2, 8, 8, 18, 18, and 32. We have seen that the populations of the orbitals depend only on the orbital quantum number ℓ and are given by $2(2\ell + 1)$. Thus

ORBITAL QUANTUM NUMBER ℓ	ORBITAL POPULATION, $2(2\ell + 1)$
0	2
1	6
2	10
3	14

We can account for each horizontal period in terms of filled orbitals in this way:

PERIOD NUMBER	ELEMENTS IN PERIOD	SUM OF ORBITALS
1	2	2
2,3	8	2 + 6
4,5	18	2 + 6 + 10
6	32	2 + 6 + 10 + 14

45-11 X RAYS AND THE NUMBERING OF THE ELEMENTS

Here we shift our attention from electrons on the outer fringes of the atom to electrons deep within the atom. We move from a region of relatively low binding energy (about 5 eV for the valence electron of sodium, for example) to a region of higher energy (about 70 keV for the binding energy of the innermost electron in tungsten, for example, over 10,000 times larger). The radiations associated with state changes shift dramatically in wavelength, from about 600 nm for the yellow light from sodium to about 20 pm for one of the characteristic tungsten radiations. We are speaking of x rays.

Our concern here is with what these rays—whose medical, dental, and industrial usefulness is so well known—can teach us about the structure of the atoms that absorb or emit them. We focus on the work of British physicist H. G. J. Moseley who, by

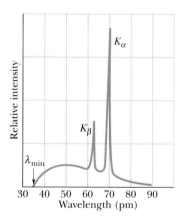

FIGURE 45-15   The distribution by wavelength of the x rays produced when 35-keV electrons strike a molybdenum target. Note the sharp peaks standing out above a continuous background.

x-ray methods, developed the concept of atomic number and gave physical meaning to the ordering of the elements in the periodic table.

As we saw in Section 41-9, x rays are produced when energetic electrons strike a solid target and are brought to rest in it. Figure 45-15 shows the wavelength spectrum of the x rays that are produced when a beam of 35-keV electrons falls on a molybdenum target. It consists of a broad spectrum of radiation, distributed continuously in wavelength, on which are superimposed peaks of sharply defined wavelengths. The broad continuous spectrum and the peaks arise in different ways, which we discuss separately.

45-12 THE CONTINUOUS X-RAY SPECTRUM

Here we examine the continuous x-ray spectrum of Fig. 45-15, ignoring for the time being the two prominent peaks that rise from it. If the incident electrons have been accelerated through a potential difference V, their kinetic energy as they strike the target will be eV. As they are being brought to rest within the target, we expect that electrons of all kinetic energies from zero to eV will be present there. Consider an electron of kinetic energy K within this range that happens to pass close to the nucleus of one of the molybdenum atoms in the target, as in Fig. 45-16. The electron may well lose an amount of energy ΔK, which will appear as the energy of an x-ray photon that is radiated away from the site of the encounter. All electrons whose kinetic energies lie in the range from 0 to eV can undergo such *bremsstrahlung** processes, and all contribute thereby to creating the continuous x-ray spectrum.

A prominent feature of the continuous spectrum of Fig. 45-15 is the sharply defined *cutoff wavelength* λ_{min}, below which the continuous spectrum does not exist. This minimum wavelength corresponds to an encounter in which one of the incident electrons, still with its initial kinetic energy eV, loses *all* this energy in a single encounter, radiating it away as a single x-ray photon. The wavelength associated with this photon, the minimum possible x-ray wavelength, is found from

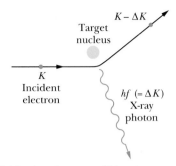

FIGURE 45-16 An electron of kinetic energy K passing near the nucleus of a target atom may generate an x-ray photon, losing part of its kinetic energy in the process.

$$eV = hf = \frac{hc}{\lambda_{min}},$$

which yields

$$\lambda_{min} = \frac{hc}{eV} \quad \text{(cutoff wavelength).} \quad (45\text{-}19)$$

The cutoff wavelength is totally independent of the target material. If you were to switch from a molybdenum to a copper target, for example, all features of the x-ray spectrum of Fig. 45-15 would change *except* the cutoff wavelength.

SAMPLE PROBLEM 45-6

A beam of 35.0-keV electrons (the kinetic energy of each is 35.0 keV) strikes a molybdenum target, generating the x rays whose spectrum is shown in Fig. 45-15. Calculate the expected cutoff wavelength λ_{min}.

SOLUTION From Eq. 45-19 we have

$$\lambda_{min} = \frac{hc}{eV} = \frac{(4.14 \times 10^{-15} \text{ eV·s})(3.00 \times 10^{8} \text{ m/s})}{35.0 \times 10^{3} \text{ eV}}$$

$$= 3.55 \times 10^{-11} \text{ m} = 35.5 \text{ pm.} \quad \text{(Answer)}$$

This agrees well with the value indicated by the vertical arrow in Fig. 45-15. Note that our calculation contains no reference to the material of which the target is made.

45-13 THE CHARACTERISTIC X-RAY SPECTRUM

We now turn our attention to the two peaks of Fig. 45-15, labeled K_α and K_β. These peaks, together with other peaks that appear at longer wavelengths, are characteristic of the target material and form what we call the **characteristic x-ray spectrum** of that material.

Here is how the x-ray photons that produce these peaks arise. (1) An energetic incoming electron strikes an atom in the target and knocks out one of its deep-lying electrons. If the electron is in the shell with $n = 1$ (called, for historical reasons, the K-shell), there remains a vacancy, or a *hole* as we shall call it, in this shell. (2) One of the outer electrons moves in to fill this hole and, in the process, the atom emits a characteristic x-ray photon. If the electron falls from the shell with $n = 2$ (called the L-shell) we have the K_α line of Fig. 45-15; if it falls from the shell with $n = 3$ (called the M-shell) we have the K_β line; and so on. Of course, such a transition will leave a hole in either the L- or the M-shell, but this will be filled in by an electron from still farther out in the atom, causing the emission of still another characteristic x-ray photon.

In studying the radiations emitted by the single electron of the hydrogen atom, we found it convenient to draw an energy level diagram in which each level corresponds to a different quantum state for that single electron. We chose as our zero-energy configuration the state of the system in which the electron is at rest and is completely removed from the atom.

In studying x rays, however, we find it much more convenient to keep track of the single hole created deep in the electron cloud, rather than of the quantum states of the many electrons that remain in the atom. We choose as our zero-energy configuration the state of the system in which the hole has been completely removed from the atom, that is, the normal neutral atom in its ground state. Recall again that the configuration to which we assign zero energy is quite arbitrary. Only differences in energy are important, and these are the same no matter what configuration we choose to represent $E = 0$.

Figure 45-17 shows an x-ray energy level diagram for molybdenum, the element to which Fig. 45-15 refers. The base line ($E = 0$) represents, as we have said, the neutral atom in its ground state. The level marked K (at $E = 20$ keV) represents the energy of the molybdenum atom with a hole in its K-shell. Similarly, the level marked L (at $E = 2.7$ keV) represents the energy of the atom with a hole in its L-shell, and so on.

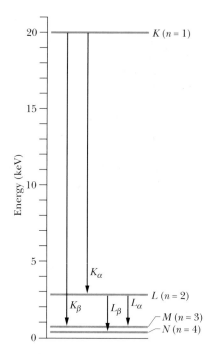

FIGURE 45-17 An atomic energy level diagram for molybdenum, showing the transitions that give rise to the characteristic x rays of that element. (All levels except the K-level consist of a number of closely lying components, not shown here.)

The transitions marked K_α and K_β in Fig. 45-17 show the origins of the two sharp x-ray lines in Fig. 45-15. The K_α line, for example, originates when an electron from the L-shell of molybdenum fills a hole in the K-shell. This corresponds to a hole moving downward on the energy level diagram of Fig. 45-17 from the K-level to the L-level.

Moseley and the X-Ray Spectrum

In his investigation of the characteristic x-ray spectrum, Moseley generated characteristic x rays by using as many elements as he could find—he found 38—as targets for electron bombardment in a special evacuated x-ray tube of his own design. By means of a trolley, manipulated by strings, he could put various targets in place in the path of the incident electron beam. He measured the wavelengths of the x rays by the crystal diffraction method described in Section 41-9.

Moseley then sought, and found, regularities in these spectra as he moved from element to element in the periodic table. In particular, he noted that if,

 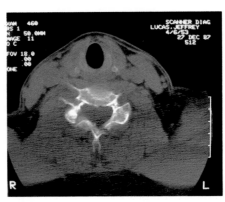

(*Left*) H. G. J. Moseley holding some of his simple x-ray apparatus. (*Right*) A computed tomography image (CAT scan) of a cross section through the broken neck of a patient. The image is produced by a computer that translates the information it receives from an x-ray detector when the detector and a source of a pencil-beam of x rays are rotated around the patient.

for a given spectral line such as K_α, he plotted the square root of the line frequency against the position of the element in the periodic table, a straight line resulted. Figure 45-18 shows a portion of his extensive data. We shall see later why it was logical to plot the data in this way and why a straight line is to be expected. Moseley's conclusion from the full body of his data was:

> **We have here a proof that there is in the atom a fundamental quantity, which increases by regular steps as we pass from one element to the next. This quantity can only be the charge on the central nucleus.**

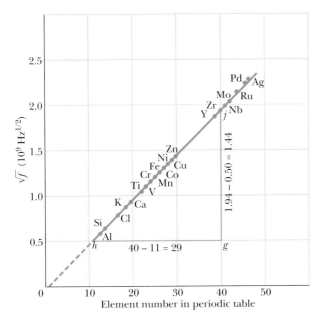

FIGURE 45-18 A Moseley plot of the K_α line of the characteristic x-ray spectra of 21 elements. The frequency is calculated from the measured wavelength.

Owing to Moseley's work, the characteristic x-ray spectrum became the universally accepted signature of an element, permitting the solution of a number of periodic-table puzzles. Prior to that time (1913) the position of an element in the table was assigned in order of atomic *weight*, although there were several cases in which it was necessary to invert this order because of compelling chemical evidence; Moseley showed that it is the nuclear charge (that is, the atomic *number*) that is the real basis for numbering the elements.

The periodic table had several empty squares, and a surprising number of claims for new elements had been advanced. The x-ray spectrum provided an indisputable test for such claims. The rare-earth elements, because of their similar chemical properties, had been only imperfectly sorted out. They were properly organized in short order. In more recent times, the identities of elements beyond uranium are pinned down without dispute when they are available in quantities large enough to permit a study of their x-ray spectra.

It is not hard to see why the characteristic x-ray spectrum shows such impressive regularities from element to element while the optical spectrum in the visible and near-visible region does not. The key to the identity of an element is the charge on its nucleus. Gold, for example, is what it is because its atoms have a nuclear charge of $+79e$. If it had one more nuclear charge it would not be gold, but mercury; if it had one less, it would be platinum. The K electrons, which play such a large role in the production of the characteristic x-ray spectrum, lie very close to the nucleus and are thus sensitive probes of its charge. The optical spectrum, on the other hand, involves transitions of the valence electrons. These outermost electrons are heavily screened from the nucleus by the remaining electrons of the atom and are not sensitive nuclear-charge probes.

Bohr Theory and the Moseley Plot

Moseley's experimental data are totally useful for numbering the elements, even if no theoretical basis for them could be established. Moseley went further, however, and showed that his results were consistent with Bohr's theory of atomic structure. We saw in Chapter 43 that the Bohr theory works quite well for the hydrogen atom but fails even for helium, the next element in the periodic sequence. How, then, can this theory work even reasonably well for such heavy atoms? We can glimpse some of the reason if we focus attention on a single hole in such a heavy atom, whose transitions generate the x-ray spectrum. From this point of view, the situation is not so much different from that of the hydrogen atom, whose spectrum is produced by the transitions of its single electron. Now for the details.

Suppose that one of the two electrons in the K-shell of a heavy atom is removed, leaving a hole in that shell. That hole will be filled by an electron moving inward from the L-shell and the atom will emit a K_α x-ray photon in the process. The effective nuclear charge "seen" by this moving electron is not Ze but something very close to $(Z - 1)e$, because the full charge of the nucleus is partially screened by the electron that has remained undisturbed in the K-shell throughout this transition.

Bohr's formula for the frequency of the radiation corresponding to a transition between any two atomic levels in hydrogenlike atoms is

$$f = \frac{me^4 Z^2}{8\epsilon_0^2 h^3}\left(\frac{1}{n_1^2} - \frac{1}{n_2^2}\right),$$

in which m is the electron mass, n_1 and n_2 are quantum numbers, and Ze is the nuclear charge (instead of e). For the K_α transition, it is appropriate to replace Z with $Z - 1$ and to put $n_1 = 1$ and $n_2 = 2$. If we do so and then take the square root of each side, we find

$$\sqrt{f} = \sqrt{\frac{3me^4}{32\epsilon_0^2 h^3}}\,(Z - 1), \qquad (45\text{-}20)$$

which we can write in the form

$$\sqrt{f} = a(Z - 1), \qquad (45\text{-}21)$$

where a is the constant identified by the square root sign in Eq. 45-20. Equation 45-21 is the equation of

a straight line, in full agreement with the experimental data of Fig. 45-18. If the plot of Fig. 45-18 is extended to higher atomic numbers, however, it departs somewhat from a straight line. The agreement with Bohr theory is, however, surprisingly good.

SAMPLE PROBLEM 45-7

A cobalt target is bombarded with electrons, and the wavelengths of its characteristic spectrum are measured. A second, fainter, characteristic spectrum is also found, due to an impurity in the target. The wavelengths of the K_α lines are 178.9 pm (cobalt) and 143.5 pm (impurity). What is the impurity?

SOLUTION Let us apply Eq. 45-21 both to cobalt and to the impurity. Substituting c/λ for f, we obtain

$$\sqrt{\frac{c}{\lambda_{Co}}} = a(Z_{Co} - 1) \quad \text{and} \quad \sqrt{\frac{c}{\lambda_x}} = a(Z_x - 1).$$

Dividing yields

$$\sqrt{\frac{\lambda_{Co}}{\lambda_x}} = \frac{Z_x - 1}{Z_{Co} - 1}.$$

Substituting gives us

$$\sqrt{\frac{178.9 \text{ pm}}{143.5 \text{ pm}}} = \frac{Z_x - 1}{27 - 1}.$$

Solving for the unknown, we find

$$Z_x = 30.0. \qquad \text{(Answer)}$$

A glance at the periodic table identifies the impurity as zinc.

SAMPLE PROBLEM 45-8

Calculate the constant a in Eq. 45-21 and compare it with the measured slope of the straight line plotted in Fig. 45-18.

SOLUTION If we compare Eqs. 45-20 and 45-21, we can write

$$a = \sqrt{\frac{3me^4}{32\epsilon_0^2 h^3}}$$

$$= \sqrt{\frac{(3)(9.11 \times 10^{-31} \text{ kg})(1.60 \times 10^{-19} \text{ C})^4}{(32)(8.85 \times 10^{-12} \text{ F/m})^2 (6.63 \times 10^{-34} \text{ J} \cdot \text{s})^3}}$$

$$= 4.95 \times 10^7 \text{ Hz}^{1/2}. \qquad \text{(Answer)}$$

Equation 45-21 shows us that a must be the slope of the straight line in Fig. 45-18. If we measure the lines hg and gj in Fig. 45-18 carefully, we find

$$a = \frac{gj}{hg} = \frac{(1.94 - 0.50) \times 10^9 \text{ Hz}^{1/2}}{40 - 11}$$

$$= 4.97 \times 10^7 \text{ Hz}^{1/2}, \qquad \text{(Answer)}$$

in good agreement with the prediction of Bohr theory. The agreement is not nearly so good for lines other than K_α in the x-ray spectrum; for them one must rely on calculations based on wave mechanics.

45-14 LASERS AND LASER LIGHT

In the late 1940s and again in the early 1960s quantum physics made two enormous contributions to technology, the transistor and the laser. The first stimulated the growth of *microelectronics*, which deals with the interaction (at the quantum level) between electrons and bulk matter. The laser is leading the way in a new field—sometimes called *photonics*—which deals with the interaction (again at the quantum level) between photons and bulk matter.

To see the importance of lasers, let us look at some of the characteristics of laser light. We shall compare it as we go along with the light emitted by such sources as a tungsten filament lamp (which emits a continuous spectrum) or a neon gas discharge tube (which emits a line spectrum). We shall thus answer this chapter's opening question.

1. *Laser Light Is Highly Monochromatic.* Tungsten light, spread over a continuous spectrum, gives us no basis for comparison. The light from selected lines in a gas discharge tube, however, can have wavelengths in the visible region that are precise to about 1 part in about 10^6. The sharpness of definition of laser light can be many times greater, as much as 1 part in 10^{15}.

2. *Laser Light Is Highly Coherent.* Wave trains for laser light may be several hundred kilometers long. This means that interference fringes can be set up by combining two separate beams whose path lengths differ by such distances. The corresponding coherence length for light from a tungsten filament lamp or a gas discharge tube is typically less than 1 meter.

3. *Laser Light Is Highly Directional.* A laser beam departs from strict parallelism only because of diffraction effects, determined (see Section 41-5) by the wavelength of the light and the diameter of the exit aperture. Light from other sources can be made into an approximately parallel beam by a lens or a mirror, but the beam divergence is much greater than from a laser. Each point on, say, a tungsten filament source forms its own separate beam; the angular divergence of the overall composite beam is determined not by diffraction but by the size of the filament.

4. *Laser Light Can Be Sharply Focused.* This property is related to the parallelism of the laser beam. As is true for starlight, the size of the focused spot for a laser beam is limited only by diffraction and not by

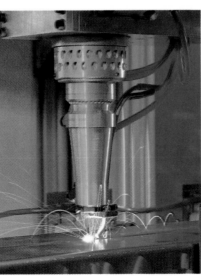

At left, a laser beam is sent into the eye of a diabetes patient to seal blood vessels in the retina. At right, a high-power laser is used for welding on an assembly line at a Porsche automobile plant.

the size of the source. Energy flux densities for laser light of about 10^{16} W/cm^2 are readily possible. An oxyacetylene flame, by contrast, has an energy flux density of only about 10^3 W/cm^2.

Laser Light Has Many Uses

The smallest lasers, used for telephone communication over optical fibers, have as their active medium a semiconducting gallium arsenide crystal about the size of a pin head. The largest lasers, used for laser fusion research and for military applications, fill a large building. They can generate pulses of laser light of about 3×10^{-9} s duration with a power level, during the pulse, of about 8×10^{13} W. This level, if maintained, would exceed the total electric power generating capacity of the United States.

Other laser uses include spot welding detached retinas; drilling tiny holes in diamonds for the drawing (stretching and shaping) of fine wires; cutting cloth (50 layers at a time, with no frayed edges) in the garment industry; precision surveying; precision length measurements via interferometry; and the generation of holograms.

45-15 EINSTEIN AND THE LASER

The word "laser" is an acronym for **l**ight **a**mplification by the **s**timulated **e**mission of **r**adiation, so you should not be surprised to learn that *stimulated emission* is the key to its operation. Einstein introduced this concept into physics in 1917; although the world had to wait until 1960 to see an operating laser, the groundwork for its development was put in place at that earlier date.

Let us consider a single isolated atom that can exist in only one of two states, of energies E_1 and E_2. We discuss below three ways in which this atom can be caused to shift from one of its two allowed states to the other.

1. *Absorption.* Figure 45-19a shows an atom initially in the lower of its two states, with energy E_1. We assume also that a continuous spectrum of radiation is present. If a photon of energy

$$hf = E_2 - E_1 \qquad (45\text{-}22)$$

in that radiation interacts with the atom, the photon will vanish and the atom will move to its upper energy state. We call this familiar process *absorption*.

2. *Spontaneous Emission.* In Fig. 45-19b the atom is in its upper state and no radiation is present. After a certain mean time τ, the atom moves of its own accord to the state of lower energy, emitting a photon of energy hf in the process. We call this familiar process *spontaneous emission*, because it was not triggered by any outside influence. The light from the glowing filament in an ordinary light bulb is generated in this way.

Normally, the mean life of excited atoms before spontaneous emission occurs is about 10^{-8} s. However, there are some states for which this mean life is much longer, perhaps as long as 10^{-3} s. We call such

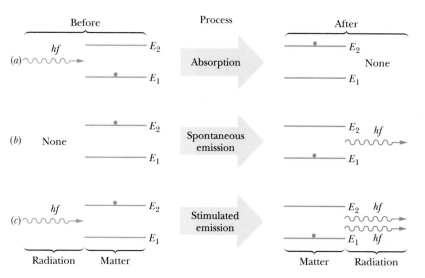

FIGURE 45-19 The interaction of matter and radiation in the processes of (a) absorption, (b) spontaneous emission, and (c) stimulated emission. It is the last process that provides the possibility of laser operation.

states *metastable;* they play an essential role in laser operation.

3. Stimulated Emission. In Fig. 45-19*c* the atom is again in its upper state, but this time a continuous spectrum of radiation is also present. As in absorption, a photon whose energy is given by Eq. 45-22 interacts with the atom. The result is that the atom moves to its lower energy state, emitting a photon of energy *hf.* There are now *two* photons where only one existed before.

The emitted photon in Fig. 45-19*c* is in every way identical to the triggering, or *stimulating;* photon. It has the same energy, direction, phase, and polarization. We can see how a chain reaction of similar processes could be triggered by one such *stimulated emission* event. Laser light is produced in this way.

Figure 45-19*c* represents the stimulated emission of a photon from a single atom. In the usual case, however, many atoms are present. Given a large number of atoms, in equilibrium at a certain temperature *T,* we may ask how many of them will be in level E_1 and how many in level E_2. Ludwig Boltzmann showed that the number n_x of atoms in any level whose energy is E_x is given by

$$n_x = Ce^{-E_x/kT}, \qquad (45\text{-}23)$$

in which C is a constant and k is Boltzmann's constant. This equation seems reasonable. The quantity kT is the mean energy of agitation of an atom at temperature *T.* The higher the temperature, the more atoms—on long-term average—will be "bumped up" by thermal agitation (that is, atom–atom collisions) to the level E_x.

If we apply Eq. 45-23 to the two levels of Fig. 45-19 and divide, the constant C cancels out and we find, for the ratio of the number of atoms in the upper level to the number in the lower level,

$$\frac{n_2}{n_1} = e^{-(E_2 - E_1)/kT}. \qquad (45\text{-}24)$$

Figure 45-20*a* illustrates this situation. Because $E_2 > E_1$, the ratio n_2/n_1 will naturally always be less than unity, which means that there will always be fewer atoms in the higher energy level than in the lower. Again, this is what we would expect if the level populations are determined only by the action of thermal agitation.

If we flood the atoms of Fig. 45-20*a* with photons of energy $E_2 - E_1$, photons will disappear by the absorption process and will be generated by the

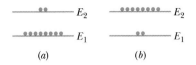

FIGURE 45-20 (*a*) The normal thermal equilibrium distribution of atoms between two states, accounted for by thermal agitation. (*b*) An inverted population, obtained by special techniques. Such an inverted population is essential for laser action.

two emission processes. However, the net effect, by sheer weight of numbers, will be absorption. To make a laser we must *generate* photons, not absorb them. The arrangement of Fig. 45-20*a* won't work.

To generate laser light we must have a situation in which stimulated emission dominates. The only way to do this is to have more atoms in the upper level than in the lower, as in Fig. 45-20*b*. A *population inversion* such as this is not consistent with simple thermal equilibrium, so we must think of clever ways to set up the inversion.

45-16 HOW A LASER WORKS

Of the many kinds of lasers, we describe two, the first being an optically pumped laser. The first operating laser, assembled by Theodore Maiman in 1960, used crystalline ruby as the lasing material and operated in this way.

An Optically Pumped Laser

Figure 45-21 shows schematically how we can set up a population inversion in a lasing material so that laser action can occur. We start with essentially all the atoms of the material in the ground state E_1. We

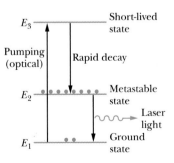

FIGURE 45-21 The basic three-level scheme for laser operation. Metastable state E_2 is more heavily populated than the ground state E_1.

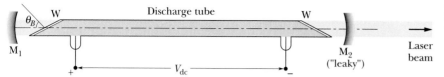

FIGURE 45-22   The elements of a helium–neon gas laser.

then supply energy to the system so that many atoms are raised to an excited state E_3. In *optical pumping*, the energy comes from an intense, continuous-spectrum light source that surrounds the lasing material.

From state E_3 many atoms decay rapidly and spontaneously to state E_2, which must be a metastable state if the material is to act as a laser; that is, state E_2 must have a relatively long mean life against spontaneous emission. If conditions are right, state E_2 can then become more heavily populated than state E_1 and we have our population inversion. A stray photon of the right energy can then trigger an avalanche of stimulated emission events from state E_2, and we have laser light.

The Helium–Neon Gas Laser

Figure 45-22 shows a type of laser commonly found in student laboratories. The glass discharge tube is filled with an 80%–20% mixture of the noble gases helium and neon, the neon being the lasing medium. In the gas laser, the necessary population inversion comes about because of collisions between helium and neon atoms.

Figure 45-23 is a simplified version of the level schemes for these two atoms. Note that four levels, labeled E_0, E_1, E_2, and E_3, are involved, rather than three levels as in the lasing scheme of Fig. 45-21. Pumping in the scheme of Fig. 45-23 is accomplished by setting up an electrical discharge in the helium–neon mixture. Electrons and ions in this discharge collide frequently enough with helium atoms to raise many to level E_3. This level is metastable, so that spontaneous emission back to the ground state (level E_0) occurs only very slowly.

Level E_3 in helium (20.61 eV) is, by chance, very close to level E_2 in neon (20.66 eV). Thus when a metastable helium atom and a ground-state neon atom collide, the excitation energy of the helium atom is often transferred to the neon atom. In this way, level E_2 in Fig. 45-23 can become more heavily populated than level E_1. This neon-atom population

inversion is maintained because (1) the metastability of level E_3 ensures a ready supply of neon atoms in level E_2 and (2) atoms in level E_1 decay rapidly (through intermediate stages not shown) to the neon ground state E_0. Stimulated emission from level E_2 to level E_1 predominates and red laser light, of wavelength 632.8 nm, is generated.

More must be done before a strong beam of laser light can be produced. Most stimulated emission photons initially produced in the discharge tube of Fig. 45-22 will not happen to be parallel to the tube axis and will quickly be stopped at the tube walls. Stimulated emission photons that *are* parallel to the axis, however, can be made to move back and forth through the discharge tube many times by successive reflections from mirrors M_1 and M_2. A chain reaction thus builds up rapidly along this axis, providing for the inherent parallelism of the laser light.

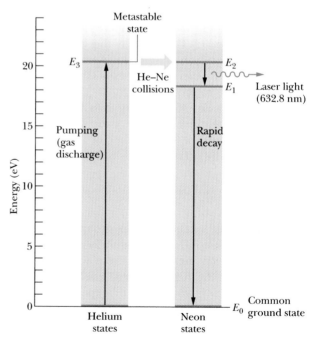

FIGURE 45-23   Four essential levels in helium and neon atoms in a He–Ne gas laser.

Rather than thinking in terms of photons bouncing back and forth between the mirrors, it is perhaps more useful to think of the entire arrangement of Fig. 45-22 as an optical resonant cavity that, like an organ pipe for sound waves, can be tuned to sharp resonance at one (or more) wavelengths.

The mirrors M_1 and M_2 are concave, with their focal points nearly coinciding at the center of the tube. Mirror M_1 is coated with a dielectric film whose thickness is carefully adjusted to make the mirror as close as possible to a perfect reflector at the wavelength of the laser light. Mirror M_2, on the other hand, is coated so as to be slightly "leaky," so that a small fraction of the laser light can escape at each reflection to form a useful beam. The windows W in Fig. 45-22, which close the ends of the discharge tube, are slanted to the Brewster angle (see Section 39-4) to minimize loss of light by reflection.

SAMPLE PROBLEM 45-9

A three-level laser of the type represented in Fig. 45-21 emits laser light at a wavelength of 550 nm, near the center of the visible band.

a. If the optical pumping mechanism is not used, what will be the equilibrium ratio of the population of the upper level (energy E_2) to that of the lower level (energy E_1)? Assume that $T = 300$ K.

SOLUTION From the Bohr frequency condition (Eq. 45-1), the energy difference between the two levels is given by

$$E_2 - E_1 = hf = \frac{hc}{\lambda}$$

$$= \frac{(6.63 \times 10^{-34}\,\text{J}\cdot\text{s})\,(3.00 \times 10^8\,\text{m/s})}{(550 \times 10^{-9}\,\text{m})\,(1.60 \times 10^{-19}\,\text{J/eV})}$$

$$= 2.26\ \text{eV}.$$

The mean energy of thermal agitation is equal to

$$kT = (8.62 \times 10^{-5}\ \text{eV/K})\,(300\ \text{K}) = 0.0259\ \text{eV}.$$

From Eq. 45-24 we have then, for the desired ratio,

$$\frac{n_2}{n_1} = e^{-(E_2 - E_1)/kT} = e^{-(2.26\ \text{eV})/(0.0259\ \text{eV})}$$

$$= e^{-87.26} \approx 1.3 \times 10^{-38}! \qquad \text{(Answer)}$$

This is an extremely small number. It is not unreasonable, however. An atom whose *mean* thermal agitation energy is only 0.0259 eV will not often impart an energy of 2.26 eV to another atom in a collision.

b. For the conditions of (a), at what temperature would the ratio n_2/n_1 be $\frac{1}{2}$?

SOLUTION Making this substitution in Eq. 45-24, taking the natural logarithm of each side, and solving for T yield

$$T = \frac{E_2 - E_1}{k(\ln 2)} = \frac{2.26\ \text{eV}}{(8.62 \times 10^{-5}\,\text{eV/K})\,(\ln 2)}$$

$$= 37{,}800\ \text{K}. \qquad \text{(Answer)}$$

This is much hotter than the surface of the sun. It is clear that, if we are to invert the populations of these two levels, some rather special pumping mechanism is needed. Without population inversion, lasing is not possible.

REVIEW & SUMMARY

Any successful theory of atomic structure must be able to explain (1) the regularities represented by the *periodic table of the elements*, (2) the details of atomic emission and absorption spectra, and (3) the fact that atoms have angular momentum and magnetic dipole moments. Modern quantum theory does this and much more.

Hydrogen Atoms
The structure of the hydrogen atom is analyzed by introducing the potential energy function of Eq. 45-2 into the mathematical machinery of wave mechanics. What

emerges is a set of quantized energies, angular momenta, magnetic moments, and wave functions that describe the allowed states of the hydrogen atom. The energies of these states are

$$E_n = -\frac{me^4}{8\epsilon_0^2 h^2}\frac{1}{n^2}$$

$$= -\frac{13.6\ \text{eV}}{n^2}, \qquad \text{for } n = 1, 2, 3, \ldots, \quad (45\text{-}3)$$

with n being the **principal quantum number.** Every hydrogen atom state has an orbital angular momentum **L** whose magnitude L is fixed by the **orbital quantum number** ℓ:

$$L = \sqrt{\ell(\ell + 1)}\hbar,$$

$$\text{for } \ell = 0, 1, 2, \ldots, n - 1. \quad (45\text{-}4, 45\text{-}5)$$

Here $\hbar = h/2\pi$ is a useful unit of angular momentum in quantum mechanics. The allowed components of **L** along the z axis depend on the **magnetic quantum number** m_ℓ and are given by

$$L_z = m_\ell \hbar, \text{ for } m_\ell = 0, \pm 1, \pm 2, \ldots, \pm \ell. \quad (45\text{-}6, 45\text{-}7)$$

The fact that only a discrete number $(= 2\ell + 1)$ of these projections is allowed is called space quantization. The allowed projections on the z axis of the magnetic moment associated with orbital angular momentum are

$$\mu_{\ell,z} = -m_\ell(eh/4\pi m) = -m_\ell \mu_B. \quad (45\text{-}9, 45\text{-}10)$$

Here $\mu_B = 9.274 \times 10^{-24}$ J/T $= 5.788 \times 10^{-5}$ eV/T is the **Bohr magneton,** a convenient unit in which to measure atomic magnetic moments.

Electron Spin and Spin Magnetic Moment

In addition to orbital angular momentum, each electron has a **spin angular momentum S** whose z component is given by

$$S_z = m_s \hbar, \quad \text{for } m_s = \pm\tfrac{1}{2}. \quad (45\text{-}11)$$

The magnetic moment associated with electron spin has possible projections given by

$$\mu_{s,z} = -2m_s \mu_B. \quad (45\text{-}12)$$

Spin angular momentum is twice as effective as orbital angular momentum in generating magnetism.

Verification of Space Quantization

The Stern–Gerlach experiment confirmed the reality of space quantization. It did so by using a nonuniform magnetic field to split a beam of neutral silver atoms into two subbeams, each characterized by the orientation in space of the magnetic moments of the atoms that make it up.

Nuclear Magnetic Resonance

Nuclear magnetic resonance provides a useful analytical technique based on the space quantization of the angular momenta of atomic nuclei, usually protons. The nuclei absorb energy from a radio-frequency beam when the frequency is just right to cause "spin flips" of the protons, that is, when

$$hf = 2\mu B. \quad (45\text{-}17)$$

Since **B** near the protons is the sum of an external field plus a local field due to the atomic environment, the absorption frequency f reveals information about chemical bonds.

Atom Building and the Periodic Table

The periodic table can be understood by assuming that electron states in multi-electron atoms are to be labeled with the hydrogen atom quantum numbers and that only one electron may be assigned to each quantum state (the **Pauli exclusion principle**).

The Continuous X-Ray Spectrum

Figure 45-15 shows the distribution by wavelength of the x rays produced when energetic electrons strike a metal target. The smooth background curve is called the **continuous x-ray spectrum.** It arises when the electrons are decelerated in the field of a nucleus. The minimum cutoff wavelength is given by

$$\lambda_{\min} = \frac{hc}{eV} \quad \text{(cutoff wavelength),} \quad (45\text{-}19)$$

where V, the only variable, is the accelerating potential of the electrons.

The Characteristic X-Ray Spectrum

The sharp peaks in Fig. 45-15 constitute the **characteristic x-ray spectrum** of the target element. The peaks arise from electron transitions deep within the atom as represented in Fig. 45-17. The frequency of any characteristic x-ray line (the K_α line, say) varies smoothly with the atomic number of the element. This fact can be used to assign atomic numbers to the elements, by way of a Moseley plot such as that of Fig. 45-18. The straight-line shape of the plot can be understood in terms of the quantum mechanics of a one-electron atom, which predicts that the frequency f of any x-ray transition should vary with atomic number Z according to the equation

$$\sqrt{f} = \sqrt{\frac{3me^4}{32\epsilon_0^2 h^3}} \, (Z - 1) = a(Z - 1). \quad (45\text{-}20, 45\text{-}21)$$

Lasers

The special properties of laser light are directly traceable to a chain reaction of stimulated emission events in a lasing medium. Figure 45-19c shows the basic process. Lasing always involves a pair of levels and, for stimulated emission to be the controlling process, there are two requirements. (1) The upper level (E_2) must be more heavily populated than the lower level (E_1); see Fig. 45-20b. (2) The upper level must be metastable; that is, it must have a relatively long mean lifetime against depletion by spontaneous emission. Figures 45-21 and 45-23 show how the necessary population inversion can be brought about, by optical pumping in the first case and by gas collision excitation processes in the second.

QUESTIONS

1. If Bohr's theory and wave mechanics predict the same result for the energies of the hydrogen atom states, then why do we need wave mechanics, with its greater complexity?

2. Compare Bohr's theory and wave mechanics. In what respects do they agree? In what respects do they differ?

3. In the laboratory, how would you show that an atom has angular momentum? That it has a magnetic dipole moment?

4. Why don't we observe space quantization for a spinning top?

5. How can we arrive at the conclusion that the spin magnetic quantum number m_s can have only the values $\pm \frac{1}{2}$? What kinds of experiments support this conclusion?

6. Why is the magnetic moment of the electron directed opposite its spin angular momentum?

7. An atom in a state with zero angular momentum has spherical symmetry as far as its interaction with other atoms is concerned. It is sometimes called a "billiard-ball atom." Explain.

8. Discuss how good an analogy the rotating Earth revolving about the sun is to a spinning electron moving about a proton in the hydrogen atom.

9. Angular momentum is a vector, and you might expect that it would take three quantum numbers to describe it, corresponding to the three space components of a vector. Instead, in an atom, only two quantum numbers characterize the angular momentum. Explain why.

10. The angular momentum of the electron in the hydrogen atom is quantized. Why isn't the linear momentum also quantized? (*Hint:* Consider the implications of the uncertainty principle.)

11. Does the Einstein–de Haas experiment (see Fig. 45-3) provide any evidence that angular momentum is quantized?

12. What are the dimensions and SI unit of a hydrogen atom wave function?

13. Define and distinguish among the terms "wave function," "probability density function," and "radial probability density function."

14. What determines the number of subbeams into which a beam of neutral atoms is split in a Stern–Gerlach experiment?

15. If in a Stern–Gerlach experiment an ion beam is resolved into five component beams, then what angular momentum quantum number does each ion have?

16. In a Stern–Gerlach apparatus, is it possible to have a magnetic field configuration in which the magnetic field

itself is zero along the beam path but the field gradient is not? If your answer is yes, can you design an electromagnet that will produce such a field configuration?

17. A beam of neutral silver atoms is used in a Stern–Gerlach experiment. What is the origin of both the force and the torque that act on the atom? How is the atom affected by each?

18. On what quantum numbers does the energy of an electron in (a) a hydrogen atom and (b) a vanadium atom depend?

19. The periodic table of the elements was based originally on atomic weight, rather than on atomic number, the latter concept having not yet been developed. Why were such early tables as successful as they proved to be? In other words, why is the atomic weight of an element (roughly) proportional to its atomic number?

20. How does the structure of the periodic table support the need for a fourth quantum number, corresponding to electron spin?

21. Why does it take more energy to remove an electron from neon ($Z = 10$) than from sodium ($Z = 11$)?

22. Why do the lanthanide elements (see Appendix E) have such similar chemical properties? How can we justify putting them all into a single square of the periodic table? Why is it that, in spite of their similar chemical properties, they can be so easily sorted out by measuring their characteristic x-ray spectra?

23. How would the properties of helium differ if the electron had no spin, that is, if the only operative quantum numbers were n, ℓ, and m_ℓ?

24. What is the origin of the cutoff wavelength $\lambda_{\min}$ of Fig. 45-15? Why is it an important clue to the photon nature of x rays?

25. Why do you expect the wavelengths of radiations generated by transitions deep within an atom to be shorter than those generated by transitions occurring in the outer fringes of the atom?

26. What are the characteristic x rays of an element? How can they be used to determine the atomic number of the element?

27. Compare Figs. 45-15 and 45-17. How can you be sure that the two prominent peaks in the former figure do indeed correspond numerically with the two transitions similarly labeled in the latter figure?

28. Can atomic hydrogen be caused to emit x rays? If so, describe how. If not, why not?

29. How does an x-ray energy level diagram differ from the energy level diagram for hydrogen? In what respects are the two diagrams similar?

30. When extended to higher atomic numbers, the Moseley plot of Fig. 45-18 is not a straight line but is slightly concave upward. Does this affect the ability to assign atomic numbers to the elements?

31. Why is it that Bohr's theory, which otherwise does not work very well even for helium ($Z = 2$), gives such a good account of the characteristic x-ray spectra of the elements?

32. Why is focused laser light inherently better than focused light from a tiny incandescent lamp filament for such delicate surgical jobs as spot-welding detached retinas?

33. Laser light forms an almost parallel beam. Does the intensity of such light fall off as the inverse square of the distance from the source?

34. In what ways are laser light and starlight similar? In what ways are they different?

35. Arthur Schawlow, one of the laser pioneers, invented a typewriter eraser based on focusing laser light on the unwanted character. What is its principle of operation?

36. We have spontaneous emission and stimulated emission. From symmetry, why don't we also have spontaneous and stimulated absorption?

37. Why is a population inversion between two atomic levels necessary for laser action to occur?

38. Comment on this statement: "Other things being equal, a four-level laser scheme such as that of Fig. 45-23 is preferable to a three-level scheme such as that of Fig. 45-21 because, in the latter scheme, half of the population of atoms in level E_1 must be moved to state E_2 before a population inversion can even begin to occur."

39. A beam of light emerges from an aperture in a "black box" and moves across your laboratory bench. How could you test this beam to find out the extent to which it is coherent over its cross section? How could you tell (without opening the box) whether or not the concealed light source is a laser?

40. Why is it difficult to build an x-ray laser?

EXERCISES & PROBLEMS

SECTION 45-5 ORBITAL ANGULAR MOMENTUM AND MAGNETIC MOMENT

1E. Use the value of Planck's constant in Appendix B to show that, to three significant figures, $\hbar = 1.06 \times 10^{-34}$ J·s $= 6.59 \times 10^{-16}$ eV·s.

2E. Verify that $\mu_B = 9.274 \times 10^{-24}$ J/T $= 5.788 \times 10^{-5}$ eV/T, as reported in Eq. 45-10.

3E. Calculate the magnitude of the orbital angular momentum of an electron in a state with $\ell = 3$.

4E. (a) What number of possible ℓ values are associated with $n = 3$? (b) What number of possible m_ℓ values are associated with $\ell = 1$?

5E. If an electron in a hydrogen atom is in a state with $\ell = 5$, what is the minimum possible angle between **L** and L_z?

6E. Write down the quantum numbers for all the hydrogen-atom states for which $n = 4$ and $\ell = 3$.

7E. A hydrogen-atom state is known to have the quantum number $\ell = 3$. What are the possible n, m_ℓ, and m_s quantum numbers?

8E. A hydrogen-atom state has a maximum m_ℓ value of $+4$. What can you say about the rest of its quantum numbers?

9E. How many hydrogen-atom states have $n = 5$?

10P. (a) Show that the magnetic moment of the electron in hydrogen, according to the Bohr model, is given by $\mu = n\mu_B$, where μ_B is the Bohr magneton and $n = 1, 2, \ldots$. (b) This result disagrees with the correct expression given in Eq. 45-9. Why does the Bohr model fail to give the correct orbital magnetic moment?

11P. Calculate and tabulate, for a hydrogen atom in a state with $\ell = 3$, the allowed values of L_z, μ_z, and θ. Find also the magnitudes of **L** and $\boldsymbol{\mu}$.

12P. Estimate (a) the quantum number ℓ for the orbital motion of the Earth around the sun and (b) the number of quantized orientations of the plane of the Earth's orbit, according to the rules of space quantization. (c) Also find θ_{min}, the half-angle of the smallest cone that can be swept out by a perpendicular to the Earth's orbit as the Earth revolves around the sun. Discuss from the point of view of the correspondence principle.

13P. Show that Eq. 45-8 is a plausible version of the uncertainty principle $\Delta p \cdot \Delta x = h$. (*Hint:* Multiply by r/r; associate p with mv, and L with mvr.)

14P. Consider the relation

$$\cos \theta_{min} = \left(1 + \frac{1}{\ell}\right)^{-1/2}$$

in Sample Problem 45-1 for the limiting case of large ℓ. (a) By expanding both sides of this equation in series form (see Appendix G), show that, to a good approximation for large ℓ,

$$\theta_{min} \approx \frac{1}{\sqrt{\ell}},$$

where θ_{min} is to be expressed in radian measure. (b) Test the validity of this approximate formula for the five values of ℓ given in Sample Problem 45-1. (c) Fill in the table in Sample Problem 45-1 by finding θ_{min} for $\ell = 10^5$, 10^6, and 10^7. (d) What computational difficulties do you encounter if you do *not* use the above approximate formula to calculate θ_{min} for large values of ℓ?

15P. Of the three scalar components of **L**, only L_z is quantized, according to Eq. 45-6. Show that the most that can be said about the other two components of **L** is

$$(L_x^2 + L_y^2)^{1/2} = [\ell(\ell + 1) - m_\ell^2]^{1/2}\hbar.$$

Note that these two components are not separately quantized. Show also that

$$\sqrt{\ell}\,\hbar \le (L_x^2 + L_y^2)^{1/2} \le [\ell(\ell + 1)]^{1/2}\hbar.$$

Correlate these results with Fig. 45-5.

16P*. An unmagnetized iron cylinder, whose radius is 5.0 mm, hangs from a frictionless bearing inside a solenoid, so that the cylinder can rotate freely about its axis; see Fig. 45-3. By passing a current through the solenoid windings, a magnetic field is suddenly applied parallel to the axis, causing the magnetic dipole moments of the atoms to align themselves parallel to the field. The atomic angular momentum vectors, which are coupled back to back with the magnetic dipole moment vectors, also become aligned and the cylinder will start to rotate. This is the Einstein–de Haas effect (see Section 45-2). Find the period of rotation of the cylinder. Assume that each iron atom has an angular momentum of $\hbar$ and that the alignment is perfect. (*Hint:* Apply conservation of angular momentum; the answer is independent of the length of the cylinder.)

SECTION 45-7 THE HYDROGEN-ATOM WAVE FUNCTIONS

17E. For a hydrogen atom in its ground state, what is the value, at $r = 2.00r_B$ of the radial probability density $P(r)$?

18E. A small sphere of radius $0.10r_B$ is located a distance r_B from the nucleus of a hydrogen atom in its ground state. What is the probability that the electron will be found inside this sphere? (Assume that ψ is constant inside the sphere.)

19E. For a hydrogen atom in its ground state, what is the probability of finding the electron between two spheres of radii $r = 1.00r_B$ and $r = 1.01r_B$?

20E. From Eq. 45-14, which gives the radial probability density for the state with $n = 2$ and $\ell = 0$, show that the corresponding wave function is

$$\psi = \frac{1}{\sqrt{32\pi r_B^3}}\left(2 - \frac{r}{r_B}\right)e^{-r/2r_B}.$$

21E. (a) Sketch the wave function given in Exercise 20 for the state with $n = 2$ and $\ell = 0$. (b) What is the value of the wave function at the center of the nucleus?

22E. For a hydrogen atom in a state with $n = 2$ and $\ell = 0$, what are the values, at $r = 5.00r_B$, of (a) the wave function $\psi(r)$, (b) the probability density $\psi^2(r)$, and (c) the radial probability density $P(r)$?

23E. Using Eq. 45-14, show that, for the hydrogen-atom state with $n = 2$ and $\ell = 0$,

$$\int_0^\infty P(r)\ dr = 1.$$

What is the physical interpretation of this result?

24P. Calculate the two maxima in the radial probability density curve of Fig. 45-7*b*.

25P. Use the results of Problem 24 to calculate the values of the radial probability density for a state with $n = 2$ and $\ell = 0$ at the two maxima; compare with Fig. 45-7*b*.

26P. Repeat Problem 36 of Chapter 44 for an electron in a state with $n = 2$ and $\ell = 0$; that is, calculate the probability that the electron will be found inside the proton, of radius = 1.1 fm, that constitutes the nucleus of the hydrogen atom.

27P. For a hydrogen atom in a state with $n = 2$ and $\ell = 0$, what is the probability of finding the electron between two spheres of radii $r = 5.00r_B$ and $r = 5.01r_B$?

28P. For a hydrogen atom in a state with $n = 2$ and $\ell = 0$, what is the probability of finding the electron somewhere within the smaller of the two bulges of its radial probability density function in Fig. 45-7*b*.

SECTION 45-8 THE STERN–GERLACH EXPERIMENT

29E. Calculate the two possible angles between the electron spin angular momentum vector and the magnetic field in Sample Problem 45-3. Bear in mind that the *orbital* angular momentum of the valence electron is zero.

30E. What is the acceleration of the silver atom as it passes through the deflecting magnet in the Stern–Gerlach experiment of Sample Problem 45-3?

31E. Assume that in the Stern–Gerlach experiment described for neutral silver atoms the magnetic field **B** has a magnitude of 0.50 T. (a) What is the energy difference between the orientations of the silver atoms in the two subbeams? (b) What is the frequency of the radiation that would induce a transition between these two states? (c) What is its wavelength, and to what part of the electromagnetic spectrum does it belong? The magnetic moment of a neutral silver atom is 1 Bohr magneton.

32P. Suppose a hydrogen atom (in its ground state) moves 80 cm in a direction perpendicular to a magnetic field that has a gradient, in the vertical direction, of 1.6×10^2 T/m. (a) What is the force on the atom due to the magnetic moment of the electron, which we take to be 1 Bohr magneton? (b) What is its vertical displacement in the 80 cm of travel if its speed is 1.2×10^5 m/s?

SECTION 45-9 SCIENCE, TECHNOLOGY, AND SPIN: AN ASIDE

33E. What is the wavelength of a photon that will induce a transition of an electron spin from parallel to antiparallel orientation in a magnetic field of magnitude 0.200 T? Assume that $\ell = 0$.

34E. The proton as well as the electron has spin $\frac{1}{2}$. In the hydrogen atom in its ground state, with $n = 1$ and $\ell = 0$, there are two energy levels, depending on whether the electron and proton spins are in the same direction or in opposite directions. The state with the spins in opposite directions has the higher energy. If an atom is in this state and one of the spins "flips over," the small energy difference is released as a photon of wavelength 21 cm. This spontaneous spin-flip process is very slow, the mean life for the process being about 10^7 y. However, radio astronomers observe this 21-cm radiation in deep space, where the density of hydrogen is so small that an atom can flip before being disturbed by collisions with other atoms. What is the effective magnetic field (due to the magnetic dipole moment of the proton) experienced by the electron in the emission of this 21-cm radiation?

35E. An external magnetic field of frequency 34 MHz is applied to molecules of a certain material that contains hydrogen atoms. Resonance is observed when the strength of this applied field equals 0.78 T. Calculate the strength of the local magnetic field at the site of the protons undergoing spin flips.

36E. Excited sodium atoms emit two closely spaced lines (the sodium doublet; see Fig. 45-24) with wavelengths 588.995 nm and 589.592 nm. (a) What is the difference in energy between the two upper energy levels? (b) This energy difference occurs because the electron's spin magnetic dipole moment (1 Bohr magneton) can be oriented either parallel or antiparallel to the internal magnetic field associated with the electron's orbital motion. Use your result in (a) to find the strength of this internal magnetic field.

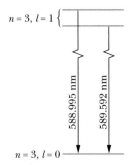

FIGURE 45-24 Exercise 36. The three energy levels that account for the two lines of the familiar sodium doublet.

SECTION 45-10 MULTI-ELECTRON ATOMS AND THE PERIODIC TABLE

37E. Label as true or false these statements involving the quantum numbers n, ℓ, m_ℓ. (a) One of these orbitals cannot exist: $n = 2$, $\ell = 1$; $n = 4$, $\ell = 3$; $n = 3$, $\ell = 2$; $n = 1$, $\ell = 1$. (b) The number of values of m_ℓ that are allowed depends only on ℓ and not on n. (c) There are four orbitals with $n = 4$. (d) The smallest value of n that can be associated with a given ℓ is $\ell + 1$. (e) All states with $\ell = 0$ also have $m_\ell = 0$, regardless of the value of n. (f) There are n orbitals for each value of n.

38E. What are the quantum numbers n, ℓ, m_ℓ, m_s for the two electrons of the helium atom in its ground state?

39E. Two electrons in lithium ($Z = 3$) have, for their quantum numbers n, ℓ, m_ℓ, m_s, the values 1, 0, 0, $\pm\frac{1}{2}$. (a) What quantum numbers can the third electron have if the atom is to be in its ground state? (b) If the atom is to be in its first excited state?

40P. Suppose there are two electrons in the same atom, both of which have $n = 2$ and $\ell = 1$. (a) If the exclusion principle did not apply, how many combinations of states would conceivably be possible? (b) How many states does the exclusion principle forbid? Which ones are they?

41P. Show that the number of states with the same n is given by $2n^2$.

42P. If the electron had no spin, and if the Pauli exclusion principle still held, how would the periodic table be affected? In particular, which of the present elements would be noble gases?

SECTION 45-12 THE CONTINUOUS X-RAY SPECTRUM

43E. Show that the cutoff wavelength in the continuous x-ray spectrum is given by

$$\lambda_{\min} = 1240 \text{ pm}/V,$$

where V is the applied potential difference in kilovolts.

44E. Determine Planck's constant from the fact that the minimum x-ray wavelength produced by 40.0-keV electrons is 31.1 pm.

45E. What is the minimum potential difference across an x-ray tube that will produce x rays with a wavelength of 0.10 nm?

46P. A 20-keV electron is brought to rest by undergoing two successive bremsstrahlung events, thus transferring its kinetic energy into the energy of two photons. The wavelength of the second photon to be emitted is 130 pm greater than the wavelength of the first photon to be emitted. (a) Find the energy of the electron after its first deceleration. (b) Calculate the wavelengths and energies of the two photons.

47P. X rays are produced in an x-ray tube by a target potential of 50.0 kV. If an electron makes three collisions in the target before coming to rest and loses one-half of its remaining kinetic energy in each of the first two collisions, determine the wavelengths of the resulting photons. Neglect the recoil of the heavy target atoms.

48P. Show that a moving electron cannot spontaneously change into an x-ray photon in free space. A third body (atom or nucleus) must be present. Why is it needed? (*Hint:* Examine the conservation of total energy and of momentum.)

SECTION 45-13 THE CHARACTERISTIC X-RAY SPECTRUM

49E. When electrons bombard a molybdenum target, they produce both continuous and characteristic x rays as shown in Fig. 45-15. In that figure the energy of the incident electrons is 35.0 keV. If the accelerating potential applied to the x-ray tube is increased to 50.0 kV, what new values of (a) λ_{min}, (b) λ for the K_α line, and (c) λ for the K_β line result?

50E. In Fig. 45-15, the x-rays shown are produced when 35.0-keV electrons fall on a molybdenum target. If the accelerating potential is maintained at 35.0 kV but a silver target ($Z = 47$) is substituted for the molybdenum target, what values of (a) λ_{min}, (b) λ for the K_α line, and (c) λ for the K_β line result? The K, L, and M atomic x-ray photon energy levels for silver (compare Fig. 45-17) are 25.51, 3.56, and 0.53 keV.

51E. The wavelength of the K_α line from iron is 193 pm. What is the energy difference between the two states of the iron atom (see Fig. 45-17) that give rise to this transition? What is the corresponding energy difference for the hydrogen atom? Why is the difference so much greater for iron than for hydrogen? (*Hint:* In the hydrogen atom the K-shell corresponds to $n = 1$ and the L-shell to $n = 2$.)

52E. From Fig. 45-15, calculate approximately the energy difference $E_L - E_M$ for the x-ray atomic energy levels of molybdenum. Compare with the result that may be found from Fig. 45-17.

53E. Calculate the ratio of the wavelengths of the K_α line for niobium (Nb) to that for gallium (Ga). Take needed data from the periodic table of Appendix E.

54P. Here are the K_α wavelengths of a few elements:

Ti	275 pm	Co	179 pm
V	250	Ni	166
Cr	229	Cu	154
Mn	210	Zn	143
Fe	193	Ga	134

Make a Moseley plot (see Fig. 45-18) and verify that its slope agrees with the value calculated in Sample Problem 45-8.

55P. A tungsten target ($Z = 74$) is bombarded by electrons in an x-ray tube. (a) What is the minimum value of the accelerating potential that will permit the production of the characteristic K_β and K_α lines of tungsten? (b) For this same accelerating potential, what is λ_{min}? (c) What are the K_β and K_α wavelengths? The K, L, and M atomic x-ray energy levels for tungsten (see Fig. 45-17) are 69.5, 11.3, and 2.30 keV, respectively.

56P. A molybdenum target ($Z = 42$) is bombarded with 35.0-keV electrons and the x-ray spectrum of Fig. 45-15 results. Here the K_β and K_α wavelengths are 63.0 pm and 71.0 pm, respectively. (a) What are the corresponding photon energies? (b) It is desired to filter these radiations through a material that will absorb the K_β line much more strongly than it will absorb the K_α line. What substance would you use? The K ionization energies for molybdenum and for four neighboring elements are as follows:

Z	40	41	42	43	44
Element	Zr	Nb	Mo	Tc	Ru
E_K (keV)	18.00	18.99	20.00	21.04	22.12

(*Hint:* A substance will selectively absorb one of two x radiations more strongly if the photons of one have enough energy to eject a K electron from the atoms of the substance but the photons of the other do not.)

57P. The binding energies of K-shell and L-shell electrons in copper are 8.979 and 0.951 keV, respectively. If a K_α x ray from copper is incident on a sodium chloride crystal and gives a first-order Bragg reflection at 74.1° when reflected from the alternating planes of sodium atoms, what is the spacing between these planes?

58P. (a) Using Bohr's theory, estimate the ratio of energies of photons due to K_α transitions in two atoms whose atomic numbers are Z and Z'. (b) How much more energetic is a K_α x ray from uranium expected to be than one from aluminum? (c) Than one from lithium?

59P. Determine how close the theoretical K_α x-ray photon energies, as obtained from Eq. 45-20, are to the measured energies in the light elements from lithium to magnesium. To do this you must first (a) determine the constant in Eq. 45-20 to five significant figures, using data in Appendix B. Next, (b) calculate the percentage deviation of the theoretical from the measured energies. (c) Plot the deviation and comment on the trend. The measured energies of the K_α x rays for these elements are

Li	54.3 eV	O	524.9 eV
Be	108.5	F	676.8
B	183.3	Ne	848.6
C	277	Na	1041
N	392.4	Mg	1254

(There is actually more than one K_α ray because of splitting of the L energy level, but that effect is negligible in the elements considered.)

SECTION 45-16 HOW A LASER WORKS

60E. A hypothetical atom has energy levels evenly spaced by 1.2 eV in energy. For a temperature of 2000 K, calculate the ratio of the number of atoms in the 13th excited state to the number in the 11th excited state.

61E. A particular (hypothetical) atom has only two atomic levels, separated in energy by 3.2 eV. In the atmosphere of a star there are 6.1×10^{13} of these atoms in the excited (upper) state per cm³ and 2.5×10^{15} per cm³ in the ground (lower) state. Calculate the temperature of the star's atmosphere.

62E. A population inversion for two levels is often described by assigning a negative Kelvin temperature to the system. Show that such a negative temperature would indeed correspond to an inversion. What negative temperature would describe the system of Sample Problem 45-9 if the population of the upper level exceeds that of the lower by 10.0%

63E. A He–Ne laser emits light at a wavelength of 632.8 nm and has an output power of 2.3 mW. How many photons are emitted each minute by this laser when it is operating?

64E. A ruby laser emits light at wavelength 694.4 nm. If a laser pulse is emitted for 1.20×10^{-11} s and the energy release per pulse is 0.150 J, (a) what is the length of the pulse, and (b) how many photons are in each pulse?

65E. Lasers have become very small as well as very large. The active volume of a laser constructed of the semiconductor GaAlAs has a volume of only 200 μm^3 (smaller than a grain of sand) and yet it can continuously deliver 5.0 mW of power at 0.80-μm wavelength. Calculate the rate at which it produces photons.

66E. It is entirely possible that techniques for modulating the frequency or amplitude of a laser beam will be developed so that such a beam can serve as a carrier for television signals, much as microwave beams do now. Assume also that laser systems will be available whose wavelengths can be precisely "tuned" to anywhere in the visible range, that is, in the range 450 nm $< \lambda <$ 650 nm. If each television channel occupies a bandwidth of 10 MHz, how many channels could be accommodated with this laser technology? Comment on the intrinsic superiority of visible light to microwaves as carriers of information.

67E. A high-powered laser beam ($\lambda = 600$ nm) with a beam diameter of 12 cm is aimed at the moon, 3.8×10^5 km distant. The spreading of the beam is caused only by diffraction effects. The angular location of the edge of the central diffraction disk (see Eq. 41-9) is given by

$$\sin \theta = \frac{1.22 \, \lambda}{d},$$

where d is the diameter of the beam aperture. What is the diameter of the central diffraction disk at the moon's surface?

68P. The active medium in a particular ruby laser ($\lambda = 694$ nm) is a synthetic ruby crystal 6.00 cm long and 1.00 cm in diameter. The crystal is silvered at one end and —to permit the formation of an external beam—only partially silvered at the other. (a) Treat the crystal as an optical resonant cavity analogous to a closed organ pipe, and calculate the number of standing wave nodes there are along the crystal axis. (b) By what amount Δf would the beam frequency have to shift to increase this number by one? Show that Δf is just the inverse of the travel time of light for one round-trip back and forth along the crystal axis. (c) What is the corresponding fractional frequency shift $\Delta f/f$? The appropriate index of refraction is 1.75.

69P. The mirrors in the laser of Fig. 45-22 form a cavity in which standing waves of laser light are set up. In the vicinity of 533 nm, how far apart in wavelength are the adjacent allowed operating modes? The mirrors are separated by 8.0 cm.

70P. An atom has two energy levels with a transition wavelength of 580 nm. At 300 K, 4.0×10^{20} atoms are in the lower state. (a) How many occupy the upper state, under conditions of thermal equilibrium? (b) Suppose, instead, that 7.0×10^{20} atoms are pumped into the upper state, with 4.0×10^{20} in the lower state. How much energy could be released in a single laser pulse?

71P. The beam from an argon laser ($\lambda = 515$ nm) has a diameter d of 3.00 mm and a continuous-wave power output of 5.00 W. The beam is focused onto a diffuse surface by a lens whose focal length f is 3.50 cm. A diffraction pattern such as that of Fig. 41-9 is formed, the radius of the central disk being given by

$$R = \frac{1.22 f \lambda}{d}$$

(see Eq. 41-11). The central disk can be shown to contain 84% of the incident power. Calculate (a) the radius R of the central disk, (b) the average power flux density in the incident beam, and (c) the average power flux density in the central disk.

72P. The use of lasers for missile defense has been proposed as part of the Strategic Defense Initiative ("Star Wars"). A beam of intensity 10^8 W/m² would probably burn into and destroy a hardened (nonspinning) missile in 1 s. (a) If the laser had 5.0-MW power, 3.0-μm wavelength, and 4.0-m beam diameter (a very powerful laser indeed), would it destroy a missile at a distance of 3000 km? (b) If the wavelength could be changed, what maximum value would work? Use the equation for the central disk given in Exercise 67, and take the focal length to be the distance to the target.

CONDUCTION OF ELECTRICITY IN SOLIDS

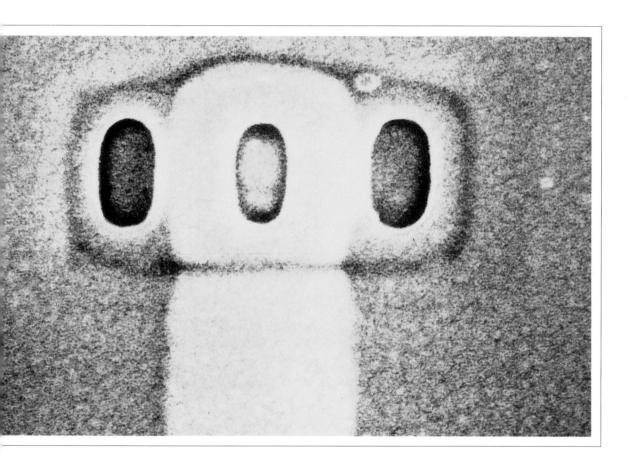

This experimental transistor, shown in a scanning electron microscope image and made by scientists of the IBM Research Division, is the world's smallest transistor: the ovals you see are only 150 nm long. Today's memory chips have a capacity of 16 megabits. With this new transistor, memory chips may soon have a capacity of 4 gigabits or even more, which will dramatically increase the capabilities of computers of all sizes. But what is a transistor and how does it function?

46-1 THE PROPERTIES OF SOLIDS

We have seen how well quantum theory works when we apply it to individual atoms. In this chapter we hope to show, by a single broad example, that this powerful theory works just as well when we apply it to aggregates of atoms in the form of solids.

Every solid has an enormous range of properties that we can choose to examine. Is it transparent? Can you hammer it out into a flat sheet? What kinds of waves travel through it and at what speeds? Does it have interesting magnetic properties? Is it a good heat conductor? What is its crystal surface? Does it have special surface properties? . . . The list goes on. We choose here to focus on a single question: "What are the mechanisms by which a solid conducts, or does not conduct, electricity?" As we shall see, the laws that govern electrical conduction are quantum laws.

46-2 ELECTRICAL CONDUCTIVITY

In studying electrical conductivity, we choose to examine only solids whose atoms are arranged to form a periodic three-dimensional lattice. Figure 46-1 shows such lattice structures for carbon (in the form of diamond), silicon, and copper. We shall not consider such materials as plastic, glass, or rubber, whose atoms are not arranged in any such regular way.

The basic electrical measurement that we can make on a sample is its *electrical resistivity* ρ at room temperature; see Section 28-4. By measuring ρ at various temperatures, we can also obtain a value for α, the *temperature coefficient of resistivity*. Finally, by making Hall effect measurements (see Section 30-4) we can find a value for *n, the number of charge carriers per unit volume* in the material being tested.

From measurements of the room temperature resistivity alone, we quickly discover that there are some materials—we call them **insulators**—that for all practical purposes do not conduct electricity at all. Diamond, for example, has a resistivity of about 10^{16} $\Omega \cdot$ m, greater than that of copper by a factor of about 10^{24}.

We can use our measured values of ρ, α, and n to divide most noninsulators into two major categories: **metals,** such as copper, and **semiconductors,** such as silicon. As we see from Table 46-1, a typical semiconductor (silicon), compared with a typical conductor (copper), (1) has far fewer charge carriers, (2) has a considerably larger resistivity, and (3) has a temper-

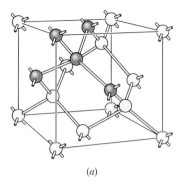

(a)

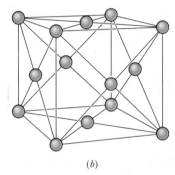

(b)

FIGURE 46-1   (*a*) The crystal structures for carbon (in the form of diamond) and for silicon, which happen to be identical. In this structure, as the darkened spheres show, each atom is bonded to four of its neighbors. (*b*) The crystal structure for copper, an arrangement called *face-centered cubic.*

ature coefficient of resistivity that is both large and negative. That is, although the resistivity of a metal *increases* with temperature, that of a semiconductor *decreases.*

We have now established an experimental basis for framing our central question about the conduc-

TABLE 46-1
SOME ELECTRIC PROPERTIES OF TWO MATERIALS[a]

	UNIT	COPPER	SILICON
Type of conductor		Metal	Semiconductor
Density of charge carriers,[b] n	m^{-3}	9×10^{28}	1×10^{16}
Resistivity, ρ	$\Omega \cdot$ m	2×10^{-8}	3×10^{3}
Temperature coefficient of resistivity, α	K^{-1}	$+4 \times 10^{-3}$	-70×10^{-3}

[a]All values are for room temperature.

[b]The value for the semiconductor includes both electrons and holes.

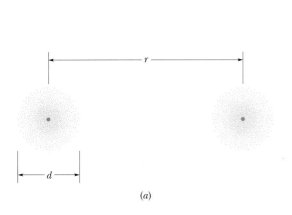

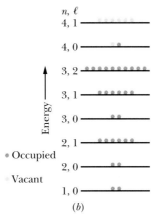

n, ℓ

4, 1

4, 0

3, 2

3, 1

3, 0

2, 1

2, 0

1, 0

• Occupied

Vacant

Energy

(a)

(b)

FIGURE 46-2 (a) Two isolated copper atoms, their electron clouds represented by dot plots. (b) Each atom has 29 electrons distributed over a set of energy levels as shown. The levels are identified by the notation n,ℓ, where n is the principal quantum number and ℓ the orbital quantum number. Each energy level contains $2(2\ell + 1)$ quantum states, defined by the quantum numbers m_ℓ and m_s. For simplicity, the levels are shown here as being uniformly spaced in energy.

tion of electricity in solids. We pose it in specific terms:

> **What is there about diamond that makes it an insulator, about copper that makes it a metal, and about silicon that makes it a semiconductor?**

As we shall see, quantum physics provides the answers to this question.

46-3 ENERGY LEVELS IN A SOLID

The distance between adjacent copper atoms in solid copper is 260 pm. Consider, as in Fig. 46-2a, two copper atoms that are separated by a much greater distance than this. As Fig. 46-2b shows, each of these isolated atoms has associated with it an array of discrete quantum states, each state defined by its unique set of quantum numbers. In the ground state of the neutral copper atom, its 29 electrons occupy the 29 states of this array that are lowest in energy, each state containing but a single electron as the Pauli exclusion principle requires.

If we bring the atoms of Fig. 46-2a closer together, they will—speaking loosely—gradually begin to sense each other's presence. In the formal language of quantum physics, their wave functions will begin to overlap. This overlap will occur first for the wave functions of the valence electrons, which, because they spend most of their time in the outer regions of the electron cloud of the isolated atom, are the first to make contact.

When the wave functions overlap, we no longer speak of two independent and isolated systems but of a single two-atom system containing $2 \times 29 = 58$ electrons. The Pauli principle requires that each of these electrons must occupy a *different* quantum

state. The only way that this can happen is for each energy level of the isolated atom to split into *two* levels for the two-atom system.

We can bring up further atoms and in this way gradually construct a lattice of solid copper. If our specimen contains N atoms, each level of the isolated copper atom must be split into N levels. In this way, each *level* of the isolated atom becomes a **band of levels** in the solid. In a typical solid, an energy band is a few electron-volts wide. Since N is of the order of the Avogadro number, we can see that the individual energy levels within a band are very close together indeed.

Figure 46-3 suggests the band structure of the levels in a hypothetical solid in which we have assumed, for simplicity, that the bands do not overlap. The gaps between the bands represent ranges of energy that no electron may possess. In much the same

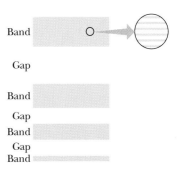

Band

Gap

Band

Gap

Band

Gap

Band

FIGURE 46-3 An idealized representation of the band–gap pattern for the energy levels of a solid. As the magnified view of the upper band suggests, each band consists of a very large number of energy levels that lie very close together. The levels of this idealized plot are, as yet, unoccupied by electrons.

way, electrons in individual atoms cannot possess energies that lie between the discrete allowed levels of the atom.

Note in Fig. 46-3 that the bands that lie lower in energy are narrower than those that lie higher. This is because these low-energy bands correspond to levels in the isolated atom that are occupied by electrons that spend most of their time deep within the electron cloud of the atom. The wave functions of these core electrons thus do not overlap as much as do the wave functions of the outer, or valence, electrons and for that reason the splitting of the levels —although it must occur—is not as great as it is for the levels normally occupied by the outer electrons.

Now that we have established the pattern of levels for a solid, we are ready to consider how these levels are filled with electrons. We shall see how this will lead us in a convincing way to the answers to the question that we raised at the end of Section 46-2.

46-4 INSULATORS

The feature that defines an *insulator* is that, as Fig. 46-4 shows, the highest occupied level coincides with the top of a band. In addition, this band must be separated from the unoccupied band above it by a substantial energy gap E_g. For diamond, $E_g =$ 5.4 eV, a value about 140 times larger than the average thermal energy of a free particle at room temperature.

By definition, an insulator is a solid through which electrons cannot flow as a directed drift current. Let us see why. If you apply an electric field **E** to an insulator, it will exert a force $- e\mathbf{E}$ on each

electron. Classically, this force will cause the electron to increase the component of its velocity in the direction $- \mathbf{E}$, which in turn means that its kinetic energy will change. In quantum terms, if the energy of an electron changes, the electron must move to a different energy level within the solid. In an insulator, however, the Pauli principle prevents the electron from doing so because all other levels within the band into which the electron might move are already occupied. These electrons are in total gridlock. It is as if a child tries to climb a ladder on which other children are standing, one every rung; since there are no vacant rungs, no one can move.

There are plenty of vacant levels in the band above the filled band in Fig. 46-4 but, if an electron is to occupy one of these levels, it must somehow jump across the gap that separates the two bands. It cannot pause at a way station within the gap because all energies in this range are strictly forbidden. In diamond, the gap is simply too wide for any detectable number of electrons to make it to the vacant band, either by the action of an external electric field or by thermal agitation.

46-5 METALS: QUALITATIVE

The feature that defines a *metal* is that, as Fig. 46-5 shows, the highest occupied level (at the absolute zero of temperature) falls somewhere in the middle of a band. The electrons that occupy this partially filled band are the valence electrons of the atoms,

Insulator

FIGURE 46-4   An idealized representation of the band–gap pattern for an insulator. Note that the highest filled level (gold) lies at the top of a band and that the next highest vacant band (gray) is separated by a relatively large energy gap E_g.

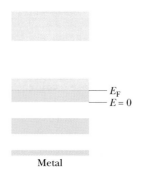

Metal

FIGURE 46-5   An idealized representation of the band–gap pattern for a metal at the absolute zero of temperature. The conduction electrons of the metal occupy the highest partially filled band. Note that vacant levels are available within the band so that these electrons can change their energies and conduction can take place. Lower lying bands are completely filled by the core electrons, that is, those electrons held close to the lattice sites and not free to move through the solid.

which, being free to move throughout the solid, become the *conduction electrons* of the solid.

At the absolute zero of temperature, thermal agitation plays no role and all electrons occupy states of the lowest possible energy. We can assume with little error that the potential energy of the conduction electrons remains constant as they move about within the solid. If we set this constant equal to zero —as we are always permitted to do—the total energy E associated with any level is equal to the kinetic energy of the electron that occupies that level.

The level at the bottom of the partially filled band of Fig. 46-5 thus corresponds to $E = 0$. The highest occupied level in this band (at the absolute zero of temperature) is called the **Fermi level** and the energy corresponding to it is called the **Fermi energy** E_F; for copper, $E_F = 7.0$ eV. The electron speed corresponding to the Fermi energy is called the **Fermi speed** v_F; for copper, $v_F = 1.6 \times 10^6$ m/s.

A glance at Fig. 46-5 should be enough to shatter the popular misconception that all motion ceases at the absolute zero of temperature. We see that, entirely because of Pauli's exclusion principle, the electrons are stacked up in the partially filled band of Fig. 46-5 with energies that range from zero up to the Fermi energy. The *average* kinetic energy for the electrons in this band for copper is about 4.2 eV. By comparison, the average translational kinetic energy of a molecule of an ideal gas at room temperature is only 0.025 eV. The conduction electrons in a metal have plenty of energy at absolute zero!

Conditions for $T > 0$

What happens to the electron distribution of Fig. 46-5 as we raise the temperature above absolute zero? The short answer is that not very much happens, although the little that does is very important. It is clear that electrons in bands below the partially filled band of Fig. 46-5 cannot be affected by thermal agitation; they are all gridlocked.

Only electrons close to the Fermi energy find vacant levels above them and it is only these electrons that are free to be boosted to higher levels by thermal agitation. Even at $T = 1000$ K, a temperature at which a metal sample would glow brightly in a dark room, the distribution of electrons among the available levels does not differ very much from the distribution at the absolute zero.

Let us see why. The quantity kT, where k is the Boltzmann constant (8.62×10^{-5} eV/K), is a convenient measure of the energy that may be given to an electron by the random thermal jiggling of the lattice. At $T = 1000$ K, $kT = 0.086$ eV; no electron can hope to have its energy changed by more than a few times this relatively small amount by thermal agitation alone. All the "action" takes place for electrons whose energies are close to the Fermi energy. It has been said, somewhat poetically, that thermal agitation normally causes only ripples on the surface of the Fermi sea; the vast depths of that sea lie undisturbed.

Electrical Conduction in a Metal

If you apply an electric field **E** to a metal, the field exerts a force $-e$**E** on each electron. This force, during time Δt, causes every conduction electron in the metal to acquire a velocity increment Δ**v** in the direction of $-$**E**. This change in velocity requires that the electrons change their energies, but there are vacant levels available so that these rearrangements can be made. To return to our previous metaphor, there are now vacant rungs on the upper half of the ladder.

The velocities of the individual conduction electrons do not increase without limit, however, because of collisions associated with the thermal vibrations of the lattice. Thus, after a certain time τ, called the *relaxation time*, the drift velocity of the conduction electrons settles down to a constant limiting value, which we associate with the constant current that is set up by the applied electric field. Note that, although *all* the conduction electrons contribute to the current, only electrons close to the Fermi energy are able to make collisions and thus play their role in establishing the limiting value of the drift velocity. It is only these electrons that have ample vacant levels nearby into which they can move after they have experienced a scattering event.

In Section 28-6, we presented the following equation for the resistivity of a metallic conductor,

$$\rho = \frac{m}{ne^2\tau}, \qquad (46\text{-}1)$$

in which m is the mass of the electron, $-e$ is its charge, and n is the number density of the charge carriers, that is, the number of conduction electrons per unit volume. Although we derived this equation on a classical basis, it holds true when the quantization of the electron energy is taken into account. The quantity τ is the relaxation time to which we referred in the preceding paragraph.

SAMPLE PROBLEM 46-1

How many conduction electrons are there in a copper cube 1.00 cm on edge?

SOLUTION In copper, there is one conduction electron per atom so that N is given by

$$N = na^3, \qquad (46\text{-}2)$$

in which n is the number of atoms per unit volume and a (= 1.00 cm) is the length of the cube edge. We can find n from

$$n = \frac{N_A d}{A}$$

in which N_A is the Avogadro constant, A is the molar mass of copper, and d is the density of copper. If we substitute values for these quantities, we find

$$n = \frac{(6.02 \times 10^{23} \text{ atoms/mol})(8900 \text{ kg/m}^3)}{0.06357 \text{ kg/mol}}$$

$$= 8.43 \times 10^{28} \text{ atoms/m}^3$$

$$= 8.43 \times 10^{28} \text{ electrons/m}^3.$$

From Eq. 46-2 we then have

$$N = na^3 = (8.43 \times 10^{28} \text{ electrons/m}^3)(1.00 \times 10^{-2} \text{ m})^3$$

$$= 8.43 \times 10^{22} \text{ electrons.} \qquad \text{(Answer)}$$

The Pauli exclusion principle requires that each of these electrons occupy a different quantum state. These states are distributed over an energy interval of only 7.0 eV (the Fermi energy) so that the average spacing between them is very small indeed.

Not all these states have different energies. Consider, for example, an electron moving along the x axis with speed v. It has the same energy as an electron moving with this same speed in any other direction in the solid but, because its motion is different, it is in a different quantum state, described by a different wave function.

SAMPLE PROBLEM 46-2

a. What is the speed of a conduction electron in copper with a kinetic energy equal to the Fermi energy (= 7.0 eV)?

SOLUTION The total energy E of the conduction electrons is all kinetic and we can write, if $E = E_F$,

$$E_F = \tfrac{1}{2}mv_F^2,$$

in which v_F is the Fermi speed. Solving for v_F yields

$$v_F = \sqrt{\frac{2E_F}{m}} = \sqrt{\frac{(2)(7.0 \text{ eV})(1.6 \times 10^{-19} \text{ J/eV})}{9.11 \times 10^{-31} \text{ kg}}}$$

$$= 1.6 \times 10^6 \text{ m/s.} \qquad \text{(Answer)}$$

You must not confuse this speed with the *drift speed* of the conduction electrons, which is typically 10^{-5} m/s and is thus smaller by about a factor of 10^{11}. As we explained more fully in Section 28-6, the drift speed is the average speed at which electrons actually drift through a conductor when an electric field is applied; the Fermi speed is the average speed of these electrons between collisions.

b. What is the average time τ between collisions for the conduction electrons in copper? The resistivity of copper at room temperature is 1.7×10^{-8} $\Omega \cdot$m.

SOLUTION Solving Eq. 46-1 for τ yields

$$\tau = \frac{m}{ne^2\rho}$$

$$= \frac{9.11 \times 10^{-31} \text{ kg}}{8.43 \times 10^{28} \text{ m}^{-3}}$$

$$\times \frac{1}{(1.6 \times 10^{-19} \text{ C})^2 (1.7 \times 10^{-8} \text{ } \Omega \cdot \text{m})}$$

$$= 2.5 \times 10^{-14} \text{ s.} \qquad \text{(Answer)}$$

c. What mean free path λ may be calculated from the results of (a) and (b) above?

SOLUTION We have

$$\lambda = v_F\tau = (1.6 \times 10^6 \text{ m/s})(2.5 \times 10^{-14} \text{ s})$$

$$= 4.0 \times 10^{-8} \text{ m} = 40 \text{ nm.} \qquad \text{(Answer)}$$

In the copper lattice the centers of neighboring atoms are 0.26 nm apart. Thus a typical conduction electron can move a substantial distance, about 150 interatomic distances, through a copper lattice at room temperature before making a collision.

46-6 METALS: QUANTITATIVE

Now let us look at the conduction of electricity in a metal quantitatively, under several headings.

Counting the Quantum States

We start by counting the number of distinct quantum states in the partially filled band of Fig. 46-5. We cannot possibly deal with this vast number of states one at a time; we must use statistical methods. Instead of asking, "What is the energy of this state?" we must ask, "How many states (per unit volume) have energies that lie in the energy range E to $E + dE$?" This number can be written as $n(E)\,dE$, where $n(E)$ is called the **density of states.**

If we assume that the conduction electrons move in a region of constant potential, $n(E)$ can be shown to be given by

$$n(E) = \frac{8\sqrt{2}\,\pi m^{3/2}}{h^3}\, E^{1/2} \quad \text{(density of states).} \quad (46\text{-}3)$$

(See Eisberg and Resnick, *Quantum Physics,* second edition, 1985, Wiley, Section 13-5.) Figure 46-6a is a

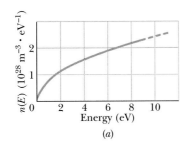

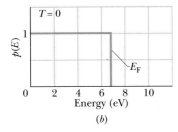

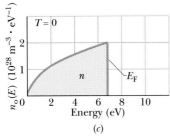

FIGURE 46-6 (*a*) The *density of states* $n(E)$ plotted as a function of energy. (*b*) The *probability function* $p(E)$ plotted as a function of energy at $T = 0$. (*c*) The *density of occupied states* $n_o(E)$, formed by multiplying the curves in (*a*) and (*b*), plotted as a function of energy. Note that all states whose energies lie below the Fermi energy are occupied but all states above that energy are vacant.

plot of Eq. 46-3. Note that there is nothing in this equation that depends on the shape or size of the sample or on the material of which it is made.

Filling the States at $T = 0$

Equation 46-3 tells us how the *unoccupied* states are distributed in energy. We must now weight each energy with a factor $p(E)$, called the **probability function,** that gives us the probability that a state with that energy will actually be occupied. At the absolute zero of temperature, all states with energies less than the Fermi energy are filled and all states with energies greater than that energy are vacant. The probability function in this case must be the simple rectangle shown in Fig. 46-6b, in which unity corresponds to a certainty that the state will be occupied and zero corresponds to a certainty that the state will *not* be occupied.

The product of these two factors gives us $n_o(E)$, the density of *occupied* states. Thus

$$n_o(E) = n(E)\,p(E). \quad (46\text{-}4)$$

Figure 46-6c is a plot of this product.

We can find the Fermi energy for a metal by adding up (integrating) the number of occupied states in Fig. 46-6c between $E = 0$ and $E = E_F$. The result must equal n, the number of conduction electrons per unit volume for the metal. In equation form we have

$$n = \int_0^{E_F} n(E)\,dE. \quad (46\text{-}5)$$

Note that n is represented by the beige area shown in Fig. 46-6c.

Calculating the Fermi Energy

If we substitute Eq. 46-3 into Eq. 46-5, we find

$$n = \frac{8\sqrt{2}\,\pi m^{3/2}}{h^3} \int_0^{E_F} E^{1/2}\,dE = \left(\frac{8\sqrt{2}\,\pi m^{3/2}}{h^3}\right)\left(\frac{2E_F^{3/2}}{3}\right).$$

Solving for E_F leads to

$$E_F = \left(\frac{3}{16\sqrt{2}\,\pi}\right)^{2/3} \frac{h^2}{m}\, n^{2/3} = \frac{0.121 h^2}{m}\, n^{2/3}. \quad (46\text{-}6)$$

Thus the Fermi energy can be calculated once n, the number of conduction electrons per unit volume, is known.

Filling the States for $T > 0$

It can be shown that the probability function for $T > 0$ is given by

$$p(E) = \frac{1}{e^{(E-E_F)/kT} + 1} \quad \begin{array}{l}\text{(probability} \\ \text{function),}\end{array} \quad (46\text{-}7)$$

in which E_F is the Fermi energy and k is the Boltzmann constant. (See Eisberg and Resnick, *Quantum Physics*, second edition, 1985, Wiley, Section 11-4.)

Note that as $T \to 0$, the exponent $(E - E_F)/kT$ in Eq. 46-7 approaches $-\infty$ if $E < E_F$ and $+\infty$ if $E > E_F$. In the first case we have $p(E) = 1$ and in the second $p(E) = 0$. Thus, at $T = 0$, Eq. 46-7 correctly yields the rectangular form shown in Fig. 46-6b. Equation 46-7 also shows us that the important quantity is not the energy E but rather $E - E_F$, the energy interval between E and the Fermi energy.

Figure 46-7b shows the probability function for $T = 1000$ K, calculated from Eq. 46-7. Note how little it differs from the rectangular form of Fig. 46-6b. Figure 46-7c, found by multiplying Figs. 46-7a and 46-7b, shows the density of occupied states for $T = 1000$ K. Note how little that differs from Fig. 46-6c, the distribution at $T = 0$.

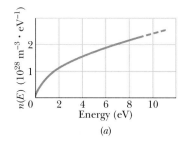

(a)

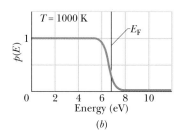

(b)

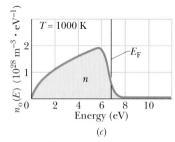

(c)

FIGURE 46-7 (a) The *density of states* $n(E)$ plotted as a function of energy; this plot is the same as that of Fig. 46-6a. (b) The *probability function* $p(E)$ plotted as a function of energy at $T = 1000$ K; note how little this plot differs from that of Fig. 46-6b. (c) The *density of occupied states* $n_o(E)$, formed by multiplying the curves in (a) and (b), plotted as a function of energy; note how little this plot differs from that of Fig. 46-6c. (This plot is idealized in that it assumes that the conduction electrons move in a region of constant potential. Measured density of states plots do not have this simple shape.)

SAMPLE PROBLEM 46-3

A cube of copper is 1.00 cm on an edge. How many quantum states lie in the energy interval between $E = 5.00$ eV and $E = 5.01$ eV?

SOLUTION These energy limits are so close together that we can say that the answer is

$$N = n(E)\, \Delta E\, V, \quad (46\text{-}8)$$

where $E = 5.00$ eV, $\Delta E = 0.01$ eV, and V is the volume of the cube. From Eq. 46-3 we have

$$n(E) = \frac{8\sqrt{2}\pi m^{3/2}}{h^3} E^{1/2}$$

$$= (8\sqrt{2}\pi)(9.11 \times 10^{-31}\, \text{kg})^{3/2}$$

$$\times \frac{(5.00\, \text{eV})^{1/2}(1.60 \times 10^{-19}\, \text{J/eV})^{1/2}}{(6.63 \times 10^{-34}\, \text{J·s})^3}$$

$$= 9.48 \times 10^{46}\, \text{m}^{-3}\text{J}^{-1} = 1.52 \times 10^{28}\, \text{m}^{-3}\text{·eV}^{-1}.$$

From Eq. 46-8 we have, putting $V = a^3$, where a is the cube edge,

$$N = n(E)\, \Delta E\, a^3$$

$$= (1.52 \times 10^{28}\, \text{m}^{-3}\, \text{eV}^{-1})(0.01\, \text{eV})(1 \times 10^{-2}\, \text{m})^3$$

$$= 1.52 \times 10^{20}. \quad \text{(Answer)}$$

SAMPLE PROBLEM 46-4

a. What is the probability that a state whose energy is 0.10 eV above the Fermi energy will be occupied? Assume a temperature of 800 K.

SOLUTION We can find $p(E)$ from Eq. 46-7. Let us first calculate the (dimensionless) exponent in that equation:

$$\frac{E - E_F}{kT} = \frac{0.10 \text{ eV}}{(8.62 \times 10^{-5} \text{ eV/K})(800 \text{ K})} = 1.45.$$

Inserting this exponent into Eq. 46-7 yields

$$p = \frac{1}{e^{1.45} + 1} = 0.19 \text{ or } 19\%. \quad \text{(Answer)}$$

b. What is the probability of occupancy for a state that is 0.10 eV *below* the Fermi energy?

SOLUTION The exponent in Eq. 46-7 has the same numerical value as above but is now negative. Thus from this equation

$$p = \frac{1}{e^{-1.45} + 1} = 0.81 \text{ or } 81\%. \quad \text{(Answer)}$$

For states whose energies lie below the Fermi energy we are often more interested in the probability that the state is *not* occupied. This is, of course, just $1 - p$, or 19% in the present case. An unfilled state in an energy range in which most of the states are filled is called a *hole*. We shall see later that this is a very useful concept.

c. What is the probability of occupancy for a state whose energy is equal to the Fermi energy?

SOLUTION For $E = E_F$ the exponent in Eq. 46-7 is zero and that equation becomes

$$p = \frac{1}{e^0 + 1} = \frac{1}{1 + 1} = 0.50 \text{ or } 50\%. \quad \text{(Answer)}$$

This result does not depend on the temperature. We can, in fact, define the Fermi energy for a metal to be that energy for which the probability of occupancy *at any temperature* is 50%.

46-7 SEMICONDUCTORS

As a comparison of Fig. 46-8 with Fig. 46-4 shows, a semiconductor is like an insulator in that its uppermost filled level (at the absolute zero of temperature) lies at the top of a band. A semiconductor differs from an insulator, however, in that the gap between this filled band and the next vacant band above it is much smaller than for an insulator, so that there is a real possibility for electrons to "jump the gap" into this empty band by thermal agitation. For semiconducting materials, the highest filled band is called the **valence band** because the elec-

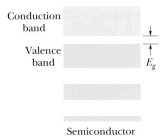

FIGURE 46-8 An idealized representation of the band–gap pattern for a semiconductor such as silicon. It resembles the pattern for an insulator (see Fig. 46-4) except that here the gap between the valence band and the conduction band is much smaller.

trons that occupy it are the valence electrons of the isolated atom. The band above the valence band, which is vacant at $T = 0$, is called the **conduction band.**

The distinction between an insulator and a semiconductor is qualitative, depending as it does on the width of the energy gap. However, there is no doubt that diamond ($E_g = 5.4$ eV) is an insulator and silicon ($E_g = 1.1$ eV) is a semiconductor. As it happens (see Fig. 46-1) these two substances have the same crystal structure.

The revolution in microelectronics that has so influenced our lives is based on semiconductors (see Fig. 46-9), so we would do well to learn more about them. Table 46-1 compares some electrical properties of silicon, our prototype semiconductor, and copper, our prototype conductor. Let us look carefully at this table, one row at a time.

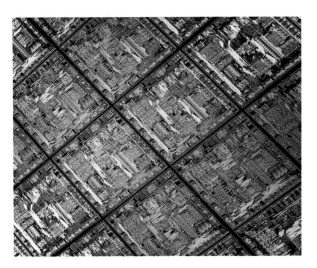

FIGURE 46-9 A photograph of an enlarged section of an integrated circuit.

The Density of Charge Carriers, n

Copper has many more charge carriers than does silicon, by a factor of about 10^{13}. For copper the carriers are the conduction electrons, present in the number of one per atom. At room temperature, to which Table 46-1 refers, charge carriers in silicon arise only because, at thermal equilibrium, thermal agitation has caused a certain (very small) number of electrons to be raised to the conduction band, leaving an equal number of vacant states (holes) in the valence band.

The holes in the valence band of a semiconductor also serve effectively as charge carriers because they permit a certain freedom of movement to the electrons in that band. If an electric field is set up in a semiconductor, the electrons in the valence band, being negatively charged, will drift opposite the direction of **E**. This causes the holes to drift in the direction of **E**. That is, the holes behave like particles carrying a charge $+e$ and, in all that follows, that is exactly how we shall regard them. Conduction by holes is an important fact of life for semiconductors.

If the concept of a migrating hole seems confusing to you, think of a vacant slot in a parking lot that is otherwise filled with cars. If one of these cars moves into the slot, it fills the slot but creates a new vacant slot in the place it just left. This vacancy, in turn, can be filled by another car, and so on. As the cars move around in this way, we can focus attention on the single migrating vacant slot as it wanders over the lot.

The Resistivity, ρ

At room temperature the resistivity of silicon is considerably higher than that of copper, by a factor of about 10^{11}. For both elements, the resistivity is determined by Eq. 46-1. The vast difference in resistivity between copper and silicon can be accounted for by the vast difference in n, the density of charge carriers. (The mean collision time τ will also be different for copper and for silicon but the effect of this on the resistivity is swamped by the enormous difference in n.)

The Temperature Coefficient of Resistivity, α

This quantity (see Eq. 28-16) is the fractional change in resistivity per unit change in temperature, or

$$\alpha = \frac{1}{\rho}\frac{d\rho}{dT}.$$

The resistivity of copper and other metals *increases* with temperature ($d\rho/dT > 0$). This happens because collisions occur more frequently the higher the temperature, thus reducing τ in Eq. 46-1. For metals, the density of charge carriers n in that equation is independent of temperature.

On the other hand, the resistivity of silicon (and other semiconductors) *decreases* with temperature ($d\rho/dT < 0$). This happens because the density of charge carriers n in Eq. 46-1 increases rapidly with temperature. The decrease in τ mentioned before for metals also occurs for semiconductors but its effect on the resistivity is swamped by the very rapid increase in the density of charge carriers.

46-8 DOPING

The versatility of semiconductors can be marvelously improved by introducing a small number of suitable

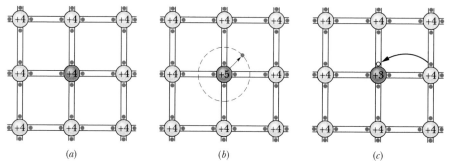

(a)	(b)	(c)

FIGURE 46-10 (a) A two-dimensional representation of a silicon lattice. Each silicon ion (core charge = $+4e$) is bonded to each of its four nearest neighbors by a shared two-electron bond. The red dots show these valence elec-

trons. (b) A phosphorus atom (valence = 5) is substituted for the central silicon atom, creating a donor site. (c) An aluminum atom (valence = 3) is substituted for the central silicon atom, creating an acceptor site.

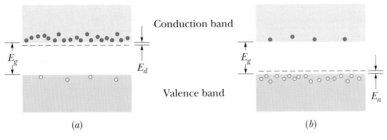

replacement atoms (it seems pejorative to call them impurities) into the semiconductor lattice, a process called **doping.** Essentially all practical semiconducting devices today are based on doped material. They are of two varieties, called *n-type* and *p-type;* we discuss each in turn.

n-Type Semiconductors

Figure 46-10*a* is a "flattened out" representation of a lattice of pure silicon; compare Fig. 46-1*a*. Each silicon atom forms a two-electron covalent bond with each of its four nearest neighbors, the electrons involved in the bonding making up the valence band of the sample. In Fig. 46-10*b* one of the silicon atoms (valence = 4) has been replaced by an atom of phosphorus (valence = 5). As Fig. 46-10*b* suggests, the "extra" electron is loosely bound to the phosphorus ion core because it is not involved in covalent bonds to neighboring ions. It is far easier for *this* electron to be thermally excited into the conduction band than it is for one of the silicon valence electrons to be so excited.

The phosphorus atom is called a **donor** atom because it so readily *donates* an electron to the conduction band. The "extra" electron in Fig. 46-10*b* can be said to lie in a localized *donor level,* as Fig.

46-11*a* shows. This level is separated from the bottom of the conduction band by an energy gap E_d, where $E_d \ll E_g$. By controlling the concentration of donor atoms it is possible to increase greatly the density of electrons in the conduction band.

Semiconductors doped with donor atoms are called *n-type* semiconductors, the "*n*" standing for "negative" because the negative charge carriers greatly outnumber the positive charge carriers. The former, called the *majority carriers,* are the electrons in the conduction band. The latter, called the *minority carriers,* are the holes in the valence band.

p-Type Semiconductors

Figure 46-10*c* shows a silicon lattice in which a silicon atom (valence = 4) has been replaced by an aluminum atom (valence = 3). Now there is a "missing" electron and it is easy for the aluminum ion to "steal" a valence electron from a nearby silicon atom, thus creating a hole in the valence band.

The aluminum atom is called an **acceptor** atom because it so readily *accepts* an electron from the valence band. The electron so accepted moves into a localized *acceptor level,* as Fig. 46-11*b* shows. This level is separated from the top of the valence band by an energy gap $E_a \ll E_g$. By controlling the concentration of acceptor atoms it is possible to greatly increase the number of holes in the valence band.

Semiconductors doped with acceptor atoms are called *p-type* semiconductors, the "*p*" standing for "positive" because the positive charge carriers in this case greatly outnumber the negative carriers. In *p*-type semiconductors the majority carriers are the holes in the valence band and the minority carriers are the electrons in the conduction band.

Table 46-2 summarizes the properties of a typical *n*-type and a typical *p*-type semiconductor. Note particularly that the donor and acceptor ion cores, although they are charged, are not charge *carriers* because, at normal temperatures, they remain fixed in their lattice sites.

TABLE 46-2
PROPERTIES OF TWO DOPED SEMICONDUCTORS

Matrix material	Silicon	Silicon
Dopant	Phosphorus	Aluminum
Type of dopant	Donor	Acceptor
Type of semiconductor	*n*-Type	*p*-Type
Dopant valence	5 (= 4 + 1)	3 (= 4 − 1)
Dopant energy gap	45 meV	57 meV
Majority carriers	Electrons	Holes
Minority carriers	Holes	Electrons
Dopant ion core charge	+ *e*	− *e*

SAMPLE PROBLEM 46-5

The number density of conduction electrons in pure silicon at room temperature is about 10^{16} m^{-3}. Assume that, by doping the lattice with phosphorus, you want to increase this number by a factor of a million (10^6). What fraction of the silicon atoms must you replace with phosphorus atoms? (Assume that, at room temperature, the thermal agitation is effective enough so that essentially every phosphorus atom donates its "extra" electron to the conduction band.)

SOLUTION The density of the phosphorus atoms must be about $(10^{16}$ m$^{-3})(10^6)$, or about 10^{22} m^{-3}. The density of silicon atoms in a pure silicon lattice may be found from

$$n_{Si} = \frac{N_A d}{A},$$

in which N_A is the Avogadro constant, d is the density of silicon, and A is the molar mass of silicon; from Appendix D we find that $d = 2330$ kg/m^3 and $A = 28.1$ g/mol. Substituting yields

$$n_{Si} = \frac{(6.02 \times 10^{23}\ \text{mol}^{-1})(2330\ \text{kg/m}^3)}{0.0281\ \text{kg/mol}}$$

$$= 5 \times 10^{28}\ \text{m}^{-3}.$$

The ratio of these two number densities is the quantity we are looking for. Thus

$$\frac{n_{Si}}{n_P} = \frac{5 \times 10^{28}\ \text{m}^{-3}}{10^{22}\ \text{m}^{-3}} = 5 \times 10^6. \quad \text{(Answer)}$$

We see that if only *one silicon atom in five million* is replaced by a phosphorus atom, the number of electrons in the conduction band will be increased by a factor of 10^6.

How can such a tiny admixture of phosphorus atoms have such a big effect? The answer is that, for pure silicon at room temperature, there were not many conduction electrons to start with. The density of conduction electrons was 10^{16} m^{-3} before doping and 10^{22} m^{-3} after doping. For copper, however, the conduction electron density (see Table 46-1) is about 10^{29} m^{-3}. Thus, even *after* doping, the conduction electron density of silicon remains much less than that of a typical metal such as copper.

46-9 THE *p-n* JUNCTION

Pass a hypothetical plane across a rod of pure silicon. Dope the rod on one side of the plane with donor atoms (thus creating *n*-type material) and on

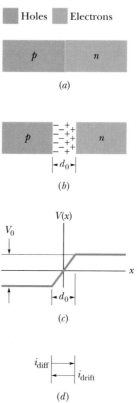

FIGURE 46-12 (*a*) A *p-n* junction at the imagined moment of its creation. Only the majority carriers are shown. (*b*) Diffusion of majority carriers across the junction plane causes a space charge of fixed donor and acceptor ions to appear. (*c*) The space charge establishes a contact potential difference V_0 across the junction plane. (*d*) In equilibrium, the diffusion of majority carriers across the junction plane is just balanced by the drift of minority carriers in the opposite direction.

the other side with acceptor atoms (thus creating *p*-type material). You have just made a ***p-n* junction;** it is at the heart of essentially all semiconducting devices.* Figure 46-12*a* represents a *p-n* junction at the imagined moment of its creation. Let us first discuss the motions of the majority carriers, which are electrons in the *n*-type material and holes in the *p*-type material.

---

*In practice, to make a *p-n* junction one usually starts with, say, *n*-type material and then diffuses acceptor atoms into the solid sample at high temperature, overcompensating the donor atoms to a certain (controllable) depth below the surface.

Motions of the Majority Carriers

Electrons close to the junction plane will tend to diffuse across it (from right to left in Fig. 46-12), for much the same reason that gas molecules will diffuse through a permeable membrane into a vacuum beyond it. In the same way, holes will tend to diffuse across the junction plane from left to right. Both motions contribute to a *diffusion current* i_{diff}, directed from left to right as in Fig. 46-12*d*.

Recall that *n*-type material is studded throughout with donor ions, fixed firmly in their lattice sites. Normally, the positive charges of these ions are compensated electrically by the majority carriers, which are electrons. When an electron diffuses from the *n*-type material and through the junction plane, however, it "uncovers" one of these donor ions, thus introducing a fixed positive charge in the *n*-type material. When this diffusing electron arrives on the other side of the barrier, it quickly finds a hole and combines with it,* thus neutralizing one of the positively charged acceptor ions that are sprinkled throughout the *p*-type material, resulting in a fixed negative charge in the *p*-type material.

Convince yourself that a hole diffusing through the barrier from left to right has exactly the same end result. Thus a region of fixed positive charge builds up on one side of the barrier and a region of fixed negative charge builds up on the other (Fig. 46-12*b*). These two regions form the so-called **depletion zone**; the fixed charges are said to be positive and negative **space charge.**

The space charge causes a *contact potential difference* to build up across the junction, as Fig. 46-12*c* shows. This potential difference is such that it serves as a barrier to limit further diffusion of both electrons and holes across the junction plane. An electron at the junction plane, for example, would be repelled back to its *n*-type home by the negative space charge in the *p*-type material that faces it across the plane. To complete the picture, let us turn our attention to the minority carriers.

Motions of the Minority Carriers

As Fig. 46-11*a* and Table 46-2 show, although the majority carriers in *n*-type material are electrons, there are nevertheless also a few holes, the minority carriers. Likewise in *p*-type material, although the majority carriers are holes, there are also a few conduction electrons.

Although the potential difference in Fig. 46-12*c* acts to retard the motions of the majority carriers—being a barrier for them—it is a downhill trip for the minority carriers, be they electrons or holes. When, by thermal agitation, an electron close to the junction plane is raised from the valence band to the conduction band of the *p*-type material in Fig. 46-12*a*, the contact potential difference causes it to drift steadily from left to right across the junction plane. Similarly, if a hole is created in the *n*-type material, it too drifts across to the other side. The space-charge region shown in Fig. 46-12*b* is effectively swept free of charge carriers by this process and, for that reason, we call it the *depletion* zone. The current represented by the motions of the minority carriers, called the *drift current* i_{drift}, is in the opposite direction to the diffusion current and just compensates it at equilibrium, as Fig. 46-12*d* shows.

Thus, at equilibrium, a *p-n* junction resting on a shelf develops a contact potential difference V_0 between its ends. The diffusion current i_{diff} that moves through the junction plane from the direction *p* to *n* is just balanced by a drift current i_{drift} that moves in the opposite direction.

46-10 THE DIODE RECTIFIER

A *p-n* junction is basically a two-terminal rectifier. If you connect this *diode rectifier* across the terminals of a battery, the current in the circuit will be very much smaller for one polarity of the battery connection than for the other, as Fig. 46-13 shows.

Figure 46-14 shows one of many applications of a diode rectifier. A sine wave input potential generates a half-wave output potential, the diode rectifier acting as essentially a short circuit (a closed switch) for one polarity of the input potential and as essentially an open circuit (an open switch) for the other. An ideal diode rectifier, in fact, has only these two modes of operation. It is either ON (zero resistance) or OFF (infinite resistance).

Figure 46-14 displays the conventional symbol for a diode rectifier. The arrowhead corresponds to the *p*-type terminal of the device and points in the direction of "easy" conventional current flow. That is, the diode is ON when the terminal with the ar-

---

*An "electron combines with a hole" when the electron drops from the conduction band to the valence band, filling a vacancy in that band.

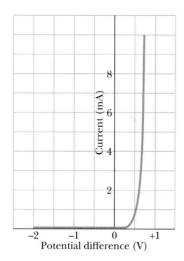

FIGURE 46-13 A current–voltage plot for a junction diode, showing that it is highly conducting in the forward direction and essentially nonconducting in the reverse direction.

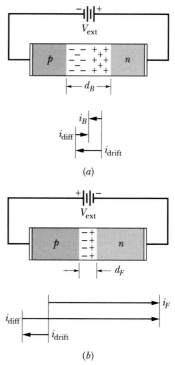

FIGURE 46-15 (a) The back-bias connection of a *p-n* junction, showing the wide depletion zone and the corresponding small back current. (b) The forward-bias connection, showing the narrowing of the depletion zone and the large forward current.

rowhead is (sufficiently) positive with respect to the other terminal.

Figure 46-15 shows details of the two connections. In Fig. 46-15a—the back-bias arrangement— the battery emf simply *adds* to the contact potential difference, thus increasing the height of the barrier that the majority carriers must surmount. Fewer of them can do so and, as a result, the diffusion current decreases markedly.

The drift current, however, senses no barrier and thus is independent of the magnitude or direction of the external potential. The nice current balance that existed at zero bias (see Fig. 46-12d) is thus upset and, as shown in Fig. 46-15a, a very small net back-current i_B appears in the circuit.

Another effect of back-bias is to widen the depletion zone, as a comparison of Figs. 46-12b and 46-15a shows. Because the depletion zone contains

very few charge carriers, it is a region of high resistivity. Thus its substantially increased width means a substantially increased resistance, consistent with the small value of the back-bias current.

Figure 46-15b shows the forward-bias connection, the positive terminal of the battery being connected to the *p*-type end of the *p-n* junction. Here the applied emf *subtracts* from the contact potential difference, the diffusion current *rises* substantially,

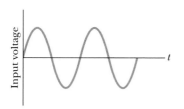

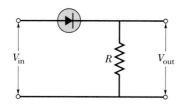

FIGURE 46-14 A *p-n* junction diode is connected as a rectifier. The action of the circuit is to pass the positive half of the input wave form but to suppress the negative

half. The average potential of the input wave form is zero; that for the output wave form is positive.

and a relatively *large* net forward current i_F results. The depletion zone *narrows,* its low resistance being consistent with the large current i_F.

46-11 THE LIGHT-EMITTING DIODE (LED)

We are all familiar with the brightly colored numbers that flash at us from cash registers and gasoline pumps. In nearly all cases this light is emitted from an assembly of *p-n* junctions operating as **light-emitting diodes** (LEDs).

Figure 46-16*a* shows the familiar seven-segment display from which the numbers are formed. Figure 46-16*b* shows that each element of this display is the end of a flat plastic lens, at the other end of which is a small LED, possibly about 1 mm² in area. Figure

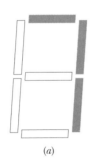

(*a*)

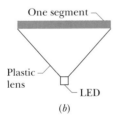

(*b*)

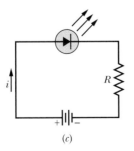

(*c*)

FIGURE 46-16 (*a*) The familiar seven-segment number display, activated to show the number ''7.'' (*b*) One segment of such a display. (*c*) An LED connected to a source of emf.

46-16*c* shows a typical circuit, in which the LED is forward-biased.

How can a *p-n* junction emit light? When an electron at the bottom of the conduction band of a semiconductor falls into a hole at the top of the valence band, an energy E_g is released, where E_g is the gap width. What happens to this energy? There are at least two possibilities. It might be transformed into thermal energy of the vibrating lattice and, with high probability, that is exactly what happens in a silicon-based semiconductor.

In some semiconducting materials, however, conditions are such that the emitted energy can also appear as electromagnetic radiation, the wavelength being given by

$$\lambda = \frac{c}{f} = \frac{c}{E_g/h} = \frac{hc}{E_g}. \qquad (46\text{-}9)$$

Commercial LEDs designed for the visible region are usually based on a semiconducting material that is a suitably chosen gallium–arsenic–phosphorus compound. By adjusting the ratio of phosphorus to arsenic, the gap width—and thus the wavelength of the emitted light—can be tailored to suit the need.

A question arises: If light is emitted when an electron falls from the conduction band to the valence band, will not light of that same wavelength be absorbed when an electron moves in the other direction, that is, from the valence band to the conduction band? It will indeed. To avoid having all the emitted photons absorbed, it is necessary to have a great surplus of both electrons and holes present in the material, in much greater numbers than would be generated by thermal agitation in the intrinsic semiconducting material.* These are precisely the conditions that result when majority carriers—be they electrons or holes—are injected across the central plane of a *p-n* junction by the action of an external potential difference. That is why a simple intrinsic semiconductor will not serve as an LED. You need a *p-n* junction! To provide lots of majority carriers (and thus lots of photons), the junction should be heavily doped and strongly forward-biased.

LEDs operating in the infrared are much used in optical communication systems based on optical fibers. The infrared region is chosen because the ab-

---

*If the surplus of electrons and holes is great enough, there may be a population inversion so that conditions for laser action are set up.

sorption per unit length of such fibers has a well-defined minimum at two different wavelengths.

In another application of the LED, the ends of a suitable *p-n* junction crystal are polished so that a slice of the crystal across the junction plane serves as a laser. Such a device is called a *laser diode;* Fig. 46-17 suggests its tiny scale.

SAMPLE PROBLEM 46-6

An LED is constructed from a *p-n* junction based on a certain Ga–As–P semiconducting material, whose energy gap is 1.9 eV. What is the wavelength of its emitted light?

SOLUTION   From Eq. 46-9 we have

$$\lambda = \frac{hc}{E_g} = \frac{(6.63 \times 10^{-34}\,\text{J}\cdot\text{s})(3.00 \times 10^8\,\text{m/s})}{(1.9\,\text{eV})(1.60 \times 10^{-19}\,\text{J/eV})}$$

$$= 6.5 \times 10^{-7}\,\text{m} = 650\,\text{nm}. \qquad \text{(Answer)}$$

Light of this wavelength is red.

46-12   THE TRANSISTOR (OPTIONAL)

The devices we have discussed so far have been diodes, that is, two-terminal devices. Here we introduce a three-terminal device, a **transistor.** As Fig. 46-18 suggests, the function of a transistor is to control a current flowing through the device from terminal D (the **drain**) to terminal S (the **source**) by varying the potential of terminal G (the **gate**).

For many applications, particularly those involving computers, we only need to be able to turn the

FIGURE 46-17   A laser diode developed at the AT&T Bell Laboratories. The cube at the right is a grain of salt.

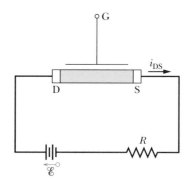

FIGURE 46-18   A representation of a transistor, showing the current i_{DS} moving through the device from the drain D to the source S. The magnitude of the current is controlled by the potential applied to G, the gate terminal.

drain–source current ON (gate open) or OFF (gate closed). One of these conditions corresponds to a "0" and the other to a "1" in the binary arithmetic on which the computer logic is based. We wish further that the gate terminal draw essentially no current from the circuit to which it is attached, thus interfering with its operation as little as possible. In more formal language, we say that we wish the transistor to have a high input impedance.

The central question proves to be: "How can we control the current in a conductor without making direct electric contact with it?" The perhaps surprising answer is that, by using the variable gate potential that is at our disposal, we can change the effective cross-sectional area of the conductor, going even so far as to reduce it to zero.

Figure 46-15 gives a clue. There we see that, by changing the bias potential of the *p-n* junction, we can control the width of the depletion zone. Simply by changing a potential we can effectively transform a conductor (*n*-type or *p*-type material) into a nonconductor (the depletion zone material). We can use the same trick in the transistor, but with a different geometrical arrangement.

Of the several types of transistor that are in common use, we choose to describe the MOSFET (Metal-Oxide-Semiconductor Field-Effect Transistor). Figure 46-19 shows its essential features.

A lightly doped *p*-type substrate has imbedded in it two "islands" of heavily doped *n*-type material, forming the drain D and the source S. These terminals are connected by a thin channel of *n*-type material, called the *n*-channel. An insulating layer of silicon dioxide (hence *O*xide in the acronym) is

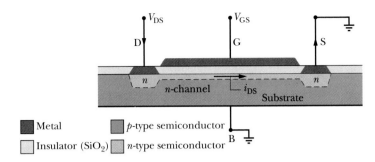

FIGURE 46-19 The construction of a MOSFET. The source S and base B are grounded, and a potential V_{DS} is applied to the drain terminal D. The magnitude of the current is controlled by the gate potential V_{GS}.

◼ Metal ◼ p-type semiconductor
◻ Insulator (SiO$_2$) ◻ n-type semiconductor

deposited on the substrate and penetrated by two metallic contacts (hence *M*etal) at D and S, so that electrical contact can be made with the drain and the source. A thin metallic layer—the gate G—is deposited opposite the n-channel. Note that the gate makes no ohmic contact with the transistor proper, being separated from it by the insulating oxide layer. Thus a MOSFET has the desired high input impedance, perhaps as high as $10^{15}\ \Omega$.

Consider first the situation with the source and the substrate grounded, the gate "floating" (that is, not connected electrically to a source of emf), and a positive potential V_{DS} applied to the drain. A drain–source current i_{DS} will be set up, as shown.

The potential difference across the boundary between the n-channel and the p-type substrate will vary from zero at the source end of the channel to V_{DS} at the drain end. The polarity is such (compare Fig. 46-15*a*) that the p-n junction that exists at this boundary is back-biased for essentially its full length. A depletion zone will exist at this boundary, increasing in thickness from the source end of the channel to the drain end. For these conditions, the n-channel will not have the same cross-sectional area along its length, being invaded by the depletion zone to a greater and greater extent as one proceeds along the channel from the source toward the drain.

The thickness of the depletion zone along its length can be influenced by the potential that we choose to apply to the gate. If we make the gate negative with respect to the source, electrons will be repelled from the n-channel into the substrate, thus

widening the depletion zone, constricting the channel and decreasing the drain–source current. Alternatively, a positive gate potential will attract electrons into the n-channel, narrow the depletion zone, widen the conducting channel, and increase the drain–source current. In this way a small change in the gate potential can generate a substantial change in the drain–source current, much as a valve controls the flow of water through a pipe.

Figure 46-20 shows a MOSFET (note the descriptive symbol) connected into a circuit as an amplifier. The input signal is applied to the gate and the output appears as a varying potential difference across a load resistor.

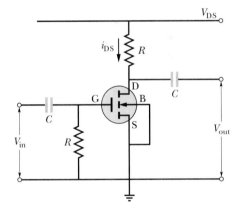

FIGURE 46-20 A MOSFET connected as an amplifier. A time-varying input signal V_{in} generates an amplified output signal V_{out}. The letters G, D, B, and S have the same meanings as in Figs. 46-18 and 46-19.

REVIEW & SUMMARY

Insulators, Metals, and Semiconductors

Quantum mechanics explains why some solids are electrical conductors, others are semiconductors, and why still others are insulators. When atoms are close to each other in a crystal lattice, their atomic energy levels become **bands** of allowed electron energies. In **insulators** the highest occupied level coincides with the top of a band; electrons are not able to accept additional kinetic energy from an applied field and so cannot conduct electricity (see Fig. 46-4). In **metals** the highest occupied level (the **Fermi energy**) falls somewhere in the middle of a band (Fig. 46-5 and Sample Problems 46-1 and 46-2). The highest occupied level in a **semiconductor** coincides, at $T = 0$, with the top of a band (the **valence band**) but the energy gap between it and the next highest band (the **conduction band**) is small enough so that charge-carrying electrons can "jump" into the conduction band because of thermal agitation (Fig. 46-8). The higher resistivity and the decrease of resistivity with increasing temperature are both easily explained by this model.

Density of States

Assuming uniform potential, the density of energy states available to electrons in the partially filled band of Fig. 46-5 is

$$n(E) = \frac{8\sqrt{2}\,\pi m^{3/2}}{h^3}\, E^{1/2} \qquad \text{(density of states)}. \qquad (46\text{-}3)$$

The Fermi Energy

At $T = 0$, electrons fill all the states up to the Fermi energy, with states above the Fermi energy being vacant. The Fermi energy corresponding to Eq. 46-3 is

$$E_{\mathrm{F}} = \frac{0.121 h^2}{m}\, n^{2/3} \qquad \text{(Fermi energy)}, \qquad (46\text{-}6)$$

in which n is the number of conduction electrons per unit volume; see Fig. 46-6.

The Probability Function

At temperatures above absolute zero the distribution of occupied states is found by multiplying the density of states by the **probability function**

$$p(E) = \frac{1}{e^{(E - E_{\mathrm{F}})/kT} + 1} \qquad \text{(probability function)} \qquad (46\text{-}7)$$

as illustrated in Fig. 46-7.

Doping: Donors and Acceptors

In practice, semiconductors are **doped** with controlled levels of selected impurities. Donor impurities contribute electrons to the conduction band and produce n-type semiconductors. Acceptor impurities contribute holes to the valence band and produce p-type semiconductors.

The p-n Junction

A $p\text{-}n$ junction (see Fig. 46-12) can serve as a diode rectifier. For forward-biasing (p positive with respect to n), the potential barrier is low, the junction is thin, and the forward current is large. For back-biasing, the potential barrier is high, the junction is thick, and the back current is small, usually negligibly so. The junction region itself, regardless of the applied potential difference, is called a **depletion layer**. It is virtually free of charge carriers and behaves like a somewhat leaky insulating slab.

The Light-Emitting Diode (LED)

A $p\text{-}n$ junction can, under certain circumstances, convert the energy lost by a charge carrier crossing the barrier into visible light whose wavelength is

$$\lambda = \frac{c}{f} = \frac{hc}{E_g}, \qquad (46\text{-}9)$$

E_g being the energy gap width.

Transistors

A **transistor** is a three-terminal solid-state device in which the current flowing from the **drain** to the **source** is controlled by varying the potential of a **gate;** see Fig. 46-18. Figure 46-19 shows a MOSFET-type transistor, in which a small variation of the potential difference V_{GS} between the gate G and the source S has a major controlling effect on the current i_{DS} between the drain D and the source S.

QUESTIONS

1. Do you think that any of the properties of solids listed in Section 46-1 are related to each other? If so, which?

2. Does the Fermi energy for a given metal depend on the volume of the sample? If, for example, you compare a sample whose volume is 1 cm^3 with one whose volume is twice that, the latter sample has just twice as many available conduction electrons; it might seem that you would have to go to higher energies to fill its available levels. Do you?

3. Why do the curves of Figs. 46-6c and 46-7c differ so little from each other?

4. The conduction electrons in a metallic sphere occupy states of quantized energy. Does the average energy interval between adjacent states depend on (a) the material of which the sphere is made, (b) the radius of the sphere, (c) the energy of the state, or (d) the temperature of the sphere?

5. What role does the Pauli exclusion principle play in accounting for the electrical conductivity of a metal?

6. Distinguish carefully among (a) the density of states $n(E)$, (b) the density of occupied states $n_o(E)$, and (c) the probability function $p(E)$, all of which appear in Eq. 46-4.

7. In what ways do the classical model and the quantum mechanical model for the electrical conductivity of a metal differ?

8. In Chapter 21 we showed that the molar specific heat of an ideal monatomic gas is $\frac{3}{2}R$. If the conduction electrons in a metal behave like such a gas, we would expect them to make a contribution of about this amount to the measured specific heat of a metal. However, this measured specific heat can be accounted for quite well in terms of energy absorbed by the vibrations of the ion cores that form the metallic lattice. The electrons do not seem to absorb much energy as the temperature of the specimen is increased. How does Fig. 46-7 provide an explanation of this prequantum-days puzzle?

9. If we compare the conduction electrons of a metal with the atoms of an ideal gas we are surprised to note that so much kinetic energy is locked into the conduction electron system at the absolute zero of temperature. Would it be better to compare the conduction electrons, not with the atoms of a gas, but with the inner electrons of a heavy atom? After all, a lot of kinetic energy is also locked up in this case, and we don't seem to find that surprising.

10. Give a physical argument to account qualitatively for the existence of allowed and forbidden energy bands in solids.

11. Is the existence of a forbidden energy gap in an insulator any harder to accept than the existence of forbidden energies for an electron in, say, the hydrogen atom?

12. In the band theory picture, what are the *essential* requirements for a solid to be (a) a metal, (b) an insulator, or (c) a semiconductor?

13. What can band theory tell us about solids that the classical model (see Section 28-6) cannot?

14. Distinguish between the drift speed and the Fermi speed of the conduction electrons in a metal.

15. Why is it that, in a solid, the allowed bands become wider as one proceeds from the inner to the outer atomic electrons?

16. Do pure (undoped) semiconductors obey Ohm's law?

17. At room temperature a given applied electric field will generate a drift speed for the conduction electrons of silicon that is about 40 times as great as that for the conduction electrons of copper. Why isn't silicon a better conductor of electricity than copper?

18. Consider these two statements. (a) At low enough temperatures silicon ceases to be a semiconductor and becomes a rather good insulator. (b) At high enough temperatures silicon ceases to be a semiconductor and becomes a rather good conductor. Discuss the extent to which each statement is either true or not true.

19. Which elements other than phosphorus are good candidates to use as donor impurities in silicon? Which elements other than aluminum are good candidates to use as acceptor impurities? Consult the periodic table given in Appendix E.

20. Identify the following as *p*-type or *n*-type semiconductors: (a) Sb in Si; (b) In in Ge; (c) Al in Ge; (d) P in Si.

21. How do you account for the fact that the resistivity of metals increases with temperature but that of semiconductors decreases?

22. The energy gaps for the semiconductors silicon and germanium are 1.14 and 0.67 eV, respectively. Which substance do you expect would have the higher density of charge carriers at room temperature? At the absolute zero of temperature?

23. Discuss this sentence: "The distinction between a metal and a semiconductor is sharp and clear-cut, but that between a semiconductor and an insulator is not."

24. What does a "hole" refer to in semiconductors?

25. Does the electrical conductivity of an intrinsic (undoped) semiconductor depend on the temperature? On the energy gap, E_g, between the full and empty bands?

26. Why does an *n*-type semiconductor have so many more electrons than holes? Why does a *p*-type semiconductor have so many more holes than electrons? Explain in your own words.

27. What does it mean to say that a *p-n* junction is biased in the forward direction?

28. A semiconductor contains equal numbers of donor and acceptor impurities. Do they cancel each other in their electrical effects? If so, what is the mechanism? If not, why not?

29. Germanium and silicon are similar semiconducting materials whose principal distinction is that the gap width E_g (see Fig. 46-8) is 0.67 eV for the former and 1.14 eV for the latter. If you wished to construct a p-n junction in which the back current is to be kept as small as possible, which material would you choose and why?

30. Consider two possible techniques for fabricating a p-n junction. (a) Prepare separately an n-type and a p-type sample and join them together, making sure that their abutting surfaces are plane and highly polished. (b) Prepare a single n-type sample and diffuse an excess acceptor impurity into it from one face, at high temperature. Which method is preferable and why?

31. In a p-n junction we have seen that electrons and holes may diffuse, in opposite directions, through the junction region. What is the eventual fate of each such particle as it diffuses into the material on the opposite side of the junction?

32. Does the diode rectifier whose characteristics are shown in Fig. 46-13 obey Ohm's law? What is your criterion for deciding?

33. We have seen that a simple intrinsic (undoped) semiconductor cannot be used as a light-emitting diode. Why not? Would a heavily doped n-type or p-type semiconductor work?

34. Explain how the MOSFET device of Fig. 46-19 works.

35. The acronym MOSFET stands for Metal-Oxide-Semiconductor Field-Effect Transistor. What is the significance of each of these terms as applied to the device shown in Fig. 46-19?

EXERCISES & PROBLEMS

SECTION 46-5 METALS: QUALITATIVE

1E. At what pressure, in atmospheres, would an ideal gas have a density of molecules equal to the density of the conduction electrons in copper ($= 8.43 \times 10^{28}$ m^{-3})? Assume $T = 300$ K.

2E. Gold is a monovalent metal with a molar mass of 197 g/mol and a density of 19.3 g/cm^3. Calculate the density of charge carriers.

3P. Calculate the number of particles per cubic meter for (a) the molecules of oxygen gas at 0°C and 1.0-atm pressure and (b) the conduction electrons in copper. (c) What is the ratio of these numbers? (d) What is the average distance between particles in each case? Assume that this distance is the edge of a cube whose volume is equal to the volume per particle. (See Sample Problem 28-3.)

4P. The density and molar mass of sodium are 971 kg/m^3 and 23 g/mol, respectively; the radius of the ion Na$^+$ is 98 pm. (a) What fraction of the volume of metallic sodium is available to its conduction electrons? (b) Carry out the same calculation for copper. Its density, molar mass, and ionic radius are, respectively, 8960 kg/m^3, 63.5 g/mol, and 135 pm. (c) For which of these two metals do you think the conduction electrons behave more like a free electron gas?

SECTION 46-6 METALS: QUANTITATIVE

5E. Use Eq. 46-6 to verify that the Fermi energy of copper is 7.0 eV. (Note, from Sample Problem 46-1, that the density of charge carriers in copper is 8.43×10^{28} m^{-3}.)

6E. (a) Show that Eq. 46-3 can be written as

$$n(E) = CE^{1/2},$$

where $C = 6.78 \times 10^{27}$ m$^{-3} \cdot$eV$^{-3/2}$. (b) Use this relation to verify a calculation of Sample Problem 46-3, namely, that for $E = 5.00$ eV, $n(E) = 1.52 \times 10^{28}$ m$^{-3} \cdot$eV^{-1}.

7E. Calculate the density $n(E)$ of conduction electron states in a metal for $E = 8.0$ eV and show that your result is consistent with the curve of Fig. 46-6a.

8E. What is the probability that a state 0.062 eV above the Fermi energy is occupied at (a) $T = 0$ K and (b) $T = 320$ K?

9E. The Fermi energy of copper is 7.0 eV. For copper at 1000 K, (a) find the energy at which the occupancy probability is 0.90. For this energy, evaluate (b) the density of states and (c) the density of occupied states.

10E. Show that Eq. 46-6 can be written as

$$E_F = An^{2/3},$$

where the constant A has the value 3.65×10^{-19} m$^2 \cdot$eV.

11E. The density of gold is 19.3 g/cm^3. Each atom contributes one conduction electron. Calculate the Fermi energy of gold.

12E. Figure 46-7c shows the density of occupied states $n_o(E)$ of the conduction electrons in a metal at 1000 K. Calculate $n_o(E)$ for copper for the energies $E = 4.00, 6.75, 7.00, 7.25,$ and 9.00 eV. The Fermi energy of copper is 7.00 eV.

13E. It can be shown that the conduction electrons in a metal behave like an ideal gas of the ordinary kind if the temperature is high enough. In particular, the temperature must be such that $kT \gg E_F$, the Fermi energy. What temperatures are required for copper ($E_F = 7.0$ eV) for this to be true? Study Fig. 46-7c in this connection and note that we have $kT \ll E_F$ for the conditions of that figure. This is just the reverse of the requirement cited above. Note also that copper boils at 2595°C.

14E. The Fermi energy of silver is 5.5 eV. (a) At $T = 0$°C, what are the probabilities that states at the following energies are occupied: 4.4 eV, 5.4 eV, 5.5 eV, 5.6 eV, 6.4 eV? (b) At what temperature will the probability that a state at 5.6 eV is occupied be 0.16?

15E. The Fermi energy of aluminum is 11.6 eV; its density is 2.70 g/cm³, and its molar mass is 27.0 g/mol (see Appendix D). From these data, determine the number of free electrons per atom.

16P. Show that the occupancy probabilities of two states whose energies are equally spaced above and below the Fermi energy add up to unity.

17P. Show that the probability p_h that a *hole* exists at a state of energy E is given by

$$p_h = \frac{1}{e^{-(E - E_F)/kT} + 1}.$$

(*Hint:* The existence of a hole means that the state is unoccupied; convince yourself that this implies that $p_h = 1 - p$.)

18P. Zinc is a bivalent metal. Calculate (a) the number of conduction electrons per cubic meter, (b) the Fermi energy E_F, (c) the Fermi speed v_F, and (d) the de Broglie wavelength corresponding to this speed. See Appendix D for needed data on zinc.

19P. Silver is a monovalent metal. Calculate (a) the number of conduction electrons per cubic meter, (b) the Fermi energy E_F, (c) the Fermi speed v_F, and (d) the de Broglie wavelength corresponding to this speed. Extract needed data from Appendix D.

20P. White dwarf stars represent a late stage in the evolution of stars like the sun. They become dense enough and hot enough that we can analyze their structure as a solid in which all Z electrons per atom are free. For a white dwarf with a mass equal to that of the sun and a radius equal to that of the Earth, calculate the Fermi energy of the electrons. Assume the atomic structure to be represented by iron atoms, and $T = 0$ K.

21P. A neutron star can be analyzed by techniques similar to those used for ordinary metals. In this case the neutrons (rather than electrons) obey the probability function, Eq. 46-7. Consider a neutron star of 2.0 solar masses with a radius of 10 km. Calculate the Fermi energy of the neutrons.

22P. Show that the density-of-states function given by Eq. 46-3 can be written in the form

$$n(E) = 1.5 n E_F^{-3/2} E^{1/2}.$$

Explain how it can be that $n(E)$ is independent of the material when the Fermi energy E_F (= 7.0 eV for copper, 9.4 eV for zinc, and so on) appears explicitly in this expression.

23P. Estimate the number N of conduction electrons in a metal that have energies greater than the Fermi energy

as follows. Strictly, N is given by

$$N = \int_{E_F}^{E_T} n(E)\, p(E)\ dE,$$

where E_T is the energy at the top of the band. By studying Fig. 46-7c, convince yourself that, to a good degree of approximation, this expression can be written as

$$N = \int_{E_F}^{E_F + 4kT} n(E_F)\left(\tfrac{1}{4}\right)\ dE.$$

By substituting the density of states function, evaluated at the Fermi energy, show that this yields for the fraction f of conduction electrons excited to energies greater than the Fermi energy,

$$f = \frac{N}{n} = \frac{3kT/2}{E_F}.$$

Why not evaluate the first integral above directly without resorting to an approximation?

24P. Use the result of Problem 23 to calculate the fraction of excited electrons in copper at temperatures of (a) absolute zero, (b) 300 K, and (c) 1000 K.

25P. At what temperature will the fraction of excited electrons in lithium equal 0.013? The Fermi energy of lithium is 4.7 eV. See Problem 23.

26P. Silver melts at 961°C. At the melting point, what fraction of the conduction electrons are in states with energies greater than the Fermi energy of 5.5 eV? See Problem 23.

27P. Show that, at the absolute zero of temperature, the average energy $\bar{E}$ of the conduction electrons in a metal is equal to $\tfrac{3}{5}E_F$, where E_F is the Fermi energy. (*Hint:* Note that, by definition of average, $\bar{E} = (1/n)\int E n_o(E)\ dE$.)

28P. Use the result of Problem 27 to calculate the total translational kinetic energy of the conduction electrons in 1.0 cm³ of copper at absolute zero.

29P. (a) Using the result of Problem 27, estimate how much energy would be released by the conduction electrons in a penny (assumed all copper; mass = 3.1 g) if we could suddenly turn off the Pauli exclusion principle. (b) For how long would this amount of energy light a 100-W lamp? Note that there is no known way to turn off the Pauli principle!

SECTION 46-8 DOPING

30E. The probability function of Section 46-6 can be applied to semiconductors as well as to metals. In semiconductors, E is the energy above the top of the valence band. The Fermi level for an intrinsic semiconductor is nearly midway between the top of the valence band and the bottom of the conduction band. For germanium these bands are separated by a gap of 0.67 eV. Calculate the probability that (a) a state at the bottom of the conduction band is occupied and (b) a state at the top of the valence band is unoccupied at 300 K.

31E. Pure silicon at room temperature has an electron density in the conduction band of approximately 1×10^{16} m^{-3} and an equal density of holes in the valence band. Suppose that one of every 10^7 silicon atoms is replaced by a phosphorus atom. (a) Which type will this doped semiconductor be, n or p? (b) What charge carrier density will the phosphorus add? (See Appendix D for needed data on silicon.) (c) What is the ratio of the charge carrier density in the doped silicon to that in the pure silicon?

32E. What mass of phosphorus would be needed to dope a 1.0-g sample of silicon to the extent described in Sample Problem 46-5?

33P. Doping changes the Fermi energy of a semiconductor. Consider silicon, with a gap of 1.11 eV between the valence and conduction bands. At 300 K the Fermi level of the pure material is nearly at the midpoint of the gap. Suppose that it is doped with donor atoms, each of which has a state 0.15 eV below the bottom of the conduction band, and suppose further that doping raises the Fermi level to 0.11 eV below the bottom of that band. (a) For both the pure and doped silicon, calculate the probability that a state at the bottom of the conduction band is occupied. (b) Also calculate the probability that a donor state in the doped material is occupied. See Fig. 46-21.

34P. A silicon sample is doped with atoms having a donor state 0.11 eV below the bottom of the conduction band. (a) If each of these states is occupied with probability 5.00×10^{-5} at $T = 300$ K, where is the Fermi level relative to the top of the valence band? (b) What then is the probability that a state at the bottom of the conduction band is occupied? The energy gap in silicon is 1.11 eV.

35P. In a simplified model of an intrinsic semiconductor (no doping), the actual distribution in energy of states is replaced by one in which there are N_v states in the valence band, all of these states having the same energy E_v, and N_c states in the conduction band, all of these states having the same energy E_c. The number of electrons in the conduction band equals the number of holes in the valence band. (a) Show that this last condition implies that

$$\frac{N_c}{e^{(E_c - E_F)/kT} + 1} = \frac{N_v}{e^{-(E_v - E_F)/kT} + 1}.$$

(*Hint:* See Problem 17.) (b) If the Fermi level is in the gap between the two bands and is far from both bands compared to kT, then the exponentials dominate in the denominators. Under these conditions, show that

$$E_F = \tfrac{1}{2}(E_c + E_v) + \tfrac{1}{2}kT \ln(N_v/N_c),$$

and therefore that, if $N_v \approx N_c$, the Fermi level is close to the center of the gap.

SECTION 46-9   THE p-n JUNCTION

36E. When a photon enters the depletion region of a p-n junction, electron–hole pairs can be created as electrons absorb part of the photon's energy and are excited from the valence band to the conduction band. These junctions are thus often used as detectors for photons, especially for x rays and nuclear gamma rays. When a 662-keV gamma-ray photon is totally absorbed by a semiconductor with an energy gap of 1.1 eV, on the average how many electron–hole pairs are created?

37P. For an ideal p-n junction diode, with a sharp boundary between the two semiconducting materials, the current i is related to the potential difference V across the diode by

$$i = i_0(e^{eV/kT} - 1),$$

where i_0, which depends on the materials but not on the current or potential difference, is called the *reverse saturation current*. V is positive if the junction is forward-biased and negative if it is back-biased. (a) Verify that this expression predicts the behavior expected of a diode by sketching i as a function of V over the range -0.12 V $< V <$ $+0.12$ V. Take $T = 300$ K and $i_0 = 5.0$ nA. (b) For the same temperature, calculate the ratio of the current for a 0.50-V forward-bias to the current for a 0.50-V back-bias.

SECTION 46-11   THE LIGHT-EMITTING DIODE (LED)

38E. (a) Calculate the maximum wavelength that will produce photoconduction in diamond, which has a band gap of 7.0 eV. (b) In what part of the electromagnetic spectrum does this wavelength lie?

39E. In a particular crystal, the highest occupied band of states is full. The crystal is transparent to light of wavelengths longer than 295 nm but opaque at shorter wavelengths. Calculate, in electron-volts, the gap between the highest occupied band and the next (empty) band.

40E. The KCl crystal has a band gap of 7.6 eV above the topmost occupied band, which is full. Is this crystal opaque or transparent to light of wavelength 140 nm?

41P. Fill in the seven-segment display shown in Fig. 46-16a to show how all 10 numbers may be generated. (b) If the numbers are displayed randomly, in what fraction of the displays will each of the seven segments be used?

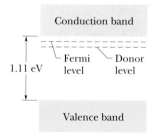

FIGURE 46-21   Problem 33.

THE LIQUID CRYSTAL STATE OF MATERIALS, OR, HOW DOES THE LIQUID CRYSTAL DISPLAY (LCD) IN MY WRISTWATCH WORK?

P. E. Cladis
AT&T Bell Laboratories

Patricia Elizabeth Cladis was born in Shanghai, China, and grew up in Vancouver, British Columbia. She received her Ph.D. in physics from the University of Rochester with a thesis on the dc superconducting transformer. Before joining AT&T Bell Laboratories, she did postdoctoral research at the University of Paris, Orsay, France, where she first learned about liquid crystals and discovered "escape into the third dimension" and point defects in nematics. At Bell Labs she discovered the "reentrant nematic" phase. Currently she uses liquid crystals to study general problems in nonlinear physics. She has published nearly 100 scientific papers and is on the editorial board of the journal Liquid Crystals.

More than 20 years ago, while at the Liquid Crystal Institute at Kent State University, Kent, Ohio, James Fergason realized that the large optical response of liquid crystals to small voltages could be used as a key component in electronic devices. A new technology was born as complex organic materials, the building blocks of nature, made their debut in the electronics industry as flat panel displays known as LCDs—liquid crystal displays. The first applications for LCDs were as low-information-density, alphanumeric displays for electronic products such as wristwatches, calculators, and "smart boxes" like the surveillance system shown in Fig. 1.

With increased understanding of their optoelectronic properties, LCDs moved toward higher-information-density applications involving pictures, as well as numbers and letters, required, for example, by lap-top PCs (Fig. 2) and color TVs. Liquid crystal color displays are awesome and need much less space than conventional displays. The LCD display package is flat and lightweight, and consumes little electrical power even when backlighting is added, as in the display shown in Fig. 2.

The purpose of this essay is to outline the physics involved in the operation of one picture element, or *pixel,* in the simplest LCD, the twisted nematic display. While the basic phys-

FIGURE 1 Alphanumeric LCDs easily transfer information between fast "smart" electronic services and human customers.

FIGURE 2   High-information-density LCDs showing pictures, as well as numbers and letters, are used in lap-top PCs.

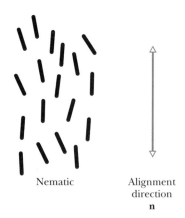

Nematic   Alignment
                 direction
                 n

FIGURE 3   Structure of the nematic liquid crystal phase: rodlike molecules align. The alignment is not perfect because molecules in the liquid state are in rapid, random thermal motion.

ics in the operation of one pixel is, roughly speaking, the same as in high-information-density displays, the passage from low-density to high-density displays required considerable R&D effort involving chemists, optoelectronic engineers, physicists,* materials scientists, and mathematicians, as well as the innovation of sophisticated manufacturing processes. Because of the expanding role computers play in our daily life, it is expected that before the turn of the century, LCDs will dominate the high-information-density display market.

Molecules That Align

In a crystal, molecules and atoms are organized in a three-dimensional structure where each atom or molecule does not move far from its position in the structure. Liquid phases have no structural order, so, while keeping close to each other, molecules or atoms move freely in the liquid state. Molecules are made up of many atoms and can exhibit a new state of matter called *liquid crystal* that is not as ordered as a solid crystal but also not as disordered as the usual liquid state, referred to as the *isotropic liquid* state.

Molecular shape plays an important role in determining liquid crystal phases. A molecule can be spherical, rodlike or cigar-shaped, disklike, bowl-shaped, or some other complex shape. Liquid crystal phases form when rod-shaped molecules spontaneously align while remaining liquid. In the liquid crystal state called *nematic,* which is used in the displays shown in Figs. 1 and 2, a degree of alignment is maintained despite

rapid, random thermal motion of individual molecules characteristic of the liquid state. The property of certain molecules to align is called *long-range orientational order.* It is the property that characterizes liquid crystals.

Molecules forming the nematic liquid crystal used in displays are 20 Å long and about 5 Å in diameter. The nematic state is a result of many molecules aligning cooperatively in the same direction. In discussing liquid crystal phases, one thinks in terms of many molecules, not one. Instead of referring to a single molecule, we talk about the collection of molecules in the nematic state and refer to the direction of alignment as the *director* or, in short-hand form, **n**, where **n** is a unit vector oriented in the direction chosen by many molecules for alignment. It is illustrated by lines, such as the ones shown in Fig. 3. The direction of alignment, **n**, defines an *optic axis* for materials in the nematic state.

Important physical principles that determine how LCDs work are as follows:

1. The orientation of **n** can be determined by small forces such as weak electric fields or surface forces.

2. When two different forces compete for the orientation of **n**, the response of **n** can be tuned by varying one of the forces.

---

*Indeed, P. G. de Gennes was awarded the 1991 Nobel prize in physics for his work on liquid crystals and polymers.

3. The behavior of polarized light traveling parallel to **n** is different from that of polarized light traveling perpendicular to **n**.

Determining the Orientation of n

Surface forces provide a useful way to select **n**. For example, when microscope slides are buffed in a single direction many times on a piece of white filter paper, a nematic liquid crystal in contact with the buffed surface orients with **n** following the buffing direction. Conventional wisdom is not that this is magic but rather that oil from the fingers is transferred to the glass surfaces. The buffing process produces microscopic grooves in the oil, providing an easy direction for the alignment of **n**.

To make a liquid crystal sample with a uniform **n**, the nematic liquid crystal is sandwiched between two buffed glass plates with the buffed surfaces facing the liquid and the directions of buffing on the two plates parallel to each other. In such a sample, **n** is parallel to the glass surfaces and parallel to the buffing direction throughout the whole sample.

Another way to orient **n** is with an *electric field* **E**. In some materials, **n** aligns parallel to **E** while in others, **n** aligns perpendicular to **E**. Materials for which **n** aligns parallel to **E** are called *positive materials*. The vector **n** orients in both ac and dc fields as there is no energy difference between the state **n** parallel to **E** and the state **n** antiparallel to **E**. Typically, LCDs use ac fields and positive materials.

When surface forces at electrodes pin **n** perpendicular to **E**, then a large-enough voltage has to be applied to a positive material before **n** responds: a *threshold voltage* $V_c = E_c/d$ (d is the distance between the electrodes) has to be applied for **E** to successfully compete with surface forces in the reorientation of **n**. For $V > V_c$, the component of **n** parallel to **E** is determined by the relative strengths of the applied electric field and the pinning force. A value of $V_c \approx 2$ V is a typical threshold voltage for nematics. As this is small by industry standards, relatively cheap electronics are needed to drive LCDs.

The Twisted Nematic

This is the configuration widely used in LCDs. A positive nematic liquid crystal is sandwiched between buffed glass plates that have transparent electrodes evaporated onto the buffed sides. The two plates are oriented with their respective buffing directions nearly perpendicular to each other. The director **n** twists nearly 90° smoothly between the two plates. A typical distance between electrodes is 6 μm. The polarization direction of incoming polarized light follows the gentle twist in **n**. When $V < V_c$, a pixel in the twisted state looks bright when viewed between crossed polar-

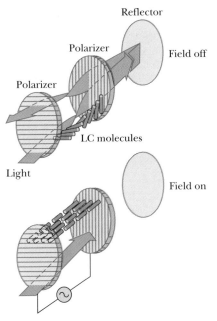

FIGURE 4 The simplest liquid crystal display uses the twisted nematic configuration. In the fully off state, shown in the upper part of the figure, the director twists nearly 90° between electrodes. The twisted structure is destroyed when **n** aligns parallel to **E**. In LCDs, the twisted nematic is sandwiched between crossed polarizers. A mirror reflects light back through the twisted structure to form the white background of the display. To display a character, an applied electric field destroys the twisted structure and no light reaches the mirror to be reflected back. The fully on state is shown in the bottom part of the figure.

izers. The upper part of Fig. 4 shows this.

When $V \gg V_c$, the optic axis is parallel to the electric field, destroying the twisted structure. The polarization direction of incoming polarized light does not rotate as the light traverses the liquid crystal and is extinguished by the second polarizer. When $V \gg V_c$, the pixel is as dark as allowed by the crossed polarizers. This is shown in the lower part of Fig. 4: the state used to show characters on a monochrome display. Intermediate shades of grey are observed for intermediate values of V. When the electric field is turned off, the orientation of **n** is again determined by surface forces and **n** relaxes back to the initially twisted configuration. The relaxation time depends on the distance between the electrodes and on an orientational diffusion constant.

Liquid crystal displays require little power because ambient light is used. Even with additional power consumption from backlighting used in higher-information-density displays, such as Fig. 2, the total power to run LCDs is smaller than conventional displays. Another LCD feature is that it works in white light, and so color displays can be made by adding color filters to the glass substrates. Indeed, liquid crystals make stunning color displays. Typical LCD switching times are about 10 ms. For some applications this is too slow. A major part of LCD research is to find ways to reduce this time.

The Effect of Temperature

Liquid crystals are useful for studying the general problem of how order in matter is created and destroyed. Indeed, some scientists are using nematic liquid crystals to model the birth of the universe immediately after the Big Bang! Still others are using liquid crystal materials to learn about nonlinear, nonequilibrium dynamic processes and transitions to chaos.

When rodlike molecules align, they are more densely packed than when randomly oriented. For example, matches arranged in a box are more closely packed than when thrown at random onto a table. At some low range of temperatures, rod-shaped molecules prefer to pack in the denser aligned state. As temperature is increased, the system becomes less dense and more energetic. The chance that all the molecules choose the same alignment is reduced. At a special temperature, called the *transition temperature*, there is a sudden change to another liquid state—to another phase—without long-range orientational order. A phase transition takes place from the nematic liquid crystal phase to the isotropic liquid.

This transition is easily observed when a nematic liquid crystal* is heated. In samples a few millimeters thick, the nematic state is translucent, like frosted glass, because of fluctuations in the alignment caused by thermal motion of the molecules. These fluctuations cause *local variations* in the index of refraction that scatter light. When heated into the isotropic liquid state, a thick sample becomes transparent because the molecules are now *uniformly* disordered.

Thus the LCD operates in a certain temperature range. If the temperature is too high, the material transforms to the isotropic liquid, losing orientational order, and the display loses contrast. If the temperature is too low, the material transforms to a more ordered liquid crystal state or, perhaps, even the crystalline state, and its orientation cannot be easily changed. Although materials are known that have a nematic state from $-50°C$ to $+400°C$, the nematic temperature range in any one compound is typically between 1 K and 20–30 K. To obtain the wide temperature range needed for applications, several different compounds are mixed together.

---

*Liquid crystals may be purchased from many chemical companies, such as EM Industries, 5 Skyline Drive, Hawthorne, NY 10532 and Roche Vitamins and Fine Chemicals, 340 Kingsland Street, Nutley, NJ 07110.

In a flat panel display such as used in today's lap-top computers, picture information reaches the screen through rows of electrodes on one substrate and columns of electrodes on the other. The intersection of the electrodes forms a grid of pixels. In a simple matrix drive system, electric signals are applied to the row and column electrodes with the proper timing to select the target pixel. Because a pixel responds to the rms voltage on a line, an *iron law* prevails that limits the number of lines that can be addressed. In the case of the twisted nematic, this is about 200 lines. When the nematic is twisted even more (between 180° and 270°, now called a *super-twisted* nematic, STN, display), the electro-optic response becomes steeper, making 768 lines available. But to make an STN display requires a liquid crystal that spontaneously twists: a chiral liquid crystal.

Chiral Liquid Crystals

In the twisted nematic display, the twist in the optic axis is determined by the surface treatment of the two glass plates. If a twist of more than 90° is applied, say, 120°, it costs less energy for **n** in nematic liquid crystals to satisfy this boundary condition by twisting only 60°. A nematic does not spontaneously twist and will always minimize the amount of twist applied by boundary conditions. When the display used is an STN display, where the twist is between 180° and 270°, the display contrast is excellent for a much wider viewing angle.

As **n** in a nematic will not twist through angles larger than 90°, a liquid crystal phase that spontaneously twists is used in STN displays. These materials are known as cholesteric liquid crystals because they were first derived from natural products like butter and cheese. Unlike the nematic display, where the director twists nearly 90° over a distance determined by the electrode separation, in cholesteric liquid crystals, **n** spontaneously twists 360° in a distance called the

pitch that is a characteristic of the material. Cholesteric liquid crystals are one example of *chiral* liquid crystal phases. The nematic is the special case of a cholesteric where the pitch is infinite.

The two ways of twisting are identified as left-handed and right-handed. In the schematic shown in Fig. 4, the director twists in a left-handed manner. To see this, extend the thumb of the left hand in the direction of twist, that is, perpendicular to the glass plates with thumb tip toward the second plate. On going through the liquid from the plate called ''one'' to the plate called ''two,'' fingers of the left hand naturally curl to maintain alignment with the director. Does it matter which plate is called ''one''? Fingers of the right hand curl in the opposite sense and follow a right-handed twist. It costs the same energy for a nematic liquid crystal to twist in a left-handed sense as in a right-handed sense. It costs no energy for a cholesteric liquid crystal to twist in one sense and a great deal to twist in the other.

Figure 5 shows half a pitch of the left-handed cholesteric structure. To make clearer the three-dimensional property of this helical structure, cylinders, rather than the simple lines used in Fig. 3, are used to depict **n** in Fig. 5. As cholesterics are also three-dimensional liquids, the cylinders are not on perfect rows or columns. A left hand is also shown in Fig. 5 to illustrate how handedness is determined. In the same way that the mirror image of a right hand is a left hand, the mirror image of a right-handed twisted structure is a left-handed one.

Cholesteric liquid crystals have an interesting way of interacting with light. They are *optically active*. While it is nontrivial to synthesize optically active materials in the laboratory, nature does it all the time. Indeed, in a Dorothy Sayers detective story, a deadly case of murder by mushroom poisoning was solved when it was discovered that the mushroom poison ingested by the victim was not opti-

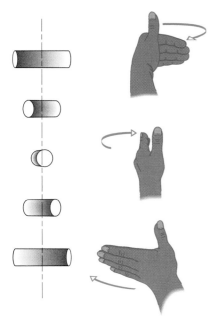

FIGURE 5 The structure of cholesteric liquid crystals. To capture the three-dimensional property of its helical structure, cylinders rather than lines are used to show the director orientation in the structure. The director **n** is perpendicular to the twist direction and rotates 360° uniformly in a fixed distance called the pitch. In the figure, half a pitch is shown. There are no layers in the cholesteric phase. The rotation sense of **n** is the same as the direction the fingers of one hand curl, shown by the arrows, when the thumb is extended in the direction **n** twists.

cally active! It had been synthesized in a university chemistry laboratory— not by the murderer, it turned out.

Unpolarized white light can be thought of as composed of both left and right *circular* polarizations rotating at all the frequencies corresponding to all the wavelengths constituting white light. A cholesteric liquid crystal with a pitch comparable to the wavelength of light transmits all the incident light *except* circularly polarized light with the *opposite* hand to its

twisted structure and of a particular wavelength corresponding to its pitch and the angle of incidence of the light. When light travels in the direction of twist, the wavelength equal to the pitch is rejected. Light traveling obliquely to the twist direction "sees" a shorter pitch, and so a shorter wavelength is scattered back. Viewed in ordinary diffuse daylight, the scattering of different wavelengths in different directions produces a striking display of vivid colors "recalling the appearance of a peacock's feathers."

In some materials the pitch depends sensitively on temperature. These materials are used as thermometers to translate small temperature differences into different colors. In some countries, newborns in a nursery have a small adhesive patch of encapsulated temperature-sensitive cholesteric on their foreheads. A single nurse can identify at a glance when one of them has a fever. A temperature map of complicated surfaces —for example, parts of the human body—is obtained by spraying a thin cholesteric liquid crystal layer onto the surface in question. In the medical field, temperature-sensitive cholesterics are an inexpensive, noninvasive way to detect at an early stage tumors, cancerous growth, and, in some Third World countries, leprosy, as the temperature of these tissues is usually different from the temperature of healthy tissue.

Discussion Topics

1. How would the two states of the LCD in Fig. 4 look if the polarizers were parallel to each other?

2. An idealized LCD is shown in Fig. 4 with the director exactly parallel to the glass plates. This is not practical for a display because there are two ways for **n** to turn toward **E**. Why is this bad for a display? Draw the picture with **n** rotated by a small angle θ toward **E** in one region and $-\theta$ in

another. How do the two regions connect? This problem is overcome by treating the glass substrates so that **n** is inclined a little to them. Why does this work? How do you think a small angle affects the sharpness of the optical response above threshold?

3. Harder question. A little bit of cholesteric material is added to most twisted nematic displays to ensure that only one hand is preferred in the twisted structure. Why is this a good idea? Draw a picture of a left-handed twist in one part of the sample and a right-handed one in the other. How do they connect up?

4. While V_c does not depend on the distance between electrodes, how does the time to turn off a pixel depend on this distance? *Hint*: The dimensions of a diffusion constant, D_0, are square centimeters per second and the distance between electrodes, d, is in centimeters. Put D_0 and d together to get a quantity that has units of time.

References for Further Reading
To learn more about:
The physics of liquid crystals: P. G. de Gennes, *The Physics of Liquid Crystals,* Clarendon Press, 1986. S. Chandrasekhar, *Liquid Crystals,* Cambridge University Press, 1977.
The twisted nematic and supertwisted displays: T. J. Scheffer and J. Nehring, in *Liquid Crystals, Applications and Uses,* Vol. I, B. Bahadur (ed.), World Scientific, 1990, pp. 232–274.
Video and color displays: E. Kaneko, *Liquid Crystal TV Displays: Principles and Applications of Liquid Crystal Displays,* Reidel Publishing, 1987.
For a popular article on another liquid crystal display concept: J. W. Doane, Polymer Dispersed Liquid Crystals: Boojums at Work, *MRS Bulletin,* January 1991, pp. 22–28.
Using nematic liquid crystals to model the birth of the universe: I. Chuang, N. Turok, and B. Yurke, *Physical Review Letters* **66**, 2472 (1991).
Using cholesteric liquid crystals to make patterns with a frequency: P. E. Cladis, J. T. Gleeson, P. L. Finn, and H. R. Brand, *Physical Review Letters* **67**, 3239 (1991).

NUCLEAR PHYSICS | 47

Radioactive nuclei, injected into a patient, collect in certain sites within the patient's body, undergo radioactive decay, and emit gamma rays. These gamma rays are recorded by a detector, and a color-coded image of the patient's body is produced on a monitor's screen. In the images reproduced here (the left one is a front view of a patient and the right one is a back view), you can tell just where the radioactive nuclei have collected (spine, pelvis, and ribs) by the color-coding of brown and orange. Why and how do nuclei undergo decay, and what exactly does "decay" mean?

47-1 DISCOVERING THE NUCLEUS

In the first years of this century not much was known about the structure of atoms beyond the fact that they contain electrons. The electron had been discovered (by J. J. Thomson) in 1897, and its mass was unknown in those early days. Thus it was not possible even to say just how many of these negatively charged electrons a given atom contained. Atoms were electrically neutral so they must also contain some positive charge, but nobody knew what form this compensating positive charge took.

In 1911 Ernest Rutherford proposed that the positive charge of the atom is densely concentrated at the center of the atom (in the *nucleus*) and that, furthermore, it is responsible for most of the mass of the atom. Rutherford's proposal was no mere conjecture but was based firmly on the results of an experiment suggested by him and carried out by his collaborators, Hans Geiger (of Geiger counter fame) and Ernest Marsden, a 20-year-old student who had not yet earned his bachelor's degree.

Rutherford's idea was to fire energetic alpha (α) particles at a thin target foil and measure the extent to which they were deflected as they passed through the foil. Alpha particles, which are about 7300 times more massive than electrons, have a charge of $+2e$ and are emitted spontaneously (with energies of a few MeV) by many radioactive materials. We now know that these useful projectiles are the nuclei of the atoms of ordinary helium. Figure 47-1 shows the experimental arrangement of Geiger and Marsden.

The experiment involves counting the number of α particles that are deflected through various scattering angles ϕ.

Figure 47-2 shows their results. Note especially that the vertical scale is logarithmic. We see that most of the α particles are scattered through rather small angles but—and this was the big surprise—a very small fraction of them are scattered through very large angles, approaching 180°. In Rutherford's words: "It was quite the most incredible event that ever happened to me in my life. It was almost as incredible as if you had fired a 15-inch shell at a piece of tissue paper and it came back and hit you."

Why was Rutherford so surprised? At the time of these experiments, most physicists believed in the so-called plum pudding model of the atom, which had been advanced by J. J. Thomson. In this view the positive charge of the atom was thought to be spread out through the entire volume of the atom. The electrons (the "plums") were thought to vibrate about fixed points within this sphere of charge (the "pudding").

The maximum deflecting force that could act on an α particle as it passed through such a large

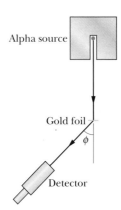

FIGURE 47-1   An arrangement (top view) used in Rutherford's laboratory in 1911–1913 to study the scattering of α particles by thin metal foils. The detector can be rotated to various values of the scattering angle ϕ. With this simple "tabletop" apparatus, the nucleus was discovered.

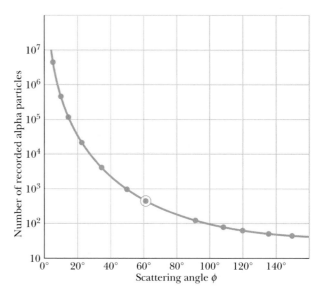

FIGURE 47-2   The dots are α-particle scattering data for a gold foil, obtained by Geiger and Marsden using the apparatus of Fig. 47-1. The solid curve is the theoretical prediction, based on the assumption that the atom has a small, massive, positively charged nucleus. Note that the vertical scale is logarithmic, covering six orders of magnitude. The data have been adjusted to fit the theoretical curve at the experimental point that is enclosed in a circle.

positive sphere of charge would be far too small to deflect the α particle by even as much as 1°. (The expected deflection has been compared to what you would observe if you fired a bullet through a sack of snowballs.) The electrons in the atom would also have very little effect on the massive, energetic α particle. They would, in fact, be themselves strongly deflected, much as a swarm of gnats would be brushed aside by a stone thrown through them.

Rutherford saw that, to deflect the α particle backward, there must be a large force; this force could be provided if the positive charge, instead of being spread throughout the atom, were concentrated tightly at its center. Then the incoming α particle could get very close to the center of the positive charge without penetrating it; such a close encounter would result in a large deflecting force.

Figure 47-3 shows possible paths taken by typical α particles as they pass through the atoms of the target foil. As we see, most are only slightly deflected, but a few (those whose extended incoming paths pass, by chance, very close to a nucleus) are deflected through large angles. From an analysis of the data, Rutherford concluded that the radius of the nucleus must be smaller than the radius of an atom

by a factor of about 10^4. In other words, the atom is mostly empty space! It is not often that the piercing insight of a gifted scientist, buttressed by a few simple calculations, leads to results of such importance.

SAMPLE PROBLEM 47-1

A 5.30-MeV α particle happens, by chance, to be headed directly toward the nucleus of an atom of gold ($Z = 79$). How close does the α particle get to the center of the gold nucleus before it comes momentarily to rest and reverses its course? Neglect the recoil of the relatively massive nucleus.

SOLUTION Initially the total mechanical energy of these two interacting bodies is just equal to K_α ($= 5.30$ MeV), the initial kinetic energy of the α particle. At the moment the α particle comes to rest, the total energy is the electrostatic potential energy of the two-body system. Because energy must be conserved, these two energies must be equal; so

$$K_\alpha = \frac{1}{4\pi\epsilon_0} \frac{Q_\alpha Q_{\text{Au}}}{d},$$

in which Q_α ($= 2e$) is the charge of the α particle, Q_{Au} ($= 79e$) is the charge of the gold nucleus, and d is the distance between the centers of the two bodies.

Substituting for the charges and solving for d yield

$$d = \frac{(2e)(79e)}{4\pi\epsilon_0 K_\alpha}$$

$$= \frac{(2 \times 79)(1.60 \times 10^{-19}\ \text{C})^2}{(4\pi)(8.85 \times 10^{-12}\ \text{F/m})(5.30\ \text{MeV})}$$

$$\times \frac{1}{(1.60 \times 10^{-13}\ \text{J/MeV})}$$

$$= 4.29 \times 10^{-14}\ \text{m} = 42.9\ \text{fm}. \qquad \text{(Answer)}$$

This is a small distance by atomic standards but not by nuclear standards. It is, in fact, considerably larger than the sum of the radii of the gold nucleus and the α particle. Thus the α particle reverses its course without ever "touching" the gold nucleus.

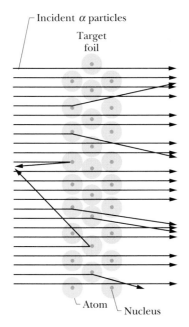

Incident α particles

Target foil

Atom   Nucleus

FIGURE 47-3 The angle through which an α particle is scattered depends on how close its extended incident path lies to an atomic nucleus. Large deflections result only from very close encounters.

47-2 SOME NUCLEAR PROPERTIES

Table 47-1 shows some properties of a few selected types of nuclei; such types are called **nuclides.** We discuss the table's several entries under separate headings.

TABLE 47-1
SOME PROPERTIES OF SELECTED NUCLIDES

NUCLIDE	Z	N	A	STABILITY[a]	MASS[b] (u)	RADIUS (fm)	BINDING ENERGY (MeV/nucleon)
^{1}H	1	0	1	99.985%	1.007825	—	—
^{7}Li	3	4	7	92.5%	7.016003	2.1	5.60
^{31}P	15	16	31	100%	30.973762	3.36	8.48
^{81}Br	35	46	81	49.3%	80.916289	4.63	8.69
^{120}Sn	50	70	120	32.4%	119.902199	5.28	8.51
^{157}Gd	64	93	157	15.7%	156.923956	5.77	8.21
^{197}Au	79	118	197	100%	196.966543	6.23	7.91
^{227}Ac	89	138	227	21.8 y	227.027750	6.53	7.65
^{239}Pu	94	145	239	24,100 y	239.052158	6.64	7.56

[a]For stable nuclides, the *isotopic abundance* is given; this is the fraction of atoms of this type found in a typical sample of the element. For radioactive nuclides, the half-life is given.

[b]Following standard practice, the reported mass is that of the neutral atom, not that of the bare nucleus.

Some Nuclear Terminology

Nuclei are made up of protons and neutrons. The number of protons in a nucleus (called the **atomic number** or **proton number** of the nucleus) is represented by the symbol Z; the number of neutrons (the **neutron number**) is represented by the symbol N. The total number of neutrons and protons in a nucleus is called its **mass number** A, so

$$A = Z + N. \qquad (47\text{-}1)$$

Neutrons and protons, when considered collectively, are called **nucleons.**

We represent nuclides with symbols such as those displayed in the first column of Table 47-1. Consider ^{197}Au, for example. The superscript (197) is the mass number A. The chemical symbol tells us that this element is gold, whose atomic number (see Appendix E) is 79. From Eq. 47-1 we see that the neutron number of this nuclide is 197 − 79 or 118.

Nuclides with the same atomic number Z but different neutron numbers N are called **isotopes.** As it happens, the element gold has 30 isotopes, ranging from ^{175}Au to ^{204}Au. Only one of them (^{197}Au) is stable, the remaining 29 being radioactive. Such *radionuclides* decay, by the spontaneous emission of a particle, with average lives (in this case) ranging from a few seconds to a few months.

Organizing the Nuclides

The neutral atoms of all isotopes for a given Z have the same number of electrons, the same chemical properties, and fit into the same box in the chemist's

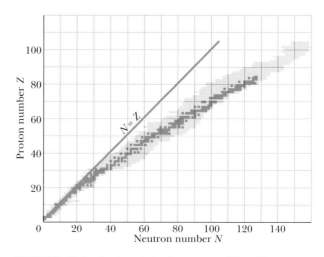

FIGURE 47-4 A plot of the known nuclides. The green shading identifies the band of stable nuclides, the beige shading the radionuclides. Light, stable nuclides have essentially equal numbers of neutrons and protons, but heavier nuclides have an increasingly larger excess of neutrons. The figure shows that there are no stable nuclides with $Z > 83$ (bismuth).

periodic table of the elements. The nuclear properties of the various isotopes, however, are very different. Thus the periodic table is of limited use to the nuclear physicist, the nuclear chemist, or the nuclear engineer.

We organize the nuclides on a *nuclidic chart* like that in Fig. 47-4, in which a nuclide is represented by plotting its proton number against its neutron number. The stable nuclides in this figure are represented by the green, the radionuclides by the beige. As you can see, the radionuclides tend to lie on either side of—and at the upper end of—a well-defined band of stable nuclides. Note too that light stable nuclides tend to lie close to the line $N = Z$, which means that they have the same numbers of neutrons and protons. Heavier nuclides, however, tend to have many more neutrons than protons. As an example, we saw that ^{197}Au has 118 neutrons and only 79 protons, a *neutron excess* of 39 neutrons.

Nuclidic charts are available as wall charts, in which each small box on the chart is filled with data about the nuclide it represents. Figure 47-5 shows a section of such a chart, centered on ^{197}Au. Relative abundances are shown for stable nuclides, half-lives (a measure of time) for radionuclides. The green line represents a line of *isobars*—nuclides of the same mass number, $A = 198$ in this case.

Nuclear Radii

A convenient unit for measuring distances on the scale of nuclei is the *femtometer* ($= 10^{-15}$ m), the prefix *femto* coming from the Danish word for 15. This unit is often called the *fermi;* the two names share the same abbreviation. Thus

$$1 \text{ femtometer} = 1 \text{ fermi} = 1 \text{ fm} = 10^{-15} \text{ m}. \quad (47\text{-}2)$$

We can learn about the size and structure of nuclei —among other ways—by bombarding them with a beam of high-energy electrons and observing the way the nuclei deflect the incident electrons. The energy of the electrons must be high enough (around 200 MeV) so that their de Broglie wavelengths will be small enough for them to act as structure-sensitive nuclear probes.

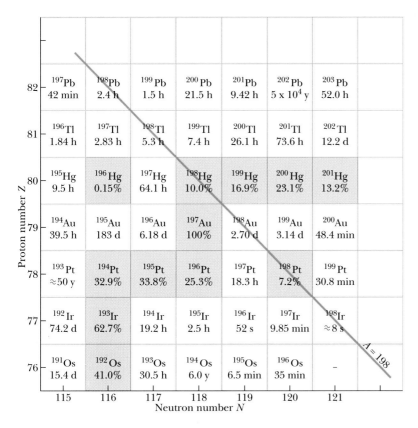

FIGURE 47-5 An enlarged section of the nuclidic chart of Fig. 47-4, centered on ^{197}Au. Shaded squares represent stable nuclides, for which relative isotopic abundances are given. Unshaded squares represent radionuclides, for which half-lives are given. Isobaric lines of constant mass number A slope as shown by the example line for $A = 198$.

Such experiments show that the nucleus (assumed to be spherical) has a characteristic mean radius R given by

$$R = R_0 A^{1/3}, \qquad (47-3)$$

in which A is the mass number and $R_0 \approx 1.2$ fm. We see that the volume of a nucleus, which is proportional to R^3, is directly proportional to the mass number A, being independent of the separate values of Z and N.

Nuclear Masses

Atomic masses can be measured with great precision using modern mass-spectrometer and nuclear-reaction techniques. Recall from Section 1-6 that such masses are reported in atomic mass units u, chosen so that the atomic mass (not the nuclear mass) of ^{12}C is exactly 12 u. The relation of this unit to the SI mass unit is, approximately,

$$1 \text{ u} = 1.661 \times 10^{-27} \text{ kg}. \qquad (47-4)$$

The mass number A of a nuclide is so named because the number represents the mass of the nuclide, expressed in atomic mass units and rounded off to the nearest integer. Thus the atomic mass of ^{197}Au is 196.966573 u, which we round to 197.

In nuclear reactions, Einstein's relation $E = \Delta m\, c^2$ is an indispensable work-a-day tool. The energy equivalent of 1 atomic mass unit can easily be shown to be 932 MeV. Thus c^2 can be written as 932 MeV/u, and we can use it to easily find the energy equivalent (in MeV) of any mass or mass difference (in u), or conversely.

Nuclear Binding Energies*

The total energy required to tear a nucleus apart into its constituent protons and neutrons can be calculated from $E = \Delta m\, c^2$ and is called the *nuclear binding energy*. If we divide the binding energy of a nucleus by its mass number we get the *binding energy*

per nucleon. Figure 47-6 shows a plot of this quantity as a function of mass number. The fact that this *binding energy curve* "droops" at both high and low mass numbers has practical consequences of the greatest importance.

The drooping of the binding energy curve at high mass numbers tells us that nucleons are more tightly bound when they are assembled into two middle-mass nuclides rather than into a single high-mass nuclide. In other words, energy can be released by the *nuclear fission*, or splitting, of a single massive nucleus into two smaller fragments.

The drooping of the binding energy curve at low mass numbers, on the other hand, tells us that energy will be released if two nuclides of small mass number combine to form a single middle-mass nuclide. This process, the reverse of fission, is called *nuclear fusion*. It occurs inside our sun and other stars and in thermonuclear explosions. Controlled nuclear fusion as a practical energy source is the subject of much current attention.

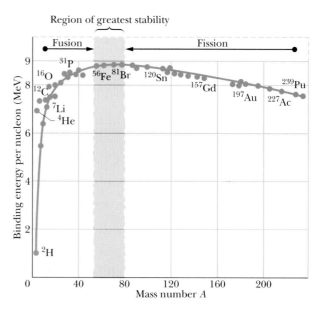

FIGURE 47-6 The curve of binding energy. The nuclides of Table 47-1 are shown, along with a few other nuclides of interest. Note the region of greatest stability. Fission can occur for heavier nuclides, and fusion for lighter nuclides. Note that the α particle (^4He) lies above the binding energy curve of its neighbors and is thus particularly stable.

*This subject is also treated in Section 8-8, which you may wish to reread at this time.

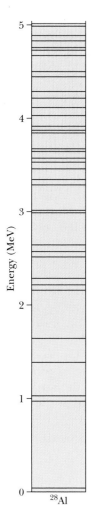

FIGURE 47-7 Energy levels for the nuclide ^{28}Al, deduced from nuclear reaction experiments.

Nuclear Energy Levels

Nuclei, like atoms, are governed by the laws of quantum physics and exist in discrete states of well-defined energy. Figure 47-7 shows these energy levels for a typical light nuclide, ^{28}Al. Note that the energy scale is in millions of electron-volts, rather than in electron-volts as for atoms. When a nucleus makes a transition from one level to a level of lower energy, the emitted photon is typically in the gamma-ray region of the electromagnetic spectrum.

Nuclear Spin and Magnetism

Many nuclides have an intrinsic *nuclear angular momentum* and an associated intrinsic *nuclear magnetic moment*. Although nuclear angular momenta are roughly of the same magnitude as the angular momenta of atomic electrons, nuclear magnetic moments are much smaller than typical atomic magnetic moments, by a factor of about 1000.

The Nuclear Force

The force that controls the motions of the atomic electrons is the familiar electromagnetic force. To bind the nucleus together, however, there must be a strong attractive nuclear force of a totally different kind, strong enough to overcome the repulsive force of the (positively charged) nuclear protons and to bind both protons and neutrons into the tiny nuclear volume. The nuclear force must also be of short range because its influence does not extend very far beyond the nuclear "surface."

The present view is that the nuclear force is not a fundamental force of nature but is due to the **strong force** that binds quarks together to form neutrons and protons (reread Section 2-9). In much the same way, certain electrically neutral molecules are held together to form solids by a secondary effect of the electromagnetic force that acts within the individual molecules.

SAMPLE PROBLEM 47-2

We can think of all nuclides as made up of a neutron–proton mixture that we can call *nuclear matter*. What is its density?

SOLUTION We know that this density will be high because virtually all the mass of the atom is found in its tiny central nucleus. The volume of a nucleus (assumed spherical) of mass number A and radius R is

$$V = (4/3)\pi R^3 = (4/3)\pi(R_0 A^{1/3})^3 = (4/3)\pi R_0^3 A.$$

Such a nucleus contains A nucleons so that its nucleon density, expressed in nucleons per unit volume, is

$$\rho_n = \frac{A}{V} = \frac{A}{(4/3)\pi R_0^3 A}$$

$$= \frac{3}{(4\pi)(1.2 \text{ fm})^3} = 0.138 \text{ nucleons/fm}^3.$$

The fact that we can speak of nuclear matter of constant density for all nuclides comes about because A cancels in the above equation. Nucleons seem to be packed into the nucleus like marbles in a sack.

The mass of a nucleon (neutron *or* proton) is about 1.67×10^{-27} kg. The mass density of nuclear matter in SI units is then

$$\rho = \left(0.138 \frac{\text{nucleons}}{\text{fm}^3}\right)\left(1.67 \times 10^{-27} \frac{\text{kg}}{\text{nucleon}}\right)$$
$$\times \left(10^{15} \frac{\text{fm}}{\text{m}}\right)^3$$
$$\approx 2 \times 10^{17} \text{ kg/m}^3. \qquad \text{(Answer)}$$

This is about 2×10^{14} times the density of water!

SAMPLE PROBLEM 47-3

a. How much energy is required to separate a typical middle-mass nucleus such as ^{120}Sn into its constituent nucleons?

SOLUTION We can find this energy from $E = \Delta m\, c^2$. Following standard practice, we carry out such calculations in terms of the masses of the neutral atoms involved, not those of the bare nuclei. As Table 47-1 shows, one *atom* of ^{120}Sn (nucleus + 50 electrons) has a mass of 119.902199 u. This atom can be separated into 50 hydrogen atoms (50 protons + 50 electrons) and 70 neutrons. Each hydrogen atom has a mass of 1.007825 u and each neutron a mass of 1.008665 u. The combined mass of the constituent particles is

$$m = 50 \times 1.007825 \text{ u} + 70 \times 1.008665 \text{ u}$$
$$= 120.99780 \text{ u}.$$

This exceeds the atomic mass of ^{120}Sn by

$$\Delta m = 120.99780 \text{ u} - 119.902199 \text{ u}$$
$$= 1.095601 \text{ u} \approx 1.096 \text{ u}.$$

Note that the masses of the 50 electrons cancel out, so that this same mass difference applies to separating a bare ^{120}Sn nucleus into 50 (bare) protons and 70 neutrons. In energy terms this mass difference becomes

$$E = \Delta m\, c^2 = (1.096 \text{ u})(932 \text{ MeV/u})$$
$$= 1021 \text{ MeV}. \qquad \text{(Answer)}$$

b. What is the binding energy per nucleon for this nuclide?

SOLUTION We have

$$E_n = \frac{E}{A} = \frac{1021 \text{ MeV}}{120} = 8.51 \text{ MeV/nucleon}, \quad \text{(Answer)}$$

in agreement with the value shown in Table 47-1.

47-3 RADIOACTIVE DECAY

As Fig. 47-4 shows, most of the nuclides that have been identified are radioactive. That is, a given nucleus spontaneously emits a particle, transforming itself in the process into a different nuclide, occupying a different square on the nuclidic chart.

Radioactive decay provided the first evidence that the laws that govern the subatomic world are statistical. Consider, for example, a 1-mg sample of uranium metal. It contains 2.5×10^{18} atoms of the very long-lived radionuclide ^{238}U. The nuclei of these particular atoms have existed without decaying since they were created—before the formation of our solar system—in the explosion of a supernova. During any given second about 12 of the nuclei in our sample will decay by emitting an α particle, transforming themselves into nuclei of ^{234}Th. However, in spite of that statistical fact:

There is absolutely no way to predict whether any given nucleus in the sample will be among the small number of nuclei that decay during the next second. All have an equal chance, namely, $12/(2.5 \times 10^{18})$ or one chance in 2×10^{17} per second.

We can express the statistical nature of the decay process by saying that if a sample contains N radioactive nuclei, then the rate $(= -dN/dt)$ at which nuclei decay is proportional to N:

$$-\frac{dN}{dt} = \lambda N, \qquad (47\text{-}5)$$

in which λ, the **disintegration constant,** has a characteristic value for every radionuclide. Equation 47-5 may be integrated to yield

$$N = N_0 e^{-\lambda t} \quad \text{(radioactive decay)}, \qquad (47\text{-}6)$$

in which N_0 is the number of radioactive nuclei in the sample at $t = 0$ and N is the number remaining at any subsequent time t. Note that light bulbs (for one example) follow no such exponential decay law. If we life-test 1000 bulbs, we expect that they will all "decay" (that is, burn out) at more or less the same time. The decay of radionuclides follows quite a different law.

We are often more interested in the decay rate R ($= -dN/dt$) than in N itself. Differentiating Eq. 47-6, we find

$$R = -\frac{dN}{dt} = \lambda N_0 e^{-\lambda t}$$

or

$$R = R_0 e^{-\lambda t} \quad \text{(radioactive decay)}, \quad (47\text{-}7)$$

an alternative form of the law of radioactive decay. Here R_0 ($= \lambda N_0$) is the decay rate at $t = 0$, and R is the rate at any subsequent time t. (In addition, using Eqs. 47-6 and 47-7, you can show that $R = \lambda N$ at any time t.) The decay rate R is often said to be the *activity* of the radionuclide. The unit for R is disintegrations per second or counts per second, the latter in the sense that the particles emitted in the disintegrations are counted (recorded) by a monitoring device such as a Geiger counter. Another unit for R is the curie (Ci), where 1 Ci is equal to 3.70×10^{10} disintegrations per second.

A quantity of special interest is the *half-life* τ, defined as the time after which both N and R are reduced to one-half their initial values. Putting $R = \frac{1}{2}R_0$ in Eq. 47-7 and substituting τ for t, we have

$$\tfrac{1}{2}R_0 = R_0 e^{-\lambda \tau}.$$

Solving for τ yields

$$\tau = \frac{\ln 2}{\lambda}, \quad (47\text{-}8)$$

a relation between the half-life and the disintegration constant.

SAMPLE PROBLEM 47-4

The table that follows shows some measurements of the decay rate of a sample of ^{128}I, a radionuclide often used medically as a tracer to measure the rate at which iodine is absorbed by the thyroid gland.

TIME (MIN)	R (COUNTS/S)	TIME (MIN)	R (COUNTS/S)
4	392.2	132	10.9
36	161.4	164	4.56
68	65.5	196	1.86
100	26.8	218	1.00

Find the disintegration constant and the half-life for this radionuclide.

SOLUTION If we take the natural logarithm of each side of Eq. 47-7, we find

$$\ln R = \ln R_0 - \lambda t.$$

Thus if we plot $\ln R$ against t, we should obtain a straight line whose slope is $-\lambda$. This is done in Fig. 47-8, from which we find

$$-\lambda = -\frac{6.2 - 0}{225 \text{ min} - 0}$$

or

$$\lambda = 0.0275 \text{ min}^{-1}. \quad \text{(Answer)}$$

We find the half-life readily from Eq. 47-8, obtaining

$$\tau = \frac{\ln 2}{\lambda} = \frac{\ln 2}{0.0275 \text{ min}^{-1}} \approx 25 \text{ min}. \quad \text{(Answer)}$$

FIGURE 47-8 Sample Problem 47-4. A semilogarithmic plot of the decay of a sample of ^{128}I, based on the data in the table. The half-life of this radionuclide (25 min) can be found from the slope of this curve.

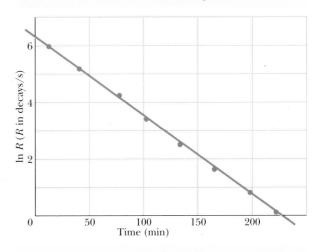

The decay rate of a given sample of ^{128}I will drop to half its initial value in 25 min, no matter what the initial value was. And the amount of ^{128}I in the sample will drop to half the initial amount in the same 25 min, no matter what that initial amount was.

SAMPLE PROBLEM 47-5

A 2.71-g sample of KCl from the chemistry stockroom is found to be radioactive and to decay at a constant rate of 4490 disintegrations/s. The decays are traced to the element potassium and in particular to the isotope ^{40}K, which constitutes 1.17% of normal potassium. Calculate the half-life of this nuclide.

SOLUTION We can find the half-life from Eq. 47-8. Since the decay rate is constant, the half-life must be very long and we cannot calculate λ by the method of Sample Problem 47-4. We must find it by determining both N and dN/dt in Eq. 47-5.

The molar mass of KCl is 74.6 g/mol, so the number of potassium atoms in the sample is

$$N_K = \frac{(6.02 \times 10^{23} \text{ mol}^{-1})(2.71 \text{ g})}{74.6 \text{ g/mol}} = 2.19 \times 10^{22}.$$

Of these, the number of ^{40}K atoms is

$$N_{40} = (2.19 \times 10^{22})(0.0117) = 2.56 \times 10^{20}.$$

From Eq. 47-5, we have

$$\lambda = \frac{-dN/dt}{N} = \frac{R_{40}}{N_{40}} = \frac{4490 \text{ s}^{-1}}{2.56 \times 10^{20}}$$

$$= 1.75 \times 10^{-17} \text{ s}^{-1}. \qquad \text{(Answer)}$$

The half-life follows from Eq. 47-8:

$$\tau = \frac{\ln 2}{\lambda} = \frac{(\ln 2)(1 \text{ y}/3.16 \times 10^7 \text{ s})}{1.75 \times 10^{-17} \text{ s}^{-1}}$$

$$= 1.25 \times 10^9 \text{ y}. \qquad \text{(Answer)}$$

This is of the order of magnitude of the age of the universe! No wonder we cannot measure the half-life of this radionuclide by waiting around for its decay rate to decrease. Interestingly, the potassium in our own bodies has its normal share of this radioisotope; we are all slightly radioactive.

47-4 ALPHA DECAY

The radionuclide ^{238}U decays by emitting an α particle, according to the scheme

$$^{238}\text{U} \rightarrow {}^{234}\text{Th} + {}^{4}\text{He}, \qquad Q = 4.25 \text{ MeV}. \quad (47\text{-}9)$$

The half-life of the decay is 4.47×10^9 y. Q is the disintegration energy of the process, that is, the amount of energy released during a single decay. We may well ask: "If energy is released in every such decay event, why did the ^{238}U nuclei not decay shortly after they were created? Why did they wait so long?" To answer this question, we must study the detailed mechanism of alpha decay.

We choose a model in which the α particle is imagined to exist (already formed) inside the nucleus before it escapes. Figure 47-9 shows the approximate potential energy function $U(r)$ for the α particle and the residual ^{234}Th nucleus as a function of their separation r. It is a combination of a potential well associated with the (attractive) strong nuclear force that acts in the nuclear interior and a Coulomb potential associated with the (repulsive) electrostatic force that acts between the two particles after the decay has occurred.

The horizontal black line marked $Q = 4.25$ MeV shows the disintegration energy for the process. If we

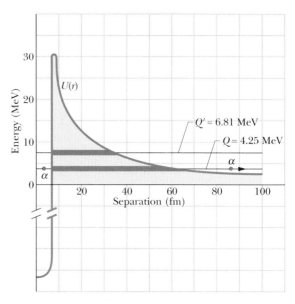

FIGURE 47-9 A potential energy function for the emission of an α particle by ^{238}U. The horizontal black line marked $Q = 4.25$ MeV shows the disintegration energy for the process. The thick green portion of this line represents separations r that are classically forbidden to the α particle. The α particle is represented by a dot, both inside the barrier (at the left) and outside it (at the right), after the particle has tunneled through. The horizontal black line marked $Q' = 6.81$ MeV shows the disintegration energy for the alpha decay of ^{228}U. (Both isotopes have the same potential energy function because they have the same number of protons.)

assume that this represents the total energy of the α particle during the decay process, then the part of the $U(r)$ curve above this line constitutes a potential energy barrier like that in Fig. 44-13. This barrier cannot be surmounted. If the α particle were able to be at a separation r within the barrier, its potential energy U would exceed its total energy E. This would mean, classically, that its kinetic energy K (which equals $E - U$) would be negative, an impossible situation.

We can see now why the α particle is not immediately emitted from the ^{238}U nucleus! That nucleus is surrounded by an impressive potential barrier, occupying—if you think of it in three dimensions—the volume lying between two spherical shells (of radii about 8 and 60 fm). This argument is so convincing that we now change our question and ask: "How can the ^{238}U nucleus ever emit an α particle? The particle seems permanently trapped inside the nucleus by the barrier." The answer is that, as you learned in Section 44-7, there is a finite probability that a particle can tunnel through an energy barrier that is classically insurmountable. And, in fact, alpha decay occurs as a result of barrier tunneling.

Since the half-life of ^{238}U is very long, the barrier must not be very "leaky." The α particle, presumed to be rattling back and forth within the nucleus, must arrive at the inner surface of the barrier about 10^{38} times before it succeeds in tunneling through the barrier. This is about 10^{21} times per second for about 4×10^9 years (the age of the Earth)! We, of course, are waiting on the outside, able to count only those α particles that *do* manage to escape.

We can test this explanation of alpha decay by examining other alpha emitters. For an extreme contrast, consider the alpha decay of another uranium isotope, ^{228}U, which has a disintegration energy Q' of 6.81 MeV, about 60% higher than that of ^{238}U. The value of Q' is also shown as a horizontal black line in Fig. 47-9. Recall from Section 44-7 that the transmission coefficient of a barrier is very sensitive to small changes in the total energy of the particle seeking to penetrate it. Thus we expect alpha

decay to occur more readily for this nuclide than for ^{238}U. Indeed it does. As Table 47-2 shows, its half-life is only 9.1 min! An increase in Q by a factor of only 1.6 produces a decrease in half-life (that is, in the effectiveness of the barrier) by a factor of 3×10^{14}. This is sensitivity indeed.

SAMPLE PROBLEM 47-6

We are given the following atomic masses:

^{238}U	238.05079 u	^{4}He	4.00260 u
^{234}Th	234.04363 u	^{1}H	1.00783 u
^{237}Pa	237.05121 u		

a. Calculate the energy released during the alpha decay of ^{238}U. The decay process is

$$^{238}\text{U} \rightarrow {}^{234}\text{Th} + {}^4\text{He}.$$

Note, incidentally, how nuclear charge is conserved in this equation: the atomic numbers of thorium (90) and helium (2) add up to the atomic number of uranium (92). The numbers of nucleons are also conserved ($238 = 234 + 4$).

SOLUTION The atomic mass of the decay products in the foregoing process ($234.04363 \text{ u} + 4.00260 \text{ u}$) is less than the atomic mass of ^{238}U by $\Delta m = 0.00456$ u, whose energy equivalent is

$$Q = \Delta m\, c^2 = (0.00456 \text{ u})(932 \text{ MeV/u})$$

$$= 4.25 \text{ MeV.} \qquad \text{(Answer)}$$

This disintegration energy appears as kinetic energy of the α particle and the recoiling ^{234}Th atom.

Note again that, following standard practice, we work with the masses of the neutral atoms rather than those of the bare nuclei; when we calculate the mass difference Δm, the masses of the extranuclear electrons cancel.

b. Show that ^{238}U cannot decay spontaneously by emitting a proton.

SOLUTION If this happened, the decay process would be

$$^{238}\text{U} \rightarrow {}^{237}\text{Pa} + {}^1\text{H}.$$

(You should verify that both nuclear charge and the numbers of nucleons are conserved in this process.) In this situation, the mass of the two decay products ($= 237.05121 \text{ u} + 1.00783 \text{ u}$) would *exceed* the mass of ^{238}U by $\Delta m = 0.00825$ u, the energy equivalence being $Q = -7.69$ MeV. The minus sign tells us that ^{238}U is stable against spontaneous proton emission.

TABLE 47-2
TWO ALPHA EMITTERS COMPARED

RADIONUCLIDE	Q	HALF-LIFE
^{238}U	4.25 MeV	4.5×10^9 y
^{228}U	6.81 MeV	9.1 min

47-5 BETA DECAY

A nucleus that decays spontaneously by emitting an electron or a positron (a positively charged particle with the mass of an electron) is said to undergo *beta decay*. This, like alpha decay, is a spontaneous process, with a definite disintegration energy and half-life. Again like alpha decay, beta decay is a statistical process, governed by Eqs. 47-6 and 47-7. Here are two examples:

$$^{32}\text{P} \rightarrow {}^{32}\text{S} + \text{e}^- + \nu \qquad (\tau = 14.3 \text{ d}) \qquad (47\text{-}10)$$

and

$$^{64}\text{Cu} \rightarrow {}^{64}\text{Ni} + \text{e}^+ + \nu \qquad (\tau = 12.7 \text{ h}). \qquad (47\text{-}11)$$

The symbol ν represents a *neutrino*, a massless, neutral particle that is emitted from the nucleus along with the electron during the decay process. Neutrinos interact only very weakly with matter and—for that reason—are so extremely difficult to detect that their presence long went unnoticed.*

Both charge and nucleon number are conserved in the above two processes. In the decay of Eq. 47-10, for example, we can write

$$(+15e) = (+16e) + (-e) + (0)$$
$$\text{(charge conservation)}$$

and

$$(32) = (32) + (0) + (0) \qquad \text{(nucleon conservation)},$$

where we have used the facts that neither the electron nor the neutrino is a nucleon and the neutrino carries no charge.

It may seem surprising that nuclei can emit electrons, positrons, and neutrinos since we have said that nuclei are made up of neutrons and protons only. However, we saw earlier that atoms emit photons, and we certainly do not say that atoms "contain" photons. We say that the photons are created during the emission process.

So it is with the electrons, positrons, and neutrinos emitted from nuclei during beta decay. They are created during the emission process, a neutron transforming itself into a proton within the nucleus (or conversely) according to

$$\text{n} \rightarrow \text{p} + \text{e}^- + \nu \qquad (47\text{-}12)$$

or

$$\text{p} \rightarrow \text{n} + \text{e}^+ + \nu. \qquad (47\text{-}13)$$

These are, in fact, the basic beta decay processes, and they provide evidence that—as was already pointed out—neutrons and protons are not truly fundamental particles. Note (as in Eqs. 47-10 and 47-11) that the mass number A of a nuclide undergoing beta decay does not change; one of its constituent nucleons simply changes its character according to Eq. 47-12 or 47-13, the total number of nucleons remaining fixed.

In both alpha decay and beta decay, the same amount of energy is released in every individual decay event. In the alpha decay of a particular radionuclide, every emitted α particle has the same sharply defined kinetic energy. However, in the beta decay of Eq. 47-12 for electron emission, the disintegration energy Q is shared—in varying proportions—between the electron and the neutrino. Sometimes the electron gets nearly all the energy, sometimes the neutrino does. In every case, however, the sum of the electron's energy and the neutrino's energy adds up to a constant value Q. Such sharing of energy, with a sum equal to Q, is also true of the beta decay of Eq. 47-13 for positron emission.

Thus in beta decay the energy of the emitted electrons or positrons may range from zero up to a

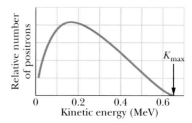

FIGURE 47-10 The distribution in kinetic energy of the positrons emitted in the beta decay of ^{64}Cu. The maximum kinetic energy of the distribution (K_{max}) is 0.653 MeV. In all the ^{64}Cu decay events, this energy is shared between the positron and the neutrino, in varying proportions.

---

*Beta decay also includes *electron capture*, in which a nucleus decays by absorbing one of its orbital electrons, emitting a neutrino in the process. We do not consider that process here. Also, the neutral particle emitted in the decay process of Eq. 47-10 is actually an *antineutrino*, a distinction that we do not wish to stress in this introductory treatment. Finally, whether the mass of the neutrino is truly zero is a subject under current investigation.

certain maximum K_{max}. Figure 47-10 shows the distribution of positron energies for the beta decay of ^{64}Cu (see Eq. 47-11). The maximum positron energy K_{max} must equal the disintegration energy Q because the neutrino carries away no energy when the positron carries away K_{max}. That is,

$$Q = K_{max}. \qquad (47-14)$$

The Neutrino

Wolfgang Pauli first suggested the existence of neutrinos in 1930. His neutrino hypothesis not only permitted an understanding of the energy distribution of electrons or positrons in beta decay but also solved another early beta decay puzzle involving "missing" angular momentum.

The neutrino is a truly elusive particle; the mean free path of an energetic neutrino in water has been calculated as no less than several thousand light years! At the same time, neutrinos left over from the Big Bang that marked the creation of the universe are the most abundant particles of physics. Billions upon billions of them pass through our bodies every second, leaving no trace.

In spite of their elusive character, neutrinos have been detected in the laboratory. The first, detected in 1953 by F. Reines and C. L. Cowan, were generated in a high-power nuclear reactor. In spite of the difficulties of detection, experimental neutrino physics is now a well-developed branch of experimental physics, with avid practitioners at several major laboratories throughout the world.

The sun emits neutrinos copiously from the nuclear furnace at its core and, at night, these messengers from the center of the sun come up at us from below, the Earth being almost totally transparent to them. In February 1987, light from an exploding star in the Large Magellanic Cloud (a nearby galaxy) reached the Earth after traveling for 170,000 years. Enormous numbers of neutrinos were generated in this explosion and about 10 of them were picked up by a sensitive neutrino detector in Japan; Fig. 47-11 shows a record of their passage.

Radioactivity and the Nuclidic Chart

Study of alpha and beta decay permits us to look at the nuclidic chart of Fig. 47-4 in a new way. Let us add a third dimension to that chart by plotting the mass excess (see the caption to Fig. 47-12) of each nuclide in a direction perpendicular to the N-Z plane. The surface so formed gives a graphic representation of nuclear stability. As Fig. 47-12 shows (for the light nuclides), this surface describes a "valley of the nuclides," the stability band of Fig. 47-4 running along its bottom. Nuclides beyond the region displayed in Fig. 47-12 decay into the valley largely by repeated alpha decay and by spontaneous fission. Nuclides on the proton-rich side of the valley decay into it by emitting positrons and those on the neutron-rich side do so by emitting electrons.

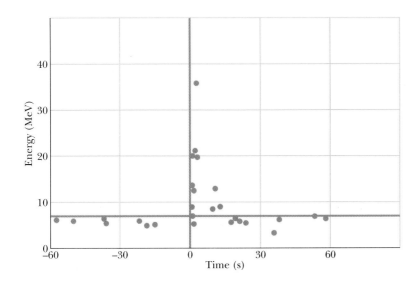

FIGURE 47-11 A burst of neutrinos from the supernova SN 1987A, which occurred at (relative) time 0, stands out from the usual reception of neutrinos. (For neutrinos, ten is a "burst!") They were detected by an elaborate detector, housed deep in a mine in Japan. The supernova was visible only in the southern hemisphere so that the neutrinos had to penetrate the Earth (a trifling barrier for them!) to reach the detector.

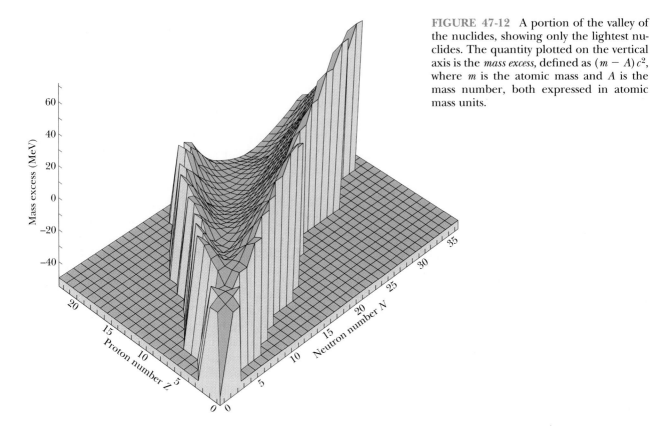

SAMPLE PROBLEM 47-7

Calculate the disintegration energy Q for the beta decay of ^{32}P, as described by Eq. 47-10. The needed atomic masses are 31.97391 u for ^{32}P and 31.97207 u for ^{32}S.

SOLUTION Because of the presence of the emitted electron, we must be especially careful to distinguish between nuclear and atomic masses. Let the boldface symbols $\mathbf{m}_P$ and $\mathbf{m}_S$ represent the nuclear masses of ^{32}P and ^{32}S and let the italic symbols m_P and m_S represent their atomic masses. We take the disintegration energy Q to be $\Delta m \, c^2$, where, from Eq. 47-10,

$$\Delta m = \mathbf{m}_P - (\mathbf{m}_S + m_e),$$

m_e being the mass of the electron. If we add and subtract $15 m_e$ on the right side of this equation, we obtain

$$\Delta m = (\mathbf{m}_P + 15 m_e) - (\mathbf{m}_S + 16 m_e).$$

The quantities in parentheses are the atomic masses of ^{32}P and ^{32}S, so

$$\Delta m = m_P - m_S.$$

We thus see that if we subtract the atomic masses in this way, the mass of the emitted electron is automatically taken into account.*

The disintegration energy for the ^{32}P decay is then

$$Q = \Delta m \, c^2$$
$$= (31.97391 \text{ u} - 31.97207 \text{ u})(932 \text{ MeV/u})$$
$$= 1.71 \text{ MeV}. \qquad \text{(Answer)}$$

Experimentally, this calculated quantity proves to be equal (as Eq. 47-14 requires) to K_{max}, the maximum energy of the emitted electrons. Although 1.71 MeV is released every time a ^{32}P nucleus decays, in essentially every case the electron carries away less energy than this. The neutrino gets essentially all the rest, carrying it undetected out of the laboratory.

---

*This is not, however, the case for positron emission.

47-6 RADIOACTIVE DATING

If you know the half-life of a given radionuclide, you can in principle use the decay of that radionuclide as a clock to measure a time interval. The decay of very long-lived nuclides, for example, can be used to measure the age of rocks, that is, the time that has elapsed since they were formed. Measurements for rocks from the Earth and the moon, and for meteorites, yield a consistent age of about 4.5×10^9 y for these bodies.

The radionuclide ^{40}K, for example, decays to a stable isotope ^{40}Ar of the noble gas argon. The half-life of this decay, as we saw in Sample Problem 47-5, is 1.25×10^9 y. By measuring the ratio of ^{40}K to ^{40}Ar found in the rock in question, one can then calculate the age of that rock. Other long-lived decays, such as that of ^{235}U to ^{207}Pb (involving a number of intermediate stages), can be used to verify these measurements.

For measuring shorter time intervals, in the range of historical interest, radiocarbon dating has proved invaluable. The radionuclide ^{14}C (with $\tau = 5730$ y) is produced at a constant rate in the upper atmosphere by the bombardment of atmospheric nitrogen by cosmic rays. This radiocarbon mixes with the carbon that is normally present in the atmosphere (as CO_2) so that there is about one atom of ^{14}C for every 10^{13} atoms of ordinary stable ^{12}C. The atmospheric carbon exchanges with the carbon in every living thing on Earth, including trees, tomatoes, rabbits, and humans, so that all living things contain a small fixed fraction of the ^{14}C nuclide.

This exchange persists as long as the organism is alive. After the organism dies, the exchange with the atmosphere stops and the amount of radiocarbon trapped in the organism, since it is no longer being replenished, dwindles away with a half-life of 5730 y. By measuring the amount of radiocarbon per gram of organic matter, it is possible to measure the time that has elapsed since the organism died. Charcoal from ancient camp fires, the Dead Sea scrolls, and many ancient artifacts have been dated in this way.

A fragment of the Dead Sea scrolls and the caves from which the scrolls were recovered. The age of the fragment can be determined by radiocarbon dating.

SAMPLE PROBLEM 47-8

Analysis of potassium and argon atoms in a moon rock sample by a mass spectrometer shows that the ratio of the number of (stable) ^{40}Ar atoms present to the number of (radioactive) ^{40}K atoms is 10.3. Assume that all the argon atoms were produced by the decay of potassium atoms, with a half-life of 1.25×10^9 y. How old is the rock?

SOLUTION If N_0 potassium atoms were present at the time the rock was formed by solidification from a molten form, the number of potassium atoms remaining at the time of analysis is, from Eq. 47-6,

$$N_K = N_0 e^{-\lambda t}, \qquad (47\text{-}15)$$

in which t is the age of the rock. For every potassium atom that decays, an argon atom is produced. Thus the number of argon atoms present at the time of the analysis is

$$N_{Ar} = N_0 - N_K. \qquad (47\text{-}16)$$

We cannot measure N_0. If we eliminate it from Eqs. 47-15 and 47-16, we find, after some algebra,

$$\lambda t = \ln\left(1 + \frac{N_{Ar}}{N_K}\right), \qquad (47\text{-}17)$$

in which N_{Ar}/N_K may be measured. Solving for t and replacing λ with $(\ln 2)/\tau$ yield

$$t = \frac{\tau \ln(1 + N_{Ar}/N_K)}{\ln 2}$$

$$= \frac{[1.25 \times 10^9 \text{ y}][\ln(1 + 10.3)]}{\ln 2}$$

$$= 4.37 \times 10^9 \text{ y}. \qquad \text{(Answer)}$$

Lesser ages may be found for other moon or Earth rock samples, but no substantially greater ones. This result may thus be taken as a good approximation to the age of the solar system.

47-7 MEASURING RADIATION DOSAGE

The effect of ionizing radiation such as gamma rays, electrons, and α particles on living tissue (particularly our own!) has become a matter of public interest. Such radiation is found in nature in cosmic rays and also arises from radioactive elements in the Earth's crust. Radiation associated with human activity also contributes, including diagnostic and therapeutic x rays and radiation from the radionuclides used in medicine and in industry. The disposal of radioactive waste and the evaluation of the probability of nuclear accidents continue to be dealt with at the level of national policy.

It is not our task here to explore the various sources of ionizing radiation but simply to describe the units in which the properties and effects of this radiation are expressed. There are four such units, and they are often used loosely or incorrectly in popular reporting.

1. *The curie (Ci).* This is a measure of the **activity** of a radioactive source. It is defined as

1 curie = 1 Ci

$$= 3.7 \times 10^{10} \text{ disintegrations per second.}$$

This definition says nothing about the nature of the decays. An example of the proper use of the curie is the following: "The activity of spent reactor fuel rod #5658 on January 15, 1987, was 9.5×10^4 Ci." The half-lives of the nuclides that make up the fuel rod and the types of radiation that they emit have no bearing on this activity measure.

2. *The Roentgen (R).* This is a measure of **exposure,** that is, of the ability of a beam of x rays or gamma rays or other radiation to deliver energy to a material through which they pass. Specifically, one roentgen is defined as the ability to deliver 8.78 mJ of energy to 1 kg of dry air at standard conditions. We might say, for example: "This dental x-ray beam provides an exposure of 300 mR/s." This says nothing about whether the energy actually reaches the patient or even whether there is a patient in the chair.

3. *The Rad.* This is an acronym for *radiation absorbed dose* and is a measure, as its name suggests, of the dose (energy) actually absorbed by a specific object. An object, which might be a person (whole body) or a specific part of the body (the hands, say), is said to have received an *absorbed dose* of 1 rad when 10 mJ/kg have been delivered to it by ionizing radiation. A typical statement using this unit is the following: "A whole-body short-term gamma-ray dose of 300 rad will cause death in 50% of the population exposed to it." By way of comfort we note that the present absorbed dose of radiation from sources of both natural and human origin is about 0.2 rad (= 200 mrad) per year.

4. *The Rem.* This is an acronym for *roentgen equivalent in man* and is a measure of *dose equivalent.* It takes account of the fact that, although different types of radiation (gamma rays and neutrons, say) may de-

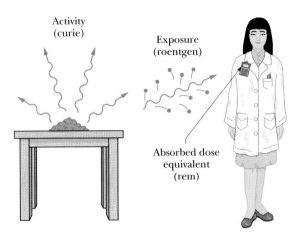

FIGURE 47-13 This sketch should help to clarify the distinction between the curie, the roentgen, the rad, and the rem.

liver the same energy per unit mass to the body, they do not have the same biological effect. The dose equivalent (in rems) is found by multiplying the absorbed dose (in rads) by a *relative biological effectiveness* (RBE) factor, which may be found tabulated in various reference sources. For x rays and electrons, RBE ≈ 1. For slow neutrons, RBE ≈ 5, and so on. Personnel-monitoring devices such as film badges are designed to register the dose equivalent in rems. An example of correct usage of the rem is the following: "The recommendation of the National Council on Radiation Protection is that no individual who is (nonoccupationally) exposed to radiations should receive a dose equivalent greater than 500 mrem (= 0.5 rem) in any one year." This includes radiation of all kinds, with the appropriate RBE being used for each kind. Figure 47-13 should help to clarify the four radiation units.

SAMPLE PROBLEM 47-9

We have seen that a gamma-ray dose of 300 rad is lethal to half of those exposed. If the equivalent energy were absorbed as heat, what rise in body temperature would result?

SOLUTION An absorbed dose of 300 rad corresponds to an absorbed energy per unit mass of

$$(300 \text{ rad}) \left(\frac{10 \times 10^{-3} \text{ J/kg}}{1 \text{ rad}} \right) = 3 \text{ J/kg}.$$

Assume that c, the specific heat of the human body, is the same as that of water, 4180 J/kg·K. The temperature rise follows from

$$\Delta T = \frac{Q/m}{c} = \frac{3 \text{ J/kg}}{4180 \text{ J/kg·K}}$$

$$= 7.2 \times 10^{-4} \text{ K} \approx 700 \text{ } \mu\text{K}. \quad \text{(Answer)}$$

Obviously the damage done by ionizing radiation has nothing to do with thermal heating. The harmful effects arise because the radiation succeeds in breaking molecular bonds and thus interfering with the normal functioning of the tissues in which it is absorbed.

47-8 NUCLEAR MODELS (OPTIONAL)

The structure of atoms is now well understood: quantum physics governs all; the force law is Coulomb's law; each atom contains a massive force center (the nucleus) that tends to dominate—and hence simplify—the physics. In principle, given enough computer time, we can calculate with confidence almost anything that we want to know about an atom.

Things are not in such a happy state for the nucleus. Quantum mechanics still governs its behavior, but the force law is complicated and cannot, in fact, be written down explicitly in full detail. Nor is there a natural force center to simplify the calculations; we are dealing with a many-body problem of great complexity.

In the absence of a comprehensive nuclear *theory*, we turn to the construction of nuclear *models*. A nuclear model is simply a way of looking at the nucleus that gives a physical insight into as wide a range of its properties as possible. The usefulness of a model is tested by its ability to provide predictions that can be verified experimentally in the laboratory.

Two models of the nucleus have proved useful. Although based on assumptions that seem flatly to exclude each other, each accounts very well for a selected group of nuclear properties. After describing them separately, we shall see how these two models may be combined to form a single coherent picture of the atomic nucleus.

The Liquid Drop Model

In the liquid drop model, formulated by Niels Bohr, the nucleons are imagined to interact strongly with each other, like the molecules in a drop of liquid. A given nucleon collides frequently with other nucleons in the nuclear interior, its mean free path as it moves about being substantially less than the nuclear radius. This constant "jiggling around" reminds us of the thermal agitation of the molecules in a drop of liquid.

The liquid drop model permits us to correlate many facts about nuclear masses and binding energies; it is useful (as we shall see later) in explaining nuclear fission. It also provides a useful model for understanding a large class of nuclear reactions.

Consider, for example, a generalized reaction of the form

$$X + a \rightarrow C^* \rightarrow Y + b. \qquad (47\text{-}18)$$

Here C^* represents an excited state of a so-called **compound nucleus** C. We imagine that projectile a enters target nucleus X, forming the compound nucleus C and conveying to it a certain amount of excitation energy. The projectile, perhaps a neutron, is at once caught up by the random motions that characterize the nuclear interior. It quickly looses its identity—so to speak—and the excitation energy it carried into the nucleus is quickly shared with all the other nucleons.

This quasistable state, represented by C^* in Eq. 47-18, may endure for as long as 10^{-16} s. By nuclear standards, this is a very long time, being one million times longer than the time required for a nucleon with a few MeV of energy to travel across a nucleus. The central feature of this compound-nucleus model is that the formation of this nucleus and its eventual decay are totally independent events. At the time of its decay, the nucleus has "forgotten" how it was formed. As an example, Fig. 47-14 shows three possible ways in which the compound nucleus ^{20}Ne* might be formed and three in which it might decay. Any of the three "formation" modes can lead to any of the three "decay" modes. All nine possible formation–decay combinations lead to the same set of energy levels for ^{20}Ne*.

The Independent Particle Model

In the liquid drop model, we assume that the nucleons move around at random and bump into each other frequently. The independent particle model, however, is based on just the opposite assumption, namely, that each nucleon moves in a well-defined orbit within the nucleus and hardly makes any collisions at all! The nucleus, unlike the atom, has no fixed center of charge; we assume in this model that each nucleon moves in a potential well that is determined by the smeared-out (time-averaged) motions of all the other nucleons.

A nucleon in a nucleus, like an electron in an atom, has a set of quantum numbers that defines its state of motion. Also, nucleons obey the Pauli exclusion principle, just as electrons do. That is, no two nucleons may occupy the same state at the same time. In this regard, the neutrons and the protons are treated separately, each having its own array of available quantized states.

The fact that nucleons obey the Pauli exclusion principle helps us to understand the relative stability of nucleon states. If two nucleons within the nucleus are to collide, the energy of each of them after the collision must correspond to the energy of an *unoccupied* state. If these states are filled, the collision simply cannot occur. In time, any given nucleon will undergo a *possible* collision, but meanwhile it will have made enough revolutions in its orbit to give meaning to the notion of a nucleon state with a quantized energy.

In the atomic realm, the repetitions of physical and chemical properties that we find in the periodic table are associated with the fact that the atomic electrons arrange themselves in shells that have a special stability when fully occupied. We can take the atomic numbers of the noble gases,

$$2, 10, 18, 36, 54, 86, \ldots,$$

as **magic electron numbers** that mark the completion (or closure) of such shells.

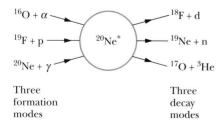

^{16}O + α ^{18}F + d

^{19}F + p ^{20}Ne* ^{19}Ne + n

^{20}Ne + γ ^{17}O + ^{3}He

Three formation modes   Three decay modes

FIGURE 47-14 The formation and decay modes of the compound nucleus ^{20}Ne*.

Nuclei also show such closed shell effects, associated with certain **magic nucleon numbers:**

$$2, 8, 20, 28, 50, 82, 126, \ldots .$$

Any nuclide whose proton number Z or neutron number N has one of these values turns out to have a special stability that may be made apparent in a variety of ways.

Examples of "magic" nuclides are ^{18}O ($Z = 8$), ^{40}Ca ($Z = 20$, $N = 20$), ^{92}Mo ($N = 50$), and ^{208}Pb ($Z = 82$, $N = 126$). Both ^{40}Ca and ^{208}Pb are said to be "doubly magic" because they contain filled shells of both protons *and* neutrons.

The magic number "2" shows up in the exceptional stability of the α particle (^{4}He), which, with $Z = N = 2$, is doubly magic. For example, the binding energy per nucleon for this nuclide stands well above that of its neighbors on the binding energy curve of Fig. 47-6. The α particle is so tightly bound, in fact, that it is impossible to add another particle to it; there is no stable nuclide with $A = 5$.

The central idea of a closed shell is that a single particle outside a closed shell can be relatively easily removed but that considerably more energy must be expended to remove a particle from the shell itself. The sodium atom, for example, has one (valence) electron outside a closed electron shell. Only about 5 eV is required to strip the valence electron away from a sodium atom; to remove a *second* electron, however (which must be plucked out of a closed shell), requires 22 eV. In a nuclear case, consider ^{121}Sb ($Z = 51$), which contains an "extra" proton outside a closed shell of 50 protons. To remove this proton requires 5.8 MeV; to remove a *second* proton, however, requires an energy of 11 MeV. Table 47-3 compares these atomic and nuclear situations. There is much additional experimental evidence that the nucleons in a nucleus form closed shells and that these shells exhibit stable properties.

We have seen that wave mechanics can account beautifully for the magic electron numbers, that is, for the populations of the orbitals into which atomic electrons are grouped. It turns out that, by making certain reasonable assumptions, wave mechanics can account equally well for the magic nucleon numbers! The 1963 Nobel prize was, in fact, awarded to Maria Mayer and Hans Jensen "for their discoveries concerning nuclear shell structure."

The Collective Model

Consider a nucleus in which a small number of neutrons (or protons) orbit outside a core of closed shells that contains a magic number of neutrons (or protons). The "extra" nucleons move in quantized orbits, in a potential well established by the central core, thus preserving the central feature of the independent particle model. These extra nucleons also interact with the core, deforming it and setting up "tidal wave" motions of rotation or vibration within it. These "liquid drop" motions of the core preserve the central feature of that model. This collective model of nuclear structure thus succeeds in combining the seemingly irreconcilable points of view of the liquid drop and independent particle models. It has been remarkably successful and perhaps represents the limits of what we can hope for in nuclear physics, given the absence of a better theory.

SAMPLE PROBLEM 47-10

Consider the neutron-capture reaction

$$^{109}\text{Ag} + \text{n} \rightarrow {}^{110}\text{Ag*} \rightarrow {}^{110}\text{Ag} + \gamma, \quad (47\text{-}19)$$

in which a compound nucleus (^{110}Ag*) is formed. Figure 47-15 shows the relative rate at which such events

TABLE 47-3
ATOMIC AND NUCLEAR SHELLS COMPARED

ENTITY	PARTICLES INVOLVED	ENERGY TO REMOVE THE FIRST PARTICLE	ENERGY TO REMOVE THE SECOND PARTICLE
Na atom	Electrons	5 eV	22 eV
^{121}Sb nucleus	Protons	5.8 MeV	11 MeV

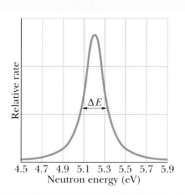

FIGURE 47-15 Sample Problem 47-10. A plot of the relative number of reaction events of the type described by Eq. 47-19, as a function of the energy of the incident neutron. The half-width ΔE of this resonance peak is about 0.20 eV.

take place, plotted against the energy of the incoming neutron. Use the uncertainty principle to find the mean lifetime of this compound nucleus.

SOLUTION We see that the relative production rate is sharply peaked at a neutron energy of about 5.2 eV. This suggests that we are dealing with a single excited

level of the compound nucleus ^{110}Ag*. When the available energy just matches the energy of this level above the ^{110}Ag ground state, we have "resonance" and the reaction really "goes."

However, the resonance peak is not infinitely sharp, having an approximate half-width (see ΔE in the figure) of about 0.20 eV. We account for this by saying that the excited level is not sharply defined in energy, having an energy uncertainty ΔE of about 0.20 eV.

We can use the uncertainty principle, written in the form

$$\Delta E \cdot \Delta t \approx h, \qquad (47\text{-}20)$$

to tell us something about any state of an atomic or nuclear system. We have seen that ΔE is a measure of the uncertainty with which the energy of the state can be defined. The quantity Δt, then, must be a measure of the time available to measure this energy. In fact, Δt is just $\bar{t}$, the mean life of the compound nucleus before it decays to its ground state.

Numerically, we have

$$\Delta t = \bar{t} \approx \frac{h}{\Delta E} = \frac{4.14 \times 10^{-15}\ \text{eV} \cdot \text{s}}{0.20\ \text{eV}}$$

$$\approx 2 \times 10^{-14}\ \text{s}. \qquad \text{(Answer)}$$

This is just the order of magnitude of the lifetime that we expect for a compound nucleus.

REVIEW & SUMMARY

The Nuclides

The nuclidic chart of Fig. 47-4 shows the approximately 2000 **nuclides** (atomic nuclei) that are known to exist. Each is characterized by an **atomic number** Z (the number of protons), a **neutron number** N, and a **mass number** A (the total number of **nucleons**—protons and neutrons). Thus $A = Z + N$. Nuclides with the same atomic number but different neutron numbers are **isotopes** of each other. Nuclei have a mean radius R given by

$$R = R_0 A^{1/3}, \qquad (47\text{-}3)$$

where $R_0 \approx 1.2$ fm (1 fm = 1 femtometer = 1 fermi = 10^{-15} m).

Mass–Energy Exchanges

Nuclear masses, universally reported as the masses of the corresponding neutral atoms, are useful in calculating disintegration energies, binding energies, and so on. The energy equivalent of one mass unit (u) is 932 MeV. The curve of binding energies (see Fig. 47-6 and Sample Prob-

lem 47-3) shows that middle-mass nuclides are the most stable and that energy can be released both by fission of heavy nuclei and by fusion of light nuclei.

The Nuclear Force

Nuclei are held together by an attractive force acting between the nucleons. It is thought to be a residual effect of the **strong force** acting between the quarks that make up the nucleons. Nuclei can exist in a number of discrete energy states (see Fig. 47-7) each with a characteristic intrinsic angular momentum and magnetic moment.

Radioactive Decay

Most known nuclides are radioactive; they spontaneously decay at a rate $R \ (= -dN/dt)$ that is proportional to the number N of radioactive atoms present, the proportionality constant being the **disintegration constant** λ. This leads to the law of exponential decay:

$$N = N_0 e^{-\lambda t}, \qquad R = \lambda N = R_0 e^{-\lambda t}$$

$$\text{(radioactive decay).} \qquad (47\text{-}6, 47\text{-}7)$$

The half-life $\tau = (\ln 2)/\lambda$ is the time required for the decay rate R (or the number N) to drop to half its initial value.

Alpha Decay

Some nuclides decay by emitting an α particle. Such decay is inhibited by a potential barrier, as in Fig. 47-9, that cannot be penetrated according to classical mechanics but is subject to tunneling according to wave mechanics. The barrier penetrability, and thus the half-life for alpha decay, is very sensitive to the energy of the α particle.

Beta Decay

In **beta decay** either an electron or a positron is emitted, along with a neutrino. They share the available disintegration energy between them. The electrons and positrons emitted in beta decay have a continuous spectrum of energies up to a limit K_{max} $(= Q = \Delta m\, c^2)$.

Radioactive Dating

Naturally occurring radioactive nuclides provide a means for estimating the dates of historic and prehistoric events. For example, the ages of organic materials can often be found by measuring their ^{14}C content; rock samples can be dated using radioactive ^{40}K.

Radiation Dosage

Four units are used to describe exposure to ionizing radiation. The **curie** (1 Ci = 3.7×10^{10} disintegrations per second) measures the **activity** of a source. The **roentgen** (an amount of radiation that can deliver 8.78 mJ of energy to 1 kg of dry air) is a unit of **exposure.** The amount of energy actually absorbed is measured in **rads,** with 1 rad corresponding to 10 mJ/kg. The estimated biological effect of the absorbed energy is quoted in **rems,** so that 1 rem of slow-neutron energy would cause the same biological effect as 1 rem of x rays even though only about one-fifth as much energy was absorbed from the neutrons.

Nuclear Models

The **liquid drop** model of nuclear structure assumes that nucleons collide constantly and that relatively long-lived **compound nuclei** are formed when a projectile is captured. The formation of a compound nucleus and the eventual decay of that nucleus are totally independent events.

The **independent particle** model of nuclear structure assumes that each nucleon moves, essentially without collisions, in a quantized orbit within the nucleus. The model predicts nucleon levels and **magic numbers** of nucleons (2, 8, 20, 28, 50, 82, and 126) associated with closed shells of nucleons; nuclides with any of these numbers of neutrons or protons are particularly stable.

The **collective** model, in which extra nucleons move in quantized orbits about a central core of closed shells, is highly successful in predicting many nuclear properties.

QUESTIONS

1. When a thin foil is bombarded with α particles, a few of them are scattered back toward the source. Rutherford concluded from this that the positive charge of the atom —and also most of its mass—must be concentrated in a very small "nucleus" within the atom. What was his line of reasoning?

2. In what basic ways do the so-called strong force and the electrostatic force differ?

3. Why does the *relative* importance of the Coulomb force compared to the strong nuclear force increase as mass numbers increase?

4. In your body, are there more neutrons than protons? More protons than electrons? Discuss.

5. Why do nuclei tend to have more neutrons than protons at high mass numbers?

6. How might the equality $1\ u = 1.661 \times 10^{-27}$ kg be arrived at in the laboratory?

7. The atoms of a given element may differ in mass, have different physical characteristics, and yet not vary chemically. Why is this?

8. The deviation of isotopic masses from integer values is due to many factors. Name some. Which is most responsible?

9. The most stable nuclides have a mass number A near 60 (see Fig. 47-6). Why don't *all* nuclides have mass numbers near 60?

10. If we neglect the very lightest nuclides, the binding energy per nucleon in Fig. 47-6 is roughly constant at 7 or 8 MeV/nucleon. Do you expect the mean electronic binding energy per electron in atoms also to be roughly constant throughout the periodic table?

11. Why is the binding energy per nucleon (Fig. 47-6) low at low mass numbers? At high mass numbers?

12. In the binding energy curve of Fig. 47-6, what is special or notable about the nuclides ^{2}H, ^{4}He, ^{56}Fe, and ^{239}Pu?

13. A particular ^{238}U nucleus was created in a massive stellar explosion, perhaps 10^{10} y ago. It suddenly decays by alpha emission while we are observing it. After all those years, why did it decide to decay just now?

14. Can you justify this statement: "In measuring half-lives by the method of Sample Problem 47-4, it is not necessary to measure the absolute decay rate R; any quantity proportional to it will suffice. However, in the method of Sample Problem 47-5 an absolute rate *is* needed."

15. Does the temperature affect the rate of decay of radioactive nuclides? If so, how?

16. You are running longevity tests on light bulbs. Do you expect their "decay" to be exponential? What is the essential difference between the decay of light bulbs and of radionuclides?

17. The half-life of ^{238}U is 4.47×10^9 y, about the age of the solar system. How can such a long half-life be measured?

18. The half-life of ^{238}U is 4.47×10^9 years. How might measurements of its activity be used to determine the age of uranium-containing rocks? How do you get around the fact that you don't know how much ^{238}U was present in the rocks to begin with? (*Hint:* What is the ultimate decay product of ^{238}U?)

19. The half-life of ^{14}C is 5730 years. This isotope is produced in the upper atmosphere at an assumed constant rate by cosmic-ray bombardment. How might measurements of its activity play a role in dating ancient carbon-containing specimens, such as wooden Egyptian artifacts or the remnants of ancient camp fires?

20. Explain why, in alpha decay, short half-lives correspond to large disintegration energies, and conversely.

21. A radioactive nucleus can emit a positron, e^+. This corresponds to a proton in the nucleus being converted to a neutron. The mass of a neutron, however, is greater than that of a proton. How then can positron emission occur?

22. In beta decay the emitted particles form a continuous energy spectrum, but in alpha decay the emitted particles form a discrete spectrum. What difficulties did this difference cause in the explanation of beta decay, and how were these difficulties finally overcome?

23. How do neutrinos differ from photons? Each has zero charge and (presumably) zero rest mass and travels at the speed of light.

24. What are the basic assumptions of the liquid drop and the independent particle models of nuclear structure? How do they differ? How does the collective model reconcile these differences?

25. Does the liquid drop model of the nucleus give us a picture of the following phenomena: (a) acceptance by the nucleus of a colliding particle; (b) loss of a particle by spontaneous emission; (c) fission; (d) dependence of stability on energy content?

26. What is so special ("magic") about the magic nucleon numbers?

27. Why aren't the magic nucleon numbers and the magic electron numbers the same? What accounts for each?

EXERCISES & PROBLEMS

SECTION 47-1 DISCOVERING THE NUCLEUS

1E. Calculate the distance of closest approach for a head-on collision between a 5.30-MeV α particle and the nucleus of a copper atom.

2E. Assume that a gold nucleus has a radius of 6.23 fm and an α particle has a radius of 1.8 fm. What minimum energy must an incident α particle have to penetrate the gold nucleus?

3P. When an α particle collides elastically with a nucleus, the nucleus recoils. Suppose a 5.00-MeV α particle has a head-on elastic collision with a gold nucleus, initially at rest. What is the kinetic energy (a) of the recoiling nucleus and (b) of the rebounding α particle?

SECTION 47-2 SOME NUCLEAR PROPERTIES

4E. A neutron star is a stellar object whose density is about that of nuclear matter, as calculated in Sample Problem 47-2. Suppose that the sun were to collapse into such a star without losing any of its present mass. What would be its radius?

5E. The nuclide ^{14}C contains how many (a) protons and (b) neutrons?

6E. The radius of a nucleus is measured, by electron-scattering methods, to be 3.6 fm. What is the likely mass number of the nucleus?

7E. Locate the nuclides displayed in Table 47-1 on the nuclidic chart of Fig. 47-4. Verify that they lie along the stability zone.

8E. Using a nuclidic chart, write the symbols for (a) all stable isotopes with $Z = 60$, (b) all radioactive nuclides with $N = 60$, and (c) all isobars with $A = 60$.

9E. The electrostatic potential energy of a uniform sphere of charge Q and radius R is

$$U = \frac{3Q^2}{20\pi\epsilon_0 R}.$$

(a) Find the electrostatic potential energy for the nuclide ^{239}Pu, assumed spherical (see Table 47-1). (b) For this nuclide, compare the electrostatic potential energy per nucleon, and also per proton, with the binding energy per nucleon of 7.56 MeV. (c) What do you conclude?

10E. The strong neutron excess of heavy nuclei is illustrated by the fact that most heavy nuclides could never fission or break up into two stable nuclei without neutrons being left over. For example, consider the spontaneous fission of a ^{235}U nucleus. If the two daughter nuclei had atomic numbers 39 and 53, and were stable, by referring to Fig. 47-4, determine the daughter nuclides and the number of neutrons left over.

11E. Arrange the 25 nuclides $^{118-122}$Te, $^{117-121}$Sb, $^{116-120}$Sn, $^{115-119}$In, and $^{114-118}$Cd in squares in a nuclidic chart similar to Fig. 47-5. Draw in and label (a) all isobaric (constant A) lines and (b) all lines of constant neutron excess, defined as $N - Z$.

12E. Calculate and compare (a) the nuclear mass density ρ_m and (b) the nuclear charge density ρ_q for the fairly light nuclide ^{55}Mn and for the fairly heavy nuclide ^{209}Bi. (c) Are the differences what you would expect?

13E. Verify the binding energy per nucleon given in Table 47-1 for ^{239}Pu, 7.56 MeV/nucleon. The needed atomic masses are 239.05216 u (^{239}Pu), 1.00783 u (^{1}H), and 1.00867 u (neutron).

14E. (a) Show that an approximate formula for the mass M of an atom is

$$M = Am_p,$$

where A is the mass number and m_p is the proton mass. (b) What percent error is committed in using this formula to calculate the masses of the atoms in Table 47-1? The mass of the bare proton is 1.007276 u. (c) Is this formula accurate enough to be used for calculations of nuclear binding energy?

15E. The characteristic nuclear time is a useful but loosely defined quantity, taken to be the time required for a nucleon with a few MeV of kinetic energy to travel a distance equal to the diameter of a middle-mass nuclide. What is the order of magnitude of this quantity? Consider 5-MeV neutrons traversing a nuclear diameter of ^{197}Au; use Table 47-1.

16E. Nuclear radii may be measured by scattering high-eneregy electrons from nuclei. (a) What is the de Broglie wavelength for 200-MeV electrons? (c) Are they suitable probes for this purpose?

17E. Because a nucleon is confined to a nucleus, we can take the uncertainty in its position to be approximately the nuclear radius R. What does the uncertainty principle say about the kinetic energy of a nucleon in a nucleus with, say, $A = 100$? (*Hint:* Take the uncertainty in momentum Δp to be the actual momentum p.)

18E. The atomic masses of ^{1}H, ^{12}C, and ^{238}U are 1.007825 u, 12.000000 u (this one is exact by definition), and 238.050785 u, respectively. (a) What would these masses be if the mass unit were defined so that the mass of ^{1}H was (exactly) 1.000000 u? (b) Use your result to suggest why this perhaps obvious choice was not made.

19P. (a) Show that the energy tied up in nuclear, or strong force, bonds is proportional to A, the mass number of the nucleus in question. (b) Show that the energy tied up in Coulomb force bonds between the protons is proportional to $Z(Z - 1)$. (c) Show that, as we move to larger and larger nuclei (see Fig. 47-4), the importance of the

Coulomb force increases more rapidly than does that of the strong force.

20P. In the periodic table, the entry for magnesium is

$$\boxed{\begin{array}{c} 12 \\ \text{Mg} \\ 24.312 \end{array}}$$

There are three isotopes:

^{24}Mg, atomic mass = 23.98504 u.

^{25}Mg, atomic mass = 24.98584 u.

^{26}Mg, atomic mass = 25.98259 u.

The abundance of ^{24}Mg is 78.99% by weight. Calculate the abundances of the other two isotopes.

21P. You are asked to pick apart an α particle (^{4}He) by removing, in sequence, a proton, a neutron, and a proton. Calculate (a) the work required for each step, (b) the total binding energy of the α particle, and (c) the binding energy per nucleon. Some needed atomic masses are

^{4}He	4.00260 u	^{2}H	2.01410 u
^{3}H	3.01605 u	^{1}H	1.00783 u
n	1.00867 u		

22P. Because the neutron has no charge, its mass must be found in some way other than by using a mass spectrometer. When a neutron and a proton meet (assume both are almost stationary), they combine and form a deuteron, emitting a gamma ray whose energy is 2.2233 MeV. The masses of the proton and the deuteron are 1.007825035 u and 2.0141019 u, respectively. Find the mass of the neutron from these data, to as many significant figures as the data warrant. (A more precise value of the mass–energy conversion factor than the one presented in the text is 931.502 MeV/u.)

23P. A penny has a mass of 3.0 g. Calculate the nuclear energy that would be required to separate all the neutrons and protons in this coin. Ignore the binding energy of the electrons, and for simplicity assume that the penny is made entirely of ^{63}Cu atoms (of mass 62.92960 u). The masses of the proton and the neutron are 1.00783 u and 1.00867 u, respectively.

24P. To simplify calculations, atomic masses are sometimes tabulated, not as the actual atomic mass m, but as $(m - A)c^2$, where A is the mass number expressed in atomic mass units. This quantity, usually reported in MeV, is called the *mass excess*, represented with symbol Δ. Using data from Sample Problem 47-3, find the mass excesses for (a) ^{1}H, (b) the neutron, and (c) ^{120}Sn.

25P. (See Problem 24.) Show that the total binding energy of a nuclide can be written as

$$E = Z\Delta_H + N\Delta_n - \Delta,$$

where Δ_H, Δ_n, and Δ are the appropriate mass excesses. Using this method calculate the binding energy per nucleon for ^{197}Au. Compare your result with the value listed in Table 47-1. The needed mass excesses, rounded to three significant figures, are $\Delta_H = +7.29$ MeV, $\Delta_n = +8.07$ MeV, and $\Delta_{197} = -31.2$ MeV. Note the economy of calculation that results when mass excesses are used in place of the actual masses.

SECTION 47-3 RADIOACTIVE DECAY

26E. The half-life of a particular radioactive isotope is 6.5 h. If there are initially 48×10^{19} atoms of this isotope, how many atoms of this isotope remain after 26 h?

27E. The half-life of a radioactive isotope is 140 d. How many days would it take for the decay rate of a sample of this isotope to fall to one-fourth of its initial value?

28E. A radioactive nuclide has a half-life of 30 y. What fraction of an initially pure sample of this nuclide will remain undecayed after (a) 60 y and (b) 90 y?

29E. Gallium ^{67}Ga has a half-life of 78 h. Consider an initially pure 3.4-g sample of this isotope. (a) What is its decay rate? (b) What is its decay rate 48 h later?

30E. A radioactive isotope of mercury, ^{197}Hg, decays into gold, ^{197}Au, with a disintegration constant of 0.0108 h^{-1}. (a) Calculate its half-life. What fraction of the original amount will remain (b) after three half-lives and (c) after 10.0 days?

31E. From data presented in the first few paragraphs of Section 47-3, deduce (a) the disintegration constant λ and (b) the half-life of ^{238}U.

32E. The plutonium isotope ^{239}Pu is produced as a by-product in nuclear reactors and hence is accumulating in our environment. It is radioactive, decaying by alpha decay with a half-life of 2.41×10^4 y. But plutonium is also one of the most toxic chemicals known; as little as 2 mg is lethal to a human. (a) How many nuclei constitute a chemically lethal dose? (b) What is the decay rate of this amount? If you were handling that quantity would you fear being poisoned or suffering radiation sickness?

33E. Cancer cells are more vulnerable to x and gamma radiation than are healthy cells. Although linear accelerators are now replacing it, in the past the standard source for radiation therapy has been radioactive ^{60}Co, which beta decays into an excited nuclear state of ^{60}Ni, which immediately drops into the ground state, emitting two gamma-ray photons, each with an approximate energy of 1.2 MeV. The controlling beta-decay half-life is 5.27 y. How many radioactive ^{60}Co nuclei are present in a 6000-Ci source used in a hospital?

34P. After long effort, in 1902, Marie and Pierre Curie succeeded in separating from uranium ore the first substantial quantity of radium, one decigram of pure RaCl$_2$. The radium was the radioactive isotope ^{226}Ra, which has a decay half-life of 1600 y. (a) How many radium nuclei had they isolated? (b) What was the decay rate of their sample, in disintegrations/s?

35P. The radionuclide ^{64}Cu has a half-life of 12.7 h. How much of an initially pure 5.50-g sample of ^{64}Cu will decay during the 2-h period beginning 14.0 h later?

36P. The radionuclide ^{32}P ($\tau = 14.28$ d) is often used as a tracer to follow the course of biochemical reactions involving phosphorus. (a) If the counting rate in a particular experimental setup is initially 3050 counts/s, after what time will it fall to 170 counts/s? (b) A solution containing ^{32}P is fed to the root system of an experimental tomato plant and the ^{32}P activity in a leaf is measured 3.48 days later. By what factor must this reading be multiplied to correct for the decay that has occurred since the experiment began?

37P. A source contains two phosphorus radionuclides, ^{32}P ($\tau = 14.3$ d) and ^{33}P ($\tau = 25.3$ d). Initially 10.0% of the decays come from ^{33}P. How long must one wait until 90.0% do so?

38P. A 1.00-g sample of samarium emits α particles at a rate of 120 particles/s. ^{147}Sm, whose natural abundance in bulk samarium is 15.0%, is the responsible isotope. Calculate the half-life for the decay process.

39P. Plutonium ^{239}Pu decays by alpha decay with a half-life of 24,100 y. How many grams of helium are produced by an initially pure 12.0-g sample of ^{239}Pu after 20,000 y? (Recall that an α particle is a helium nucleus.)

40P. Calculate the mass of a sample of (initially pure) ^{40}K with an initial decay rate of 1.70×10^5 disintegrations/s. The isotope has a half-life of 1.28×10^9 y.

41P. One of the dangers of radioactive fallout from a nuclear bomb is ^{90}Sr, which beta-decays with a 29-year half-life. Because it has chemical properties much like calcium, the strontium, if eaten by a cow, becomes concentrated in its milk and ends up in the bones of whoever drinks the milk. The energetic decay electrons damage the bone marrow and thus impair the production of red blood cells. A 1-megaton bomb produces approximately 400 g of ^{90}Sr. If the fallout spreads uniformly over a 2000-km^2 area, what area would receive radioactivity equal to the allowed bone burden for one person, which is 74,000/s?

42P. After a brief neutron irradiation of silver, two isotopes are present: ^{108}Ag ($\tau = 2.42$ min) with an initial decay rate of 3.1×10^5/s, and ^{110}Ag ($\tau = 24.6$ s) with an initial decay rate of 4.1×10^6/s. Make a semilog plot similar to Fig. 47-8 showing the total combined decay rate of the two isotopes as a function of time from $t = 0$ until $t = 10$ min. We used Fig. 47-8 to illustrate the extraction of the half-life for simple (one isotope) decays. Given only your plot of total decay rate, can you suggest a way to analyze it in order to find the half-lives of both isotopes?

43P. A certain radionuclide is being manufactured, say, in a cyclotron, at a constant rate R. It is also decaying, with a disintegration constant λ. Let the production process continue for a time that is long compared to the half-life of the radionuclide. Show that the number of radioactive nuclei present after such time will remain constant and will be given by $N = R/\lambda$. Now show that this result holds no matter how many radioactive nuclei were present initially. The nuclide is said to be in *secular equilibrium* with its source; in this state its decay rate is just equal to its production rate.

44P. (See Problem 43.) The radionuclide ^{56}Mn has a half-life of 2.58 h and is produced in a cyclotron by bombarding a manganese target with deuterons. The target contains only the stable manganese isotope ^{55}Mn and the reaction that produces ^{56}Mn is

$$^{55}\text{Mn} + \text{d} \rightarrow {}^{56}\text{Mn} + \text{p}.$$

After bombardment for a time $\gg 2.58$ h, the activity of the target, due to ^{56}Mn, is 8.88×10^{10}. (a) At what constant rate R are ^{56}Mn nuclei being produced in the cyclotron during the bombardment? (b) At what rate are they decaying (also during the bombardment)? (c) How many ^{56}Mn nuclei are present at the end of the bombardment? (d) What is their total mass?

45P. (See Problems 43 and 44.) A radium source contains 1.00 mg of ^{226}Ra, which decays with a half-life of 1600 y to produce ^{222}Rn, a noble gas. This radon isotope in turn decays by alpha emission with a half-life of 3.82 d. (a) What is the rate of disintegration of ^{226}Ra in the source? (b) How long does it take for the radon to come to secular equilibrium with its radium parent? (c) At what rate is the radon then decaying? (d) How much radon is in equilibrium with its radium parent?

SECTION 47-4 ALPHA DECAY

46E. Consider a ^{238}U nucleus to be made up of an α particle (^{4}He) and a residual nucleus (^{234}Th). Plot the electrostatic potential energy $U(r)$, where r is the distance between these particles. Cover the approximate range 10 fm $< r <$ 100 fm and compare your plot with that of Fig. 47-9.

47E. Generally, heavier nuclides tend to be more unstable to alpha decay. For example, the most stable isotope of uranium, ^{238}U, has an alpha decay half-life of 4.5×10^9 y. The most stable isotope of plutonium is ^{244}Pu with an 8.2×10^7 y half-life, and for curium we have ^{248}Cm and 3.4×10^5 y. When half of an original sample of ^{238}U has decayed, what fractions of the original isotopes of plutonium and curium are left?

48P. A ^{238}U nucleus emits an α particle of energy 4.196 MeV. Calculate the disintegration energy Q for this process, taking the recoil energy of the residual ^{234}Th nucleus into account.

49P. Consider that a ^{238}U nucleus emits (a) an α particle or (b) a sequence of neutron, proton, neutron, proton. Calculate the energy released in each case. (c) Convince yourself both by reasoned argument and by direct calculation that the difference between these two numbers is just the total binding energy of the α particle. Find that binding energy. Some needed atomic and particle masses are

^{238}U	238.05079 u	^{234}Th	234.04363 u
^{237}U	237.04873 u	^{4}He	4.00260 u
^{236}Pa	236.04891 u	^{1}H	1.00783 u
^{235}Pa	235.04544 u	n	1.00867 u

50P. Under certain circumstances, a nucleus can decay by emitting a particle heavier than an α particle. Such decays are very rare and have only recently been observed. Consider the decays

$$^{223}\text{Ra} \rightarrow \,^{209}\text{Pb} + \,^{14}\text{C}$$

and

$$^{223}\text{Ra} \rightarrow \,^{219}\text{Rn} + \,^{4}\text{He}.$$

(a) Calculate the Q values for these decays and determine that both are energetically possible. (b) The Coulomb barrier height for α particles in this decay is 30.0 MeV. What is the barrier height for ^{14}C decay? The needed atomic masses are

^{223}Ra	223.01850 u	^{14}C	14.00324 u
^{209}Pb	208.98107 u	^{4}He	4.00260 u
^{219}Rn	219.01008 u		

51P. Heavy radionuclides emit an α particle rather than other combinations of nucleons because the α particle is such a stable, tightly bound structure. To confirm this, calculate the disintegration energies for these hypothetical decay processes and discuss the meaning of your findings:

$$^{235}\text{U} \rightarrow \,^{232}\text{Th} + \,^{3}\text{He}, \qquad Q_3;$$

$$^{235}\text{U} \rightarrow \,^{231}\text{Th} + \,^{4}\text{He}, \qquad Q_4;$$

$$^{235}\text{U} \rightarrow \,^{230}\text{Th} + \,^{5}\text{He}, \qquad Q_5.$$

The needed atomic masses are

^{232}Th	232.0381 u	^{3}He	3.0160 u
^{231}Th	231.0363 u	^{4}He	4.0026 u
^{230}Th	230.0331 u	^{5}He	5.0122 u
^{235}U	235.0439 u		

SECTION 47-5 BETA DECAY

52E. A certain stable nuclide, after absorbing a neutron, emits an electron and then splits spontaneously into two α particles. Identify the nuclide.

53E. ^{137}Cs is present in the fallout from above-ground detonations of nuclear bombs. Because it beta-decays with a slow 30.2-y half-life into ^{137}Ba, releasing considerable energy in the process, it is of environmental concern. The

atomic masses of the Cs and Ba are 136.9073 u and 136.9058 u, respectively; calculate the total energy released in such a decay.

54E. Heavy radionuclides, which may be either alpha or beta emitters, belong to one of four decay chains, depending on whether their mass numbers A are of the form $4n$, $4n + 1$, $4n + 2$, or $4n + 3$, where n is a positive integer. (a) Justify this statement and show that if a nuclide belongs to one of these families, all its decay products belong to the same family. (b) Classify these nuclides as to family: ^{235}U, ^{236}U, ^{238}U, ^{239}Pu, ^{240}Pu, ^{245}Cm, ^{246}Cm, ^{249}Cf, and ^{253}Fm.

55E. A free neutron decays according to Eq. 47-12. If the neutron–hydrogen atom mass difference is 840 μu, what is the maximum kinetic energy K_{max} of the electron energy spectrum?

56E. An electron is emitted from a middle-mass nuclide ($A = 150$, say) with a kinetic energy of 1.0 MeV. (a) What is its de Broglie wavelength? (b) Calculate the radius of the emitting nucleus. (c) Can such an electron be confined as a standing wave in a "box" of such dimensions? (d) Can you use these numbers to disprove the argument (long since abandoned) that electrons actually exist in nuclei?

57P. Some radionuclides decay by capturing one of their own atomic electrons, a K-shell electron, say. An example is

$$^{49}\text{V} + \text{e}^- \rightarrow \,^{49}\text{Ti} + \nu, \qquad \tau = 331 \text{ d}.$$

Show that the disintegration energy Q for this process is given by

$$Q = (m_{\text{V}} - m_{\text{Ti}})c^2 - E_K,$$

where m_{V} and m_{Ti} are the atomic masses of ^{49}V and ^{49}Ti, respectively, and E_K is the binding energy of the vanadium K-electron. (*Hint:* Put $\mathbf{m}_{\text{V}}$ and $\mathbf{m}_{\text{Ti}}$ as the corresponding nuclear masses and proceed as in Sample Problem 47-7.)

58P. Find the disintegration energy Q for the decay of ^{49}V by K-electron capture, as described in Problem 57. The needed data are $m_{\text{V}} = 48.94852$ u, $m_{\text{Ti}} = 48.94787$ u, and $E_K = 5.47$ keV.

59P. The radionuclide ^{11}C decays according to

$$^{11}\text{C} \rightarrow \,^{11}\text{B} + \text{e}^+ + \nu, \qquad \tau = 20.3 \text{ min.}$$

The maximum energy of the positron spectrum is 0.960 MeV. (a) Show that the disintegration energy Q for this process is given by

$$Q = (m_{\text{C}} - m_{\text{B}} - 2m_{\text{e}})c^2,$$

where m_{C} and m_{B} are the atomic masses of ^{11}C and ^{11}B, respectively, and m_{e} is the mass of a positron and also an electron. (b) Given that $m_{\text{C}} = 11.011434$ u, $m_{\text{B}} = 11.009305$ u, and $m_{\text{e}} = 0.0005486$ u, calculate Q and compare it with the maximum energy of the positron spectrum, given above. (*Hint:* Let $\mathbf{m}_{\text{C}}$ and $\mathbf{m}_{\text{B}}$ be the nuclear masses and proceed as in Sample Problem 47-7 for beta decay. Note that positron decay is an exception to the general rule that, if atomic masses are used in nuclear decay processes, the mass of the emitted electron is automatically taken care of.)

60P. Two radioactive materials that are unstable with regard to alpha decay, ^{238}U and ^{232}Th, and one that is unstable with regard to beta decay, ^{40}K, are sufficiently abundant in granite to contribute significantly to the heating of the Earth through the decay energy produced. The alpha-unstable isotopes give rise to decay chains that stop when stable lead isotopes are formed. ^{40}K has a single beta decay. Decay information follows:

PARENT	DECAY MODE	HALF-LIFE (y)	STABLE ENDPOINT	Q (MeV)	f (ppm)
^{238}U	α	4.47×10^9	^{206}Pb	51.7	4
^{232}Th	α	1.41×10^{10}	^{208}Pb	42.7	13
^{40}K	β	1.25×10^9	^{40}Ca	1.31	4

Q is the total energy released in the decay of one parent nucleus to the final stable endpoint and f is the abundance of the isotope in kilograms per kilogram of granite; ppm means parts per million. (a) Show that these materials give rise to a total heat production of 9.8×10^{-10} W for each kilogram of granite. (b) Assuming that there is 2.7×10^{22} kg of granite in a 20-km-thick spherical shell at the surface of the Earth, estimate the power this decay process will produce over the whole Earth. Compare this power production with the total solar power intercepted by the Earth, 1.7×10^{17} W.

61P*. The radionuclide ^{32}P decays to ^{32}S as described by Eq. 47-10. In a particular decay event, a 1.71-MeV electron is emitted, the maximum possible value. What is the kinetic energy of the recoiling ^{32}S atom in this event? (*Hint:* For the electron it is necessary to use the relativistic expressions for kinetic energy and linear momentum. Newtonian mechanics may safely be used for the relatively slow-moving ^{32}S atom.)

SECTION 47-6 RADIOACTIVE DATING

62E. ^{238}U decays to ^{206}Pb with a half-life of 4.47×10^9 y. Although the decay occurs in many individual steps, the first step has by far the longest half-life; therefore one can often consider the decay to go directly to lead. That is,

$$^{238}\text{U} \rightarrow {}^{206}\text{Pb} + \text{various decay products.}$$

A rock is found to contain 4.20 mg of ^{238}U and 2.135 mg of ^{206}Pb. Assume that the rock contained no lead at formation, all the lead now present arising from the decay of uranium. (a) How many atoms of ^{238}U and ^{206}Pb does the rock now contain? (b) How many atoms of ^{238}U did the rock contain at formation? (c) What is the age of the rock?

63E. A 5.00-g charcoal sample from an ancient fire pit has a ^{14}C activity of 63.0 disintegrations/min. A living tree has a ^{14}C activity of 15.3 disintegrations/min per 1.00 g. The half-life of ^{14}C is 5730 y. How old is the charcoal sample?

64P. A particular rock is thought to be 260 million years old. If it contains 3.70 mg of ^{238}U, how much ^{206}Pb should it contain? See Exercise 62.

65P. A rock, recovered from far underground, is found to contain 0.86 mg of ^{238}U, 0.15 mg of ^{206}Pb, and 1.6 mg of ^{40}Ar. How much ^{40}K will it likely contain? Needed half-lives are listed in Problem 60.

SECTION 47-7 MEASURING RADIATION DOSAGE

66E. A Geiger counter records 8700 counts in 1 min. Calculate the activity of the source in curies, assuming that the counter records all decays.

67E. The nuclide ^{198}Au, with half-life = 2.70 d, is used in cancer therapy. Calculate the mass of this isotope required to produce an activity of 250 Ci.

68E. An airline pilot spends an average of 20 h per week flying at 35,000 ft, at which altitude the equivalent dose due to cosmic radiation is 0.70 mrem/h. What is the annual equivalent dose from this source alone? Note that the maximum permitted yearly equivalent dose (from all sources) for the general population is 500 mrem, and for radiation workers it is 5000 mrem.

69E. A 75-kg person receives a whole-body radiation dose of 24 mrad, delivered by α particles for which the RBE factor is 12. Calculate (a) the absorbed energy in joules and (b) the equivalent dose in rem.

70P. A typical chest x-ray radiation dose is 25 mrem, delivered by x rays with an RBE factor of 0.85. Assuming that the mass of the exposed tissue is one-half the patient's mass of 88 kg, calculate the energy absorbed in joules.

71P. An 85-kg worker at a breeder reactor plant accidentally ingests 2.5 mg of plutonium ^{239}Pu dust. ^{239}Pu has a half-life of 24,100 y, decaying by alpha decay. The energy of the emitted α particles is 5.2 MeV, with an RBE factor of 13. Assume that the plutonium resides in the worker's body for 12 h, and that 95% of the emitted α particles are stopped within the body. Calculate (a) the number of plutonium atoms ingested, (b) the number that decay during the 12 h, (c) the energy absorbed by the body, (d) the resulting physical dose in rad, and (e) the equivalent biological dose in rem.

SECTION 47-8 NUCLEAR MODELS

72E. An intermediate nucleus in a particular nuclear reaction decays within 10^{-22} s of its formation. (a) What is the uncertainty ΔE in our knowledge of this intermediate state? (b) Can this state be called a compound nucleus? (See Sample Problem 47-10.)

73E. A typical kinetic energy for a nucleon in a middle-mass nucleus may be taken as 5.00 MeV. To what effective nuclear temperature does this correspond, using the assumptions of the liquid drop model of nuclear structure? (*Hint:* See Eq. 21-16.)

74E. In the following list of nuclides, identify (a) those with filled nucleon shells, (b) those with one nucleon outside a filled shell, and (c) those with one vacancy in an otherwise filled shell: ^{13}C, ^{18}O, ^{40}K, ^{49}Ti, ^{60}Ni, ^{91}Zr, ^{92}Mo, ^{121}Sb, ^{143}Nd, ^{144}Sm, ^{205}Tl, and ^{207}Pb.

75P. Consider the three formation processes shown for the compound nucleus ^{20}Ne* in Fig. 47-14. What energy must (a) the α particle, (b) the proton, and (c) the x-ray photon have to provide 25.0 MeV of excitation energy to the compound nucleus? Some needed atomic and particle masses are

^{20}Ne	19.99244 u	α	4.00260 u
^{19}F	18.99840 u	p	1.00783 u
^{16}O	15.99491 u		

76P. Consider the three decay processes shown for the compound nucleus ^{20}Ne* in Fig. 47-14. If the compound nucleus is initially at rest and has an excitation energy of 25.0 MeV, what kinetic energy, measured in the laboratory, will (a) the deuteron, (b) the neutron, and (c) the ^{3}He nuclide have when the nucleus decays? Some needed atomic and particle masses are

^{20}Ne	19.99244 u	d	2.01410 u
^{19}Ne	19.00188 u	n	1.00867 u
^{18}F	18.00094 u	^{3}He	3.01603 u
^{17}O	16.99913 u		

77P. The nuclide ^{208}Pb is "doubly magic" in that both its proton number Z ($= 82$) and its neutron number N ($= 126$) represent filled nucleon shells. An additional proton would yield ^{209}Bi, and an additional neutron ^{209}Pb. These "extra" nucleons should be easier to remove than a proton or a neutron from the filled shells of ^{208}Pb. (a) Calculate the energy required to remove the "extra" proton from ^{209}Bi and compare it with the energy required to remove a proton from the filled proton shell of ^{208}Pb. (b) Calculate the energy required to remove the "extra" neutron from ^{209}Pb and compare it with the energy required to remove a neutron from the filled neutron shell of ^{208}Pb. Do your results agree with expectation? Use these atomic mass data:

NUCLIDE	Z	N	ATOMIC MASS (u)
^{209}Bi	82 + 1	126	208.9804
^{208}Pb	82	126	207.9767
^{207}Tl	82 − 1	126	206.9774
^{209}Pb	82	126 + 1	208.9811
^{207}Pb	82	126 − 1	206.9759

The masses of the proton and the neutron are 1.00783 u and 1.00867 u, respectively.

78P. The nucleus ^{91}Zr ($Z = 40$, $N = 51$) has a single neutron outside a filled 50-neutron core. Because 50 is a magic number, this neutron should perhaps be especially loosely bound. (a) What is its binding energy? (b) What is the binding energy of the next neutron, which would have to be extracted from the filled core? (c) What is the binding energy per nucleon for the entire nucleus? Compare these three numbers and discuss. Some needed atomic masses are

^{91}Zr	90.90564 u	n	1.00867 u
^{90}Zr	89.90471 u	p	1.00783 u
^{89}Zr	88.90890 u		

79P. Verify the data for ^{121}Sb presented in Table 47-3. That is, calculate (a) the energy needed to remove a proton from a ^{121}Sb nucleus, and (b) the energy needed to remove a proton from the resulting ^{120}Sn nucleus. Needed atomic masses are

^{121}Sb	120.9038 u
^{120}Sn	119.9022 u
^{119}In	118.9058 u

ADDITIONAL PROBLEMS

80. A nucleus of mass M decays from an excited state S to the ground state by emitting a gamma ray of energy E, which causes the nucleus to recoil. The gamma ray is then absorbed by a second, identical nucleus, which is brought from its ground state to the (same) excited state S. The absorption causes the initially stationary second nucleus to move. Estimate the lifetime Δt of state S by using the uncertainty principle $\Delta E \, \Delta t \approx h$.

81. The electrons emitted in beta decay of ^{60}Co nuclei have a maximum kinetic energy $K_{max} = 0.310$ MeV. (a) What is the de Broglie wavelength of an electron with that maximum energy? (b) What is the radius of a ^{60}Co nucleus? (c) Considering your answers to (a) and (b), can an electron exist in the nucleus prior to the emission?

ENERGY FROM THE NUCLEUS

The image that has transfixed the world since World War II. When Robert Oppenheimer, the head of the scientific team that developed the atomic bomb, witnessed the first atomic explosion, he quoted from Hindu scripture: "Now I am become Death, the destroyer of worlds." What is the physics behind this image that has so horrified the world?

48-1 THE ATOM AND ITS NUCLEUS

When we get energy from coal by burning it in a furnace, we are tinkering with atoms of carbon and oxygen, rearranging their outer *electrons* into more stable combinations. When we get energy from uranium by burning it in a nuclear reactor, we are tinkering with its nucleus, rearranging its *nucleons* into more stable combinations.

Electrons are held in atoms by the electromagnetic Coulomb force, and it takes a few electron-volts to pull one of them out. On the other hand, nucleons are held in nuclei by the strong nuclear force, and it takes a few *million* electron-volts to pull one of *them* out. This factor of a few million is reflected in the fact that we can extract about that much more energy from a kilogram of uranium than we can from a kilogram of coal.

In both atomic and nuclear burning, the release of energy is accompanied by a decrease in mass, according to Einstein's equation $E = \Delta m \, c^2$. The only difference between burning uranium and burning coal is that, in the former case, a much larger fraction of the available mass (again, by a factor of a few million) is consumed.

The different processes that can be used for atomic or nuclear burning do provide different levels of power, or rates at which the energy is delivered. In the nuclear case you can burn your kilogram of uranium slowly in a power reactor or explosively in a bomb. In the atomic case, you might consider exploding a stick of dynamite or digesting a jelly doughnut. (Surprisingly, the total energy release is greater in the second case than in the first!)

Table 48-1 shows how much energy can be extracted from 1 kg of matter by doing various things to it. Instead of reporting the energy directly, we measure it by showing how long the extracted energy could operate a 100-W light bulb. Only processes in the first three rows of the table have actually been carried out; the remaining three represent theoretical limits that may not be attainable in practice. The bottom row, the total mutual annihilation of matter and antimatter, is an ultimate energy-production goal. When you have converted all the available mass, you can do no more.

Keep in mind that the comparisons of Table 48-1 are computed on a per-unit-mass basis. Kilogram for kilogram you get several million times more energy from uranium than you do from coal or from falling water. On the other hand, there is a lot of coal in the Earth's crust and water is easily backed up behind a dam.

48-2 NUCLEAR FISSION: THE BASIC PROCESS

In 1932, English physicist James Chadwick discovered the neutron. A few years later Enrico Fermi and his collaborators in Rome discovered that, if various elements are bombarded by these new projectiles, new radioactive elements are produced. Fermi had predicted that the neutron, being uncharged, would be a useful nuclear projectile; unlike the proton or the α particle, it experiences no repulsive Coulomb force when it approaches a nuclear surface. *Thermal neutrons*, which are neutrons in thermal equilibrium with the surrounding matter at room temperature, move slowly, with a mean kinetic energy of only about 0.04 eV, but are nevertheless particularly useful projectiles.

In 1939, German chemists Otto Hahn and Fritz Strassman, following up work initiated by Fermi and his collaborators, bombarded solutions of uranium salts with such thermal neutrons. They found by chemical analysis that after the bombardment a number of new radioactive elements were present, among them one whose chemical properties were remarkably similar to those of barium. Repeated tests finally convinced these able chemists that the "new" element was not new at all; it really *was* barium. How could this middle-mass element ($Z = 56$) be produced by bombarding uranium ($Z = 92$) with neutrons?

The puzzle was solved within a few weeks by Lise Meitner and her nephew Otto Frisch. They showed

TABLE 48-1
ENERGY RELEASED BY 1 KG OF MATTER

FORM OF MATTER	PROCESS	TIME[a]
Water	A 50-m waterfall	5 s
Coal	Burning	8 h
Enriched UO_2 (3%)	Fission in a reactor	690 y
^{235}U	Complete fission	3×10^4 y
Hot deuterium gas	Complete fusion	3×10^4 y
Matter and antimatter	Complete annihilation	3×10^7 y

[a] These numbers show how long the energy generated could power a 100-W light bulb.

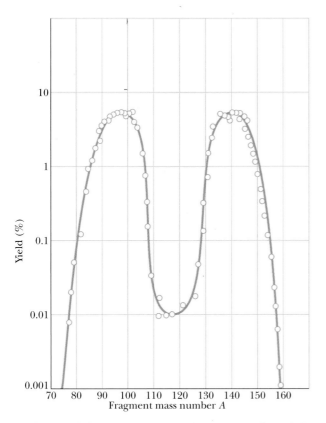

FIGURE 48-1 The distribution by mass number of the fragments that are found when many fission events of ^{235}U are examined. Note that the vertical scale is logarithmic.

that a uranium nucleus, having absorbed a thermal neutron, could split, with the release of energy, into two roughly equal parts, one of which might well be barium. Frisch named the process **fission.*** Figure 48-1 shows the distribution by mass number of the fragments produced when ^{235}U is bombarded with thermal neutrons. The most probable mass numbers, occurring in about 7% of the events, are centered around $A \approx 95$ and $A \approx 137$.

A Typical Fission Event

In a typical ^{235}U fission event, a ^{235}U nucleus absorbs a thermal neutron, producing a compound nucleus ^{236}U in a highly excited state. It is *this* nucleus that

---

actually undergoes fission, splitting into two fragments. These fragments—between them—rapidly emit two *prompt neutrons,* leaving ^{140}Xe and ^{94}Sr as fission fragments. Thus the overall fission equation for this event is

$$^{235}\text{U} + \text{n} \rightarrow {}^{236}\text{U}^* \rightarrow {}^{140}\text{Xe} + {}^{94}\text{Sr} + 2\text{n}. \quad (48\text{-}1)$$

The fragments ^{140}Xe and ^{94}Sr are both highly unstable, undergoing beta decay (with the emission of an electron) until each reaches a stable end product. Thus

$$^{140}\text{Xe} \rightarrow {}^{140}\text{Cs} \rightarrow {}^{140}\text{Ba} \rightarrow {}^{140}\text{La} \rightarrow {}^{140}\text{Ce}$$

τ	14 s	64 s	13 d	40 h	Stable
Z	54	55	56	57	58

$(48\text{-}2)$

and

$$^{94}\text{Sr} \rightarrow {}^{94}\text{Y} \rightarrow {}^{94}\text{Zr}$$

τ	75 s	19 min	Stable
Z	38	39	40

$(48\text{-}3)$

As expected, the mass numbers (140 and 94) of the fragments remain unchanged during these beta-decay processes; the atomic numbers (initially 54 and 38) increase by unity at each step.

Inspection of the stability band on the nuclidic chart of Fig. 47-4 can show us why the fission fragments are unstable. The nuclide ^{236}U, which is the fissioning nucleus in the reaction of Eq. 48-1, has 92 protons and $236 - 92$ or 144 neutrons, for a neutron/proton ratio of about 1.6. The primary fragments formed immediately after the fission reaction retain this same neutron/proton ratio. However, stable nuclides in the middle-mass region have smaller neutron/proton ratios, in the range 1.3–1.4. The primary fragments will thus be neutron rich and will "boil off" a small number of neutrons, two in the case of the reaction of Eq. 48-1. The fragments that remain are still too neutron rich to be stable. Beta decay offers a mechanism for getting rid of the excess neutrons, namely, by changing them into protons within the nucleus according to Eq. 47-12.

We can use the binding energy curve of Fig. 47-6 to estimate the energy released in fission. From this curve, we see that for heavy nuclides ($A \approx 240$) the mean binding energy per nucleon is about 7.6 MeV. For middle-mass nuclides ($A \approx 120$) it is about 8.5 MeV. The difference in total binding energy between a single large nucleus ($A = 240$) and two fragments (assumed equal) into which it may be split is then

$$Q = 2(8.5 \text{ MeV})(\tfrac{1}{2}A) - (7.6 \text{ MeV})(A)$$

$$\approx 200 \text{ MeV}. \qquad (48\text{-}4)$$

The more careful calculation of Sample Problem 48-1 agrees remarkably well with this rough estimate.

SAMPLE PROBLEM 48-1

Calculate the disintegration energy Q for the fission event of Eq. 48-1, taking into account the decay of the fission fragments as displayed in Eqs. 48-2 and 48-3.

SOLUTION   We can calculate the disintegration energy from $E = \Delta m \, c^2$. Some atomic and particle masses that we will need are

^{235}U	235.0439 u	^{140}Ce	139.9054 u
n	1.00867 u	^{94}Zr	93.9063 u

If we combine Eq. 48-1 with Eqs. 48-2 and 48-3, we see that the overall transformation is

$$^{235}\text{U} \rightarrow {}^{140}\text{Ce} + {}^{94}\text{Zr} + \text{n}. \qquad (48\text{-}5)$$

The single neutron comes about because the initiating neutron on the left side of Eq. 48-1 cancels one of the two neutrons on the right of that equation. The mass difference for the reaction of Eq. 48-5 is

$$\Delta m = (235.0439 \text{ u})$$
$$- (139.9054 \text{ u} + 93.9063 \text{ u} + 1.00867 \text{ u})$$
$$= 0.224 \text{ u},$$

and the corresponding disintegration energy is

$$Q = \Delta m \, c^2 = (0.224 \text{ u})(932 \text{ MeV/u})$$
$$= 209 \text{ MeV}, \qquad \text{(Answer)}$$

in good agreement with our estimate of Eq. 48-4.

If the fission event takes place in a bulk solid, most of this disintegration energy appears eventually as an increase in the internal energy of that body, revealing itself as a rise in temperature. Five or six percent or so of the disintegration energy, however, is associated with neutrinos that are emitted during the beta decay of the primary fission fragments. This energy is carried out of the system and is lost.

48-3 A MODEL FOR NUCLEAR FISSION

Soon after the discovery of fission, Niels Bohr and John Wheeler developed a model, based on the

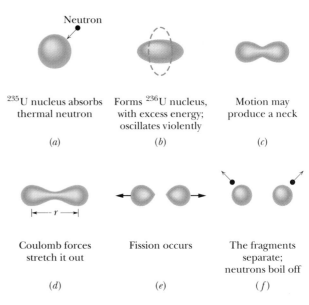

FIGURE 48-2   The stages of a typical fission process, according to the liquid drop fission model of Bohr and Wheeler.

analogy between a nucleus and a charged liquid drop, that explained its main features. Figure 48-2 suggests how the fission process proceeds from this point of view. When a heavy nucleus—let us say ^{235}U—absorbs a slow (thermal) neutron, as in Fig. 48-2a, that neutron falls into the potential well associated with the strong nuclear forces that act in the nuclear interior. The neutron's potential energy is then transformed into internal excitation energy, as Fig. 48-2b suggests.

The amount of excitation energy that a slow neutron carries into the nucleus is equal to the work required to pull a neutron out of the nucleus, that is, to the binding energy E_n of the neutron. In much the same way, the amount of excitation energy delivered to a well when a stone is dropped into it is equal to the work required to pull the stone back up out of the well, that is, to the "binding energy" E_s of the stone.

Figures 48-2c and 2d show that the nucleus, behaving like an energetically oscillating charged liquid drop, will sooner or later develop a short "neck" and will begin to separate into two charged "globs." If conditions are right, the electrostatic repulsion between these two globs will force them apart, breaking the neck. The two fragments, each still carrying some residual excitation energy, then fly apart. Fission has occurred.

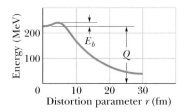

FIGURE 48-3 The potential energy at various stages in the fission process, as predicted from the liquid drop fission model of Bohr and Wheeler. The Q of the reaction (about 200 MeV) and the fission barrier height E_b are both indicated.

Thus far this model gives a good qualitative picture of the fission process. It remains to see, however, whether it can answer a hard question: "Why are some heavy nuclides (^{235}U and ^{239}Pu, say) readily fissionable by thermal neutrons but other, equally heavy, nuclides (^{238}U and ^{243}Am, say) are not?"

Bohr and Wheeler were able to answer this question. Figure 48-3 shows the potential energy curve that they derived from their model for the fission process. The horizontal axis displays the *distortion parameter r*, which is a rough measure of the extent to which the oscillating nucleus departs from a spherical shape. Figure 48-2*d* suggests how this parameter is defined before fission occurs. When the fragments are far apart, this parameter is simply the distance between their centers.

The energy difference between the initial state and the final state of the fissioning nucleus—that is, the disintegration energy Q—is displayed in Fig. 48-3. The central feature of that figure, however, is that the potential energy curve passes through a maximum at a certain value of r. There is a *potential barrier* of height E_b that must be surmounted (or tunneled!) before fission can occur. This reminds us of alpha decay (see Fig. 47-9), which is also a process that is inhibited by a potential barrier.

We see then that fission will occur only if the absorbed neutron provides an excitation energy E_n great enough to overcome the barrier. E_n need not be quite as great as the barrier height E_b because of the possibility of wave-mechanical tunneling.

Table 48-2 shows a test of fissionability by thermal neutrons applied to four heavy nuclides, chosen from dozens of possible candidates. For each nuclide both the barrier height E_b and the excitation energy E_n are given. The former was calculated from the theory of Bohr and Wheeler; the latter was computed from the known masses, using $E = \Delta m\, c^2$.

For ^{235}U and ^{239}Pu we see that $E_n > E_b$. This means that fission by absorption of a thermal neutron is predicted to occur for these nuclides. For the other two nuclides (^{238}U and ^{243}Am), we have $E_n < E_b$, so that there is not enough energy for a thermal neutron to surmount the barrier or to tunnel through it effectively. The excited nucleus (Fig. 48-2*b*) prefers to get rid of its excitation energy by emitting a gamma ray instead of by breaking into two large fragments.

^{238}U and ^{243}Am *can* be made to fission, however, if they absorb a substantially energetic (rather than a thermal) neutron. For ^{238}U, for example, the absorbed neutron must have at least 1.3 MeV of energy for this *fast fission* process to occur with meaningful probability.

48-4 THE NUCLEAR REACTOR

To make large-scale use of the energy released in fission, one fission event must trigger another, so that the process spreads throughout the nuclear fuel like flame through a log. The fact that more neutrons are produced in fission than are consumed raises just this possibility of a **chain reaction.** Such a reaction can be either rapid (as in a nuclear bomb) or controlled (as in a nuclear reactor).

TABLE 48-2
TEST OF THE FISSIONABILITY OF FOUR NUCLIDES

TARGET NUCLIDE	NUCLIDE BEING FISSIONED	E_n (MeV)	E_b (MeV)	$E_n - E_b$ (MeV)	FISSION BY THERMAL NEUTRONS?
^{235}U	^{236}U	6.5	5.2	+ 1.3	Yes
^{238}U	^{239}U	4.8	5.7	− 0.9	No
^{239}Pu	^{240}Pu	6.4	4.8	+ 1.6	Yes
^{243}Am	^{244}Am	5.5	5.8	− 0.3	No

Suppose that we wish to design a reactor based on the fission of ^{235}U by thermal neutrons. Natural uranium contains 0.7% of this isotope, the remaining 99.3% being ^{238}U, which is not fissionable by thermal neutrons. Let us give ourselves an edge by artificially enriching the uranium fuel so that it contains perhaps 3% ^{235}U in a 97% ^{238}U *matrix*. Three difficulties still stand in the way of a working reactor.

1. *The Neutron Leakage Problem.* Some of the neutrons produced by fission will leak out of the reactor and be lost to the chain reaction. Leakage is a surface effect, its magnitude being proportional to the square of a typical reactor dimension ($= 6a^2$ for a cube of edge a). Neutron production, however, occurs throughout the volume of the fuel and is thus proportional to the cube of a typical dimension ($= a^3$ for a cube). We can make the fraction of neutrons lost by leakage as small as we wish by making the reactor core large enough, thereby reducing the surface-to-volume ratio ($= 6/a$ for a cube).

2. *The Neutron Energy Problem.* The neutrons produced by fission are fast, with kinetic energies of about 2 MeV. However, fission is induced most effectively by thermal neutrons. The fast neutrons can be slowed down by mixing the uranium fuel with a substance—called a *moderator*—that has two properties: it is effective in slowing down neutrons by elastic collisions, and it does not remove neutrons

from the core by absorbing them in ways that do not result in fission. Most power reactors in this country use water as a moderator, the hydrogen nuclei (protons) being the effective component.

3. *The Neutron Capture Problem.* As the fast (2 MeV) neutrons generated by fission are slowed down in the moderator to thermal energies (about 0.04 eV), they must pass through a critical energy interval (from 1 to 100 eV) in which they are particularly susceptible to nonfission capture by ^{238}U nuclei. Such capture, which results in the emission of a gamma ray, removes the neutron from the fission chain. To minimize such *resonance capture,* the uranium fuel and the moderator are not intimately mixed but are "clumped," occupying different regions of the reactor volume. This increases the chance that a fast neutron, produced in a uranium clump, will find itself in the moderator as it passes through the critical energy interval. Once the neutron has reached thermal energies, it may *still* be captured in ways that do not result in fission (*thermal capture*). However, it is much more likely that the thermal neutron will wander into a clump of fuel and produce a fission event.

Figure 48-4 shows the neutron balance in a typical power reactor operating at constant power. Let us trace a sample of 1000 thermal neutrons through one complete cycle, or "generation," in the reactor

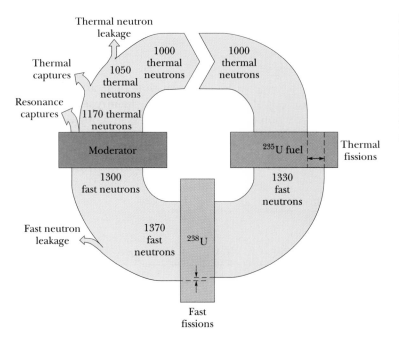

Thermal neutron leakage

Thermal captures

Resonance captures

1050 thermal neutrons

1000 thermal neutrons

1000 thermal neutrons

1170 thermal neutrons

Moderator

^{235}U fuel

Thermal fissions

1300 fast neutrons

1330 fast neutrons

Fast neutron leakage

1370 fast neutrons

^{238}U

Fast fissions

FIGURE 48-4 Neutron bookkeeping in a reactor. A generation of 1000 thermal neutrons is followed as they interact with the ^{235}U fuel, the ^{238}U matrix, and the moderator. We see that 1370 neutrons are produced by fission; 370 of these are lost, by nonfission capture or by leakage, so that exactly 1000 thermal neutrons are left to form the next generation. The figure is drawn for a reactor running at a steady power level.

core. They produce 1330 neutrons by fission in the ²³⁵U fuel and 40 neutrons by fast fission in the ²³⁸U, making a total of 370 new neutrons, all of them fast. Exactly this same number of neutrons is then lost by leakage from the core and by nonfission capture, leaving 1000 thermal neutrons to continue the chain reaction. What has been gained in this cycle, of course, is that each of the 370 neutrons produced by fission represents a deposit of about 200 MeV of energy in the reactor core, heating it up.

The *multiplication factor k*—an important reactor parameter—is the ratio of the number of neutrons present at the beginning of a particular generation to the number present at the beginning of the next generation. For the situation of Fig. 48-4, the multiplication factor is 1000/1000 or exactly unity. For $k = 1$, the operation of the reactor is said to be exactly *critical*, which is what we wish it to be for steady-power operation. Reactors are designed so that they are inherently *supercritical* ($k > 1$); the multiplication factor is then adjusted to critical operation ($k = 1$) by inserting *control rods* into the reactor core. These rods, containing a material, such as cadmium, that absorbs neutrons readily, can then be withdrawn as needed to compensate for the tendency of reactors to go subcritical as (neutron-absorbing) fission products build up in the core during continued operation.

If you pulled out one of the control rods, how fast would the reactor power level increase? This *response time* is controlled by the fascinating circumstance that a small fraction of the neutrons generated by fission are not boiled off promptly from the newly formed fission fragments but are emitted from these fragments later, as the fragments decay by beta emission. Of the 370 "new" neutrons analyzed in Fig. 48-4, for example, perhaps 16 are delayed, being emitted from fragments following beta decays whose half-lives range from 0.2 to 55 s. These delayed neutrons are few in number but they serve the useful purpose of slowing down the reactor response time to match human reaction times.

Figure 48-5 shows the broad outlines of an electric power plant based on a *pressurized-water reactor* (PWR), a type in common use in this country. In such a reactor, water is used both as the moderator and as the heat transfer medium. In the *primary loop*, water at high temperature and pressure (possibly 600 K and 150 atm) circulates through the reactor vessel and transfers heat from the reactor core to the

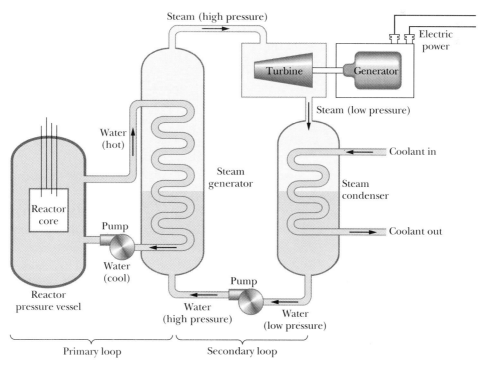

FIGURE 48-5 A simplified layout of a nuclear power plant, based on a pressurized-water reactor. Many features are omitted—among them the arrangement for cooling the reactor core in case of an emergency.

The scene 20 m from the Chernobyl reactor unit 4 (near Kiev), after it exploded and set the building on fire. Nearly all of the volatile radionuclides inside the reactor were released into the air. This reactor was not a pressurized-water reactor. Its catastrophic failure, which affected more than 600,000 people and made a wasteland out of the surrounding area, was due to poor engineering and an unbelievable disregard for safety.

steam generator; there, evaporation provides high-pressure steam to operate the turbine that drives the generator. To complete the *secondary loop*, low-pressure steam from the turbine is condensed to water and forced back into the steam generator by a pump. To give some idea of scale, a typical reactor

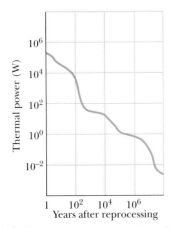

FIGURE 48-6   The thermal power released by the radioactive wastes from one year's operation of a typical large nuclear power plant, shown as a function of time. The curve is the superposition of the effects of many radionuclides, with a wide variety of half-lives. Note that both scales are logarithmic.

vessel for a 1000-MW (electric) plant may be 40 ft high and weigh 450 tons. Water flows through the primary loop at a rate of about 300,000 gal/min.

An unavoidable feature of reactor operation is the accumulation of radioactive wastes, including both fission products and heavy "transuranic" nuclides such as plutonium and americium. One measure of their radioactivity is the rate at which they release energy in thermal form. Figure 48-6 shows the theoretical variation with time of the thermal power generated by such wastes from one year's operation of a typical large nuclear plant. Note that both scales are logarithmic. The total activity of the waste 10 years after reprocessing is as high as 3×10^7 Ci.

SAMPLE PROBLEM 48-2

A large electric generating station is powered by a pressurized-water nuclear reactor. The thermal power in the reactor core is 3400 MW, and 1100 MW of electricity is generated. The fuel charge is 86,000 kg of uranium, in the form of 110 tons of uranium oxide, distributed among 57,000 fuel rods. The uranium is enriched to 3.0% ^{235}U.

a. What is the plant efficiency?

SOLUTION

$$\text{eff} = \frac{\text{useful output}}{\text{energy input}} = \frac{1100 \text{ MW (electric)}}{3400 \text{ MW (thermal)}}$$

$$= 0.32 \text{ or } 32\%. \qquad \text{(Answer)}$$

The efficiency—as for all power plants, whether based on fossil fuel or nuclear fuel—is controlled by the second law of thermodynamics. To run this plant, 3400 MW − 1100 MW or 2300 MW of power must be discharged as thermal energy to the environment.

b. At what rate R do fission events occur in the reactor core?

SOLUTION   If $P = 3400$ MW is the thermal power in the core and $Q = 200$ MeV is the average energy released per fission event, then, in steady-state operation,

$$R = \frac{P}{Q} = \left(\frac{3.4 \times 10^9 \text{ W}}{200 \text{ MeV/fission}} \right) \left(\frac{1 \text{ MeV}}{1.60 \times 10^{-13} \text{ J}} \right) \left(\frac{1 \text{ J/s}}{1 \text{ W}} \right)$$

$$= 1.06 \times 10^{20} \text{ fissions/s}$$

$$\approx 1.1 \times 10^{20} \text{ fissions/s.} \qquad \text{(Answer)}$$

c. At what rate is the ^{235}U fuel disappearing? Assume conditions at start-up.

SOLUTION ^{235}U disappears by fission at the rate calculated in (b) above. It is also consumed by (nonfission) neutron capture at a rate about one-fourth as large. The total ^{235}U consumption rate is then $(1.25)(1.06 \times 10^{20} \text{ s}^{-1})$ or $1.33 \times 10^{20} \text{ s}^{-1}$. We recast this as a mass rate as follows:

$$\frac{dM}{dt} = (1.33 \times 10^{20} \text{ s}^{-1})\left(\frac{0.235 \text{ kg/mol}}{6.02 \times 10^{23} \text{ atoms/mol}}\right)$$

$$= 5.19 \times 10^{-5} \text{ kg/s} \approx 4.5 \text{ kg/d}. \qquad \text{(Answer)}$$

d. At this rate of fuel consumption, how long would the fuel supply last?

SOLUTION From the data given, we can calculate that, at start-up, about $(0.030)(86{,}000 \text{ kg})$ or 2580 kg of ^{235}U were present. Thus a somewhat simplistic answer would be

$$T = \frac{2580 \text{ kg}}{4.5 \text{ kg/d}} \approx 570 \text{ d}. \qquad \text{(Answer)}$$

In practice, the fuel rods must be replaced (often in batches) before their ^{235}U content is entirely consumed.

e. At what rate is mass being converted to other forms of energy in the reactor core?

SOLUTION From Einstein's relation $E = \Delta mc^2$, we can write

$$\frac{dM}{dt} = \frac{dE/dt}{c^2} = \frac{3.4 \times 10^9 \text{ W}}{(3.00 \times 10^8 \text{ m/s})^2}$$

$$= 3.8 \times 10^{-8} \text{ kg/s} \quad \text{or} \quad 3.3 \text{ g/d}. \qquad \text{(Answer)}$$

We see that the mass conversion rate is about the mass of one penny every day! This rate of conversion of mass to other forms of energy is quite a different quantity from the fuel consumption rate (loss of ^{235}U) calculated in (c) above.

48-5 A NATURAL NUCLEAR REACTOR (OPTIONAL)

On December 2, 1942, when the reactor assembled by Enrico Fermi and his associates first went critical (Fig. 48-7), they had every right to expect that they had put into operation the first fission reactor that had ever existed on this planet. About 30 years later it was discovered that, if they did in fact think that, they were wrong.

FIGURE 48-7 A painting of the first nuclear reactor, assembled during World War II on a squash court at the University of Chicago by a team headed by Enrico Fermi. It went critical on December 2, 1942. This reactor—built of lumps of uranium imbedded in blocks of graphite—served as a prototype for later reactors whose purpose was to manufacture plutonium for the construction of nuclear weapons.

Some two billion years ago, in a uranium deposit now being mined in Gabon, West Africa, a natural fission reactor went into operation and ran for perhaps several hundred thousand years before shutting itself down. We can analyze this claim by considering two questions:

1. *Was There Enough Fuel?* The fuel for a uranium-based fission reactor must be the easily fissionable isotope ^{235}U, which constitutes only 0.72% of natural uranium. This isotopic ratio has been measured for terrestrial samples, in moon rocks, and in meteorites; in all cases the abundance values are the same. The clue to the discovery in West Africa was that the uranium in that deposit was deficient in ^{235}U, some samples having abundances as low as 0.44%. Investigation led to the speculation that this deficit in ^{235}U could be accounted for if, at some time in the past, the isotope was partially consumed by the operation of a natural fission reactor.

The serious problem remains that, with an isotopic abundance of only 0.72%, a reactor can be assembled (as Fermi and his team learned) only with the greatest of difficulty. There seems no chance at all that it could have happened naturally.

However, things were different in the distant past. Both ^{235}U and ^{238}U are radioactive, with half-lives of 7.04×10^8 y and 44.7×10^8 y, respectively. Thus the half-life of the readily fissionable ^{235}U is about 6.4 times shorter than that of ^{238}U. Because

^{235}U decays faster, there must have been more of it, relative to ^{238}U, in the past. Two billion years ago, in fact, this abundance was not 0.72%, as it is now, but 3.8%. This abundance happens to be just about the abundance to which natural uranium is artificially enriched to serve as fuel in modern power reactors.

With this readily fissionable fuel available, the presence of a natural reactor (providing certain other conditions are met) is much less surprising. The fuel was there. Two billion years ago, incidentally, the highest order of life forms that had evolved were the blue-green algae.

2. *What Is the Evidence?* The mere depletion of ^{235}U in an ore deposit is not enough evidence on which to hang a claim for the existence of a natural fission reactor. One looks for more convincing proof.

If there were a reactor, there must also be fission products. Of the 30 or so elements whose stable isotopes are produced in this way, some must still remain. Study of their isotopic abundances could provide the convincing evidence we need.

Of the several elements investigated, the case of neodymium is spectacularly convincing. Figure 48-8*a* shows the isotopic abundances of the seven stable neodymium isotopes as they are normally found in nature. Figure 48-8*b* shows these abundances as they appear among the ultimate stable fission products of the fission of ^{235}U. The clear differences are not surprising, considering the totally different origins of the two sets of isotopes. The isotopes shown in Fig. 48-8*a* were formed in supernova explosions that oc-

curred before the formation of our solar system. The isotopes of Fig. 48-8*b* were cooked up in a reactor by totally different processes. Note particularly that ^{142}Nd, the dominant isotope in the natural element, is totally absent from the fission products.

The big question is: "What do the neodymium isotopes found in the uranium ore body in West Africa look like?" We must expect that, if a natural reactor operated there, isotopes from *both* sources (that is, natural isotopes as well as fission-produced isotopes) should be present. Figure 48-8*c* shows the results after this and other corrections have been made to the raw data. Comparison of Figs. 48-8*b* and 48-8*c* leaves little doubt that there was indeed a natural fission reactor at work.

SAMPLE PROBLEM 48-3

The isotopic ratio of ^{235}U to ^{238}U in natural uranium deposits today is 0.0072. What was this ratio 2.0×10^9 y ago? The half-lives of the two isotopes are 7.04×10^8 y and 44.7×10^8 y, respectively.

SOLUTION Consider two samples that, at a time t in the past, contained $N_5(0)$ and $N_8(0)$ atoms of ^{235}U and ^{238}U, respectively. The numbers of atoms remaining at the present time are

$$N_5(t) = N_5(0)e^{-\lambda_5 t} \quad \text{and} \quad N_8(t) = N_8(0)e^{-\lambda_8 t},$$

respectively, in which λ_5 and λ_8 are the corresponding

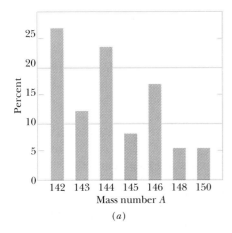
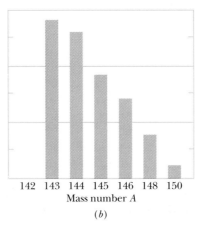
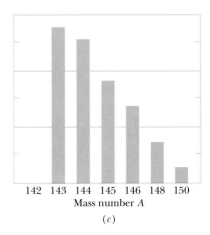

FIGURE 48-8   The distribution by mass number of the isotopes of neodymium as they occur in (*a*) natural terrestrial deposits of the ores of this element and (*b*) the spent fuel of a power reactor. (*c*) The distribution (after several corrections) found for neodymium from the uranium mine in Gabon, West Africa. Note that (*b*) and (*c*) are virtually identical and are quite different from (*a*).

disintegration constants. Dividing gives

$$\frac{N_5(t)}{N_8(t)} = \frac{N_5(0)}{N_8(0)} e^{-(\lambda_5 - \lambda_8)t}.$$

Expressed in terms of isotopic ratios, this becomes

$$R(0) = R(t) e^{(\lambda_5 - \lambda_8)t}.$$

The disintegration constants are related to the half-lives by Eq. 47-8, which yields

$$\lambda_5 = \frac{\ln 2}{\tau_5} = \frac{\ln 2}{7.04 \times 10^8 \text{ y}} = 9.85 \times 10^{-10} \text{ y}^{-1}$$

and

$$\lambda_8 = \frac{\ln 2}{\tau_8} = \frac{\ln 2}{44.7 \times 10^8 \text{ y}} = 1.55 \times 10^{-10} \text{ y}^{-1}.$$

The exponent in the expression for $R(0)$ above is then

$$(\lambda_5 - \lambda_8)t = [(9.85 - 1.55) \times 10^{-10} \text{ y}^{-1}][2 \times 10^9 \text{ y}]$$

$$= 1.66.$$

The isotopic ratio is then

$$R(0) = R(t) e^{(\lambda_5 - \lambda_8)t}$$

$$= (0.0072)(e^{1.66})$$

$$= 0.0379 \approx 3.8\%. \qquad \text{(Answer)}$$

Two billion years ago, the ratio of ^{235}U to ^{238}U in natural uranium deposits was much higher than it is today. You should be able to show that when the Earth was formed (4.5 billion years ago) this ratio was 30%.

48-6 THERMONUCLEAR FUSION: THE BASIC PROCESS

The binding energy curve of Fig. 47-6 shows that energy can be released if two light nuclei combine to form a single larger nucleus, a process called nuclear **fusion.** The process is hindered by the Coulomb repulsion that acts to prevent the two particles from getting close enough to each other to be within range of their attractive nuclear forces and "fusing." The height of the Coulomb barrier depends on the charges and the radii of the two interacting nuclei. We show in Sample Problem 48-4 that, for two deuterons ($Z = 1$), the barrier height is 200 keV. For more highly charged particles, of course, the barrier is correspondingly higher.

To generate useful amounts of power, nuclear fusion must occur in bulk matter. The best hope for bringing this about is to raise the temperature of the material so that the particles have enough energy—due to their thermal motions alone—to penetrate the Coulomb barrier. We call this process **thermonuclear fusion.**

In thermonuclear studies, temperatures are reported in terms of kinetic energy K via the relation

$$K = kT, \qquad (48\text{-}6)$$

in which k is the Boltzmann constant. Thus rather than saying, "The temperature at the center of the sun is 1.5×10^7 K," it is more common to say, "The temperature at the center of the sun is 1.3 keV."

Room temperature corresponds to $K \approx 0.03$ eV; a particle with only this amount of energy could not hope to overcome a barrier as high as, say, 200 keV. Even at the center of the sun, where $kT = 1.3$ keV, the outlook for thermonuclear fusion does not seem promising at first glance. Yet we know that thermonuclear fusion not only occurs in the core of the sun but is the dominant feature of that body and of all other stars.

The puzzle is solved when we realize two facts: (1) The energy calculated with Eq. 48-6 is that of the particles with the *most probable* speed, as defined in Section 21-7; there is a long Maxwellian tail of particles with much higher speeds and, correspondingly, much higher energies. (2) The barrier heights that we have calculated represent the *peaks* of the barriers. Barrier tunneling can occur at energies considerably lower than these peaks, as we saw in the case of alpha decay in Section 47-4.

Figure 48-9 sums things up. The curve marked $n(K)$ in this figure is a Maxwell distribution curve for

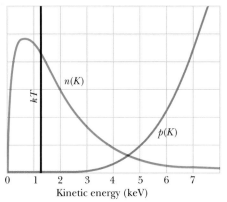

FIGURE 48-9 The curve marked $n(K)$ gives the relative distribution in energy for protons at the center of the sun. The curve marked $p(K)$ gives the relative probability of barrier penetration for proton–proton collisions at the sun's central temperature. The vertical line marks the value of kT at this temperature. Note that the two curves are drawn to (separate) arbitrary vertical scales.

the protons in the sun's core, drawn to correspond to the sun's central temperature. This curve differs from the Maxwell distribution curve of Fig. 21-8 in that it is drawn in terms of energy and not of speed. Specifically, $n(K) \, dK$ gives the probability that a proton will have a kinetic energy lying between K and $K + dK$. The value of kT in the core of the sun is marked on the figure; note that many protons have energies greater than this.

The curve marked $p(K)$ in Fig. 48-9 is the relative probability of barrier penetration for two colliding protons. Study of the two curves in Fig. 48-9 suggests that there is a particular proton energy at which proton–proton fusion events occur at a maximum rate. If the energy of the interacting protons is much higher than this value, the barrier is transparent enough but there are too few protons in the Maxwellian tail to sustain the reaction. If the energy is much lower than this value, there are plenty of protons but the barrier is now too formidable.

SAMPLE PROBLEM 48-4

The deuteron (^{2}H) has a charge $+e$ and may be taken as a sphere of effective radius $R = 2.1$ fm. Two such particles are fired at each other with the same kinetic energy K. What must K be if the particles are brought to rest by their mutual Coulomb repulsion when they are just "touching" each other? We take this value of K as a measure of the height of the Coulomb barrier.

SOLUTION   Because the two deuterons are momentarily at rest when they just touch, their initial kinetic energy has all been transformed into electrostatic potential energy. Their centers are separated by a distance $2R$ and we have

$$2K = \frac{1}{4\pi\epsilon_0} \frac{q_1 q_2}{r} = \frac{1}{4\pi\epsilon_0} \frac{e^2}{2R},$$

which yields

$$K = \frac{e^2}{16\pi\epsilon_0 R}$$

$$= \frac{(1.60 \times 10^{-19} \text{ C})^2}{(16\pi)(8.85 \times 10^{-12} \text{ F/m})(2.1 \times 10^{-15} \text{ m})}$$

$$= 2.74 \times 10^{-14} \text{ J} = 171 \text{ keV} \approx 200 \text{ keV}. \quad \text{(Answer)}$$

48-7 THERMONUCLEAR FUSION IN THE SUN AND OTHER STARS

The sun radiates energy at the rate of 3.9×10^{26} W and has been doing so for several billion years. From where does all this energy come? Chemical burning is ruled out; if the sun had been made of coal and oxygen—in the right proportions for combustion—it would have lasted for only about 1000 y. Another possibility is that the sun is slowly shrinking, under the action of its own gravitational forces. By transferring gravitational potential energy to thermal energy, the sun might maintain its temperature and continue to radiate. Calculation, however, shows that this mechanism also fails, producing a solar lifetime that is too short by a factor of at least 500. That leaves only thermonuclear fusion. The sun, as we shall see, burns not coal but hydrogen, and in a nuclear furnace, not an atomic or chemical one.

The fusion reaction in the sun is a multistep process in which hydrogen is burned into helium, hydrogen being the "fuel" and helium the "ashes." Figure 48-10 shows the **proton–proton** (p–p) **cycle** by which this is accomplished.

The p–p cycle starts with the thermal collision of two protons (^{1}H + ^{1}H) to form a deuteron (^{2}H), with the simultaneous creation of a positron (e^+)

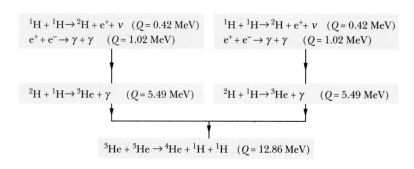

FIGURE 48-10 The proton–proton mechanism that accounts for energy production in the sun. In this process, protons fuse to form an α particle (^{4}He), with a net energy release of 26.7 MeV for each event.

and a neutrino (ν). The positron very quickly encounters a free electron (e^-) in the sun and both particles annihilate (see Section 23-6), their mass energy appearing as two gamma-ray photons (γ).

A pair of such events is shown in the top row of Fig. 48-10. These events are actually extremely rare. In fact, only once in about 10^{26} proton–proton collisions is a deuteron formed; in the vast majority of cases, the two protons simply rebound elastically from each other. It is the slowness of this "bottleneck" or "safety valve" process that regulates the rate of energy production and keeps the sun from exploding. Interestingly, in spite of this slowness, there are so very many protons in the huge and dense volume of the sun's core that deuterium is produced there in this way at the rate of 10^{12} kg/s!

Once a deuteron (^{2}H) has been produced it quickly collides with another proton and forms a ^{3}He nucleus, as the middle row of Fig. 48-10 shows. Two such ^{3}He nuclei may eventually (within 10^5 y; there is plenty of time) find each other, forming an α particle (^{4}He) and two protons, as the bottom row in the figure shows.

Taking an overall view, we see from Eq. 48-10 that the p–p cycle amounts to the combination of four protons and two electrons to form an α particle, two neutrinos, and six gamma rays. Thus

$$4\,^1\text{H} + 2e^- \rightarrow\,^4\text{He} + 2\nu + 6\gamma. \quad (48\text{-}7)$$

Now, in a formal way, let us add two electrons to each side of Eq. 48-7, yielding

$$(4\,^1\text{H} + 4e^-) \rightarrow (^4\text{He} + 2e^-) + 2\nu + 6\gamma. \quad (48\text{-}8)$$

The quantities in the first two parentheses then represent *atoms* (not bare nuclei) of hydrogen and of helium.

The energy release in the reaction of Eq. 48-8 is

$$Q = \Delta m\, c^2$$
$$= [(4)(1.007825\,\text{u}) - 4.002603\,\text{u}][932\,\text{MeV/u}]$$
$$= 26.7\,\text{MeV},$$

in which 1.007825 u is the mass of a hydrogen atom and 4.002603 u is the mass of a helium atom; neutrinos and gamma-ray photons have no mass and thus do not enter into the calculation of the disintegration energy.

This same value of Q follows (as it must) by adding up the Q values for the separate steps of the proton–proton cycle in Fig. 48-10. Thus

$$Q = (2)(0.42\,\text{MeV}) + (2)(1.02\,\text{MeV})$$
$$\quad + (2)(5.49\,\text{MeV}) + 12.86\,\text{MeV}$$
$$= 26.7\,\text{MeV}.$$

About 0.5 MeV of this energy is carried out of the sun by the two neutrinos in Eq. 48-8; the rest ($= 26.2$ MeV) is deposited in the core of the sun as thermal energy.

The burning of hydrogen in the sun's core is alchemy on a grand scale in the sense that one element is turned into another. The medieval alchemists, however, were more interested in changing lead into gold than in changing hydrogen into helium. In a sense, they were on the right track, except that their furnaces were not hot enough. Instead of being at 600 K, they should have been at least as high as 10^8 K!

Hydrogen burning has been going on in the sun for about 5×10^9 y, and calculations show that there is enough hydrogen left to keep the sun going for about the same length of time into the future. The sun's core, which by that time will be largely helium, will begin to cool and the sun will start to collapse under its own gravity. This will raise the core temperature and cause the outer envelope to expand, turning the sun into what astronomers call a *red giant*.*

If the core temperature increases to about 10^8 K again, energy can be produced once more by burning helium to make carbon. As a star evolves and becomes still hotter, other elements can be formed by other fusion reactions. However, elements more massive than those with $A \approx 56$ (^{56}Fe, ^{56}Co, ^{56}Ni) cannot be manufactured by further fusion processes. $A = 56$ marks the peak of the binding energy curve of Fig. 47-6, and fusion between nuclides beyond this point involves the consumption—not the production—of energy.

Elements with mass numbers beyond $A = 56$ are thought to be formed by neutron capture during cataclysmic stellar explosions that we call *supernovas* (Fig. 48-11). In such an event the outer shell of the star is blown outward into space where it mixes with —and becomes part of—the tenuous medium that

---

*The details of this event, which promises to be rather unpleasant, are spelled out in "When the Sun Swallows the Earth," *Sky & Telescope*, December 1987, News Notes.

(a)

(b)

FIGURE 48-11   (a) A star known as Sanduleak, as it appeared until 1987. (b) We then began to intercept light from the supernova of that star; the explosion was 100 million times brighter than our sun and could be seen with the unaided eye. It took place 155,000 light-years away and thus actually occurred 155,000 years ago.

fills the space between the stars. It is from this medium, continually enriched by debris from stellar explosions, that new stars form, by condensation under the influence of the gravitational force.

The fact that the Earth abounds in elements heavier than hydrogen and helium suggests that our solar system has condensed out of interstellar material that contained the remnants of such explosions. Thus all the elements around us—including those in our own bodies—were manufactured in the interiors of stars that no longer exist. As one scientist put it: "In truth, we are the children of the Universe."

SAMPLE PROBLEM 48-5

At what rate is hydrogen being consumed in the core of the sun by the p–p cycle of Fig. 48-10?

SOLUTION We have seen that 26.2 MeV appears as thermal energy in the sun for every four protons that are consumed, a rate of 6.6 MeV/proton. We can express this energy transfer rate as

$$\frac{dE}{dm} = (6.6 \text{ MeV/proton}) \left(\frac{1 \text{ proton}}{1.67 \times 10^{-27} \text{ kg}} \right)$$
$$\times \left(\frac{1.60 \times 10^{-13} \text{ J}}{1 \text{ MeV}} \right)$$
$$= 6.3 \times 10^{14} \text{ J/kg}.$$

This tells us that the sun radiates away 6.3×10^{14} J of

energy for every kilogram of protons consumed. The hydrogen consumption rate is then the sun's power ($= 3.9 \times 10^{26}$ W) divided by the above quantity, or

$$R = \frac{3.9 \times 10^{26} \text{ W}}{6.3 \times 10^{14} \text{ J/kg}} = 6.2 \times 10^{11} \text{ kg/s}. \quad \text{(Answer)}$$

This seems like a large mass loss per second but—to keep things in perspective—we point out that the sun's mass is 2×10^{30} kg.

48-8 CONTROLLED THERMONUCLEAR FUSION

The first thermonuclear reaction to take place on Earth occurred at Eniwetok Atoll on October 31, 1952, when the United States exploded a fusion device, generating an energy release equivalent to 10 million tons of TNT. The high temperatures and densities needed to initiate the reaction were provided by using a fission bomb as a trigger.

A sustained and controllable source of fusion power—a fusion reactor—is considerably more difficult to achieve. The goal, however, is being pursued vigorously in many countries around the world because many look to the fusion reactor as the power source of the future, at least as far as the generation of electricity is concerned.

The p–p scheme displayed in Fig. 48-10 is not suitable for an Earth-bound fusion reactor because the scheme is hopelessly slow. The reaction succeeds in the sun only because of the enormous density of protons in the center of the sun. The most attractive reactions for terrestrial use appear to be the deuteron–deuteron (d–d) and the deuteron–triton (d–t) reactions:*

$$^2H + {}^2H \rightarrow {}^3He + n \qquad (d–d)$$
$$Q = +3.27 \text{ MeV}, \qquad (48\text{-}9)$$

$$^2H + {}^2H \rightarrow {}^3H + {}^1H \qquad (d–d)$$
$$Q = +4.03 \text{ MeV}, \qquad (48\text{-}10)$$

and

$$^2H + {}^3H \rightarrow {}^4He + n \qquad (d–t)$$
$$Q = +17.59 \text{ MeV}. \qquad (48\text{-}11)$$

Deuterium, whose isotopic abundance is 1 part in 6700, is available in unlimited quantities as a component of seawater. Proponents of power from the nucleus have described our ultimate power choice—when we have burned up all our fossil fuels—as either "burning rocks" (fission of uranium extracted from ores) or "burning water" (fusion of deuterium extracted from water).

There are three requirements for a successful thermonuclear reactor:

1. *A High Particle Density n.* The density of interacting particles (deuterons, say) must be great enough to ensure that the d–d collision rate is high enough. At the high temperatures required, the deuterium would be completely ionized, forming a neutral *plasma* (ionized gas) consisting of deuterons and electrons.

2. *A High Plasma Temperature T.* The plasma must be hot. Otherwise the colliding deuterons will not be energetic enough to penetrate the Coulomb barrier that tends to keep them apart. In fusion research, temperatures are often reported by giving the value of kT. A plasma ion temperature of 20 keV, corresponding to 23×10^7 K, has been achieved in the laboratory. This is more than 15 times higher than the sun's central temperature (1.3 keV or 1.5×10^7 K).

3. *A Long Confinement Time τ.* A major problem is containing the hot plasma long enough to ensure that its density and temperature remain sufficiently high for enough of the fuel to be fused. It is clear that no solid container can withstand the high temperatures that are necessary, so clever confining techniques are called for; we discuss two in the next two sections.

It can be shown that, for the successful operation of a thermonuclear reactor, it is necessary to have

$$n\tau > 10^{20} \text{ s} \cdot \text{m}^{-3}. \qquad (48\text{-}12)$$

This condition is known as **Lawson's criterion,** and the quantity $n\tau$ is known as the **Lawson number.** Equation 48-12 tells us that we have a choice between confining a lot of particles for a short time or confining fewer particles for a longer time. Beyond meeting this criterion, it is also necessary that the plasma temperature be high enough.

48-9 THE TOKAMAK

Tokamak, a Russian-language acronym for "toroidal magnetic chamber," implies a type of thermonuclear fusion device first developed in the USSR. Large tokamaks have been built and operated in several countries, and several major new machines are in the design stage.

In a tokamak the charged particles that make up the hot plasma are confined by a magnetic field in the shape of a doughnut or torus. As Fig. 48-12a suggests, the confining magnetic field is a sheath of helical field lines—only one of which is shown in the figure—that spiral around the plasma ring. The magnetic forces acting on the moving charges of the plasma keep the hot plasma from touching the walls of the chamber. Figures 48-12b and 48-12c show how the helical confining field is made up by combining a toroidal field (Fig. 48-12b) and a so-called poloidal field (Fig. 48-12c). The currents required to produce these fields are also shown. The current that generates the poloidal field is induced in the plasma itself, and it serves also to heat the plasma.

Figure 48-13 shows a plot of Lawson number $n\tau$ versus plasma temperature for various tokamaks and other magnetic confinement devices, in various

---

*The nucleus of the hydrogen isotope ^3H is called the *triton.* It is a radionuclide, with a half-life of 12.3 y.

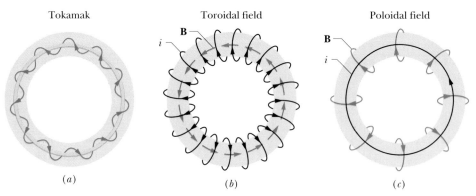

Tokamak   Toroidal field   Poloidal field

(a)   (b)   (c)

FIGURE 48-12   (a) The gold ring suggests the confined plasma in a tokamak. The green wavy line suggests the nature of the confining magnetic field. (b) The toroidal component of this magnetic field is established by currents that loop around the torus. (c) The poloidal component of this field is established by a current induced in the plasma.

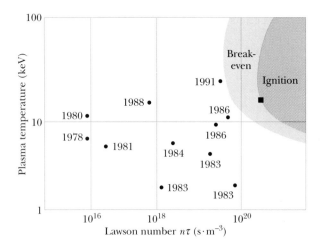

FIGURE 48-13   A plot of Lawson number versus plasma temperature for a number of magnetic confinement fusion devices. The year of first successful operation is shown for each device. The square shows the expected performance of two machines.

countries. *Break-even* corresponds to meeting the Lawson criterion with a sufficiently hot, thermalized plasma; *ignition* corresponds to a self-sustaining thermonuclear reaction. As Fig. 48-13 indicates, break-even has almost been achieved, but no device has yet achieved ignition. In spite of the rapid progress being made at present, many formidable engineering problems remain, and a practical thermonuclear power plant does not seem possible before the early decades of the next century.

SAMPLE PROBLEM 48-6

Suppose that a tokamak achieves ignition with a plasma temperature of 10 keV and a confinement time of 980 ms.

a. What would the particle density of its plasma have to be?

SOLUTION   From Fig. 48-13 we see that the 10-keV temperature line intersects the curve marked "ignition" at a value of the Lawson number of about 4×10^{20} s·m^{-3}. (In making this last estimate, bear in mind that the scale is logarithmic.) The necessary particle density is then

$$n = \frac{4 \times 10^{20} \text{ s·m}^{-3}}{980 \times 10^{-3} \text{ s}} \approx 4 \times 10^{20} \text{ m}^{-3}. \quad \text{(Answer)}$$

b. How does this number compare with the particle density of the atoms of an ideal gas at standard temperature and pressure?

SOLUTION   The number density of atoms in an ideal gas at standard conditions is given by $n = N_A/V$, where N_A is the Avogadro number and V $(= 22{,}460 \text{ cm}^3)$ is the volume occupied by one mole of an ideal gas at standard conditions. Thus

$$n = \frac{N_A}{V} = \frac{6.02 \times 10^{23} \text{ mol}^{-1}}{22{,}460 \times 10^{-6} \text{ m}^3}$$

$$\approx 2.7 \times 10^{25} \text{ m}^{-3}. \quad \text{(Answer)}$$

This is larger than the particle density of the plasma in (a) above by a factor of about 70,000.

48-10 LASER FUSION

A second technique for confining plasma so that thermonuclear reactions can take place is called **inertial confinement.** It involves compressing a fuel pellet by "zapping" it from all sides by laser beams (or particle beams), thus compressing it and increasing its temperature (to perhaps 10^8 K) and particle density (by perhaps a factor of 10^3) so that thermonuclear fusion can occur. By comparison with magnetic confinement devices such as tokamaks, inertial confinement involves working with much higher particle densities for much shorter times.

Laser fusion is being investigated in many laboratories, in the United States and elsewhere. At the Lawrence Livermore Laboratory, for example, in the NOVA laser-fusion arrangement deuterium–tritium fuel pellets, each smaller than a grain of sand (Fig. 48-14a), are to be "zapped" by 10 synchronized high-powered laser pulses, symmetrically arranged around the pellet (Fig. 48-14b). The laser pulses are designed to deliver, in total, some 200 kJ of energy to each fuel pellet in less than a nanosecond. This is a delivered power of about 2×10^{14} W during the pulse, which is roughly 100 times the total installed (sustained) electric power generating capacity of the world!

In an operating thermonuclear reactor of the laser-fusion type, it is visualized that fuel pellets would be exploded—like miniature hydrogen bombs—at a rate of perhaps 10–100 per second. The feasibility of laser fusion as the basis of a thermonuclear power reactor has not been demonstrated as of 1992, but research is continuing at a vigorous pace.

SAMPLE PROBLEM 48-7

Suppose that a fuel pellet in a laser-fusion device is made of a liquid deuterium–tritium mixture containing equal numbers of deuterium and tritium atoms. The density $d = 200$ kg/m³ of the pellet is increased by a factor of 10^3 by the action of the laser pulses.

a. How many particles per unit volume (either deuterons or tritons) does the pellet contain in its compressed state?

FIGURE 48-14   The small spheres on the quarter in (a) are deuterium–tritium fuel pellets, designed to be used in the laser fusion chamber in (b).

(a)

(b)

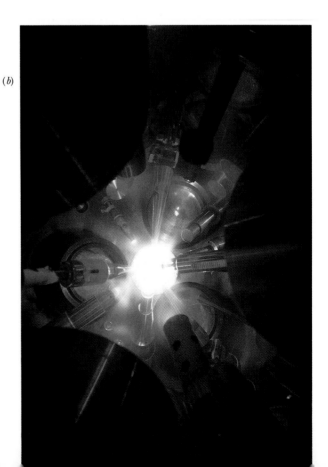

SOLUTION We can write, for the density d^* of the compressed pellet,

$$d^* = 10^3\, d = m_d\left(\frac{n}{2}\right) + m_t\left(\frac{n}{2}\right),$$

in which n is the number of particles per unit volume (either deuterons or tritons) in the compressed pellet, m_d is the mass of a deuterium atom, and m_t is the mass of a tritium atom. These atomic masses are related to the Avogadro constant N_A and to the corresponding molar masses (M_d and M_t) by

$$m_d = \frac{M_d}{N_A} \quad \text{and} \quad m_t = \frac{M_t}{N_A}.$$

Combining the foregoing equations and solving for n lead to

$$n = \frac{2000 d N_A}{M_d + M_t},$$

which gives us

$$n = \frac{(2000)\,(200\ \text{kg/m}^3)\,(6.02 \times 10^{23}\ \text{mol}^{-1})}{2.0 \times 10^{-3}\ \text{kg/mol} + 3.0 \times 10^{-3}\ \text{kg/mol}}$$

$$= 4.8 \times 10^{31}\ \text{m}^{-3}. \qquad \text{(Answer)}$$

b. According to Lawson's criterion, for how long must the pellet maintain this particle density if break-even operation is to take place?

SOLUTION From the foregoing we have, where L is the Lawson number,

$$\tau = \frac{L}{n} \approx \frac{10^{20}\ \text{s} \cdot \text{m}^{-3}}{4.8 \times 10^{31}\ \text{m}^{-3}} \approx 10^{-12}\ \text{s}. \qquad \text{(Answer)}$$

The pellet must remain compressed for at least this long if break-even operation is to occur. (It is also necessary for the effective temperature to be suitably high.)

REVIEW & SUMMARY

Energy from the Nucleus

Nuclear processes are about a million times more effective, per unit mass, than chemical processes in transforming mass into other forms of energy.

Induced Fission

Equation 48-1 shows a **fission** of ^{235}U induced by thermal neutrons. Equations 48-2 and 48-3 show the beta-decay chains of the primary fragments. The energy released in such a fission event is $Q \approx 200$ MeV.

A Model for Fission

Fission can be understood in terms of the liquid-drop model, in which a nucleus is likened to a charged liquid drop carrying a certain excitation energy: see Fig. 48-2. A potential barrier (see Fig. 48-3) must be tunneled if fission is to occur. Table 48-2 shows that fissionability depends on the relationship between the barrier height E_b and the excitation energy E_n.

Chain Reactions and Reactors

The free neutrons released during fission make possible a fission **chain reaction.** Figure 48-4 shows the neutron balance for one cycle of a typical reactor. Figure 48-5 suggests the outlines of a complete nuclear power plant.

Fusion

The release of energy by the **fusion** of two light nuclei is inhibited by their mutual Coulomb barrier. Fusion can occur in bulk matter only if the temperature is high enough (that is, if the particle energy is high enough) for

appreciable barrier tunneling to occur. In this regard, Fig. 48-9 shows (1) the energy distribution $n(K)$ for protons at the central temperature of the sun and (2) the proton–proton barrier penetrability factor $p(K)$.

The Proton–Proton Cycle

The sun's energy arises mainly from the thermonuclear "burning" of hydrogen to form helium by the **proton–proton cycle** outlined in Fig. 48-10. The overall Q is 26.7 MeV per cycle.

Element Building

Elements up to $A = 56$ (the peak of the binding energy curve of Fig. 47-6) can be built up by other fusion processes once the hydrogen fuel supply of a star has been exhausted. Heavier elements are probably formed by successive neutron captures in supernova explosions.

Controlled Fusion

Controlled **thermonuclear fusion** for power generation has not yet been achieved, even on a laboratory scale. The d–d and the d–t reactions (Eqs. 48-9 to 48-11) are the most likely mechanisms. A successful fusion reactor must satisfy **Lawson's criterion,**

$$n\tau > 10^{20}\ \text{s} \cdot \text{m}^{-3}, \qquad (48\text{-}12)$$

and must have a plasma temperature T greater than about 10^8 K ($kT \approx 9$ keV). Confining such a hot plasma is a major problem.

In a **tokamak** the plasma is confined by a magnetic field. In **laser fusion** inertial confinement is used.

QUESTIONS

1. Can you say, from examining Table 48-1, that one source of energy, or of power, is better than another? If not, what other considerations enter?

2. To which of the processes in Table 48-1 does the relationship $E = \Delta m\, c^2$ apply?

3. In the generalized equation for the fission of ^{235}U by thermal neutrons, $^{235}U + n \to X + Y + b n$, do you expect the Q of the reaction to depend on the identity of X and Y?

4. Is the fission fragment curve of Fig. 48-1 necessarily symmetrical about its central minimum? Explain.

5. In the chain decays of the primary fission fragments, see Eqs. 48-2 and 48-3, why do no e^+ decays occur?

6. The half-life of ^{235}U is 7.0×10^8 y. Discuss the assertion that if it had turned out to be shorter by a factor of 10 or so, there would not be any atomic bombs today.

7. The half-life for the decay of ^{235}U by alpha emission is 7×10^8 y; by spontaneously occurring fission, acting alone, it would be 3×10^{17} y. Both are barrier tunneling processes, as Figs. 47-9 and 48-3 reveal. Why this enormous difference in barrier tunneling probability?

8. In what sense does a chain reaction occur in both a nuclear reactor and a coal fire? What is the energy-releasing mechanism in each case?

9. Not all neutrons produced in a reactor are destined to initiate a fission event. What happens to those that do not?

10. Explain just what is meant by the statement that in a reactor core, neutron leakage is a surface effect and neutron production is a volume effect.

11. Explain the purpose of the moderator in a nuclear reactor. Is it possible to design a reactor that does not need a moderator? If so, what would be some of the advantages and disadvantages of such a reactor?

12. Describe how the control rods of a nuclear reactor should be operated (a) during initial start-up, (b) to reduce the power level, and (c) on a long-term basis, as fuel is consumed.

13. A reactor is operating at full power with its multiplication factor k adjusted to unity. If the reactor is now adjusted to operate stably at half power, what value must k now assume?

14. Separation of the two isotopes ^{238}U and ^{235}U from natural uranium requires a physical method, such as diffusion, rather than a chemical method. Explain why.

15. A piece of pure ^{235}U (or ^{239}Pu) will spontaneously explode if it is larger than a certain "critical size." A smaller piece will not explode. Explain.

16. The Earth's core is thought to be made of iron because, during the formation of the Earth, heavy elements such as iron would have sunk toward the Earth's center and lighter elements, such as silicon, would have floated upward to form the Earth's crust. However, iron is far from the heaviest element. Why isn't the Earth's core made of uranium?

17. Do you think that the thermonuclear fusion reaction controlled by the two curves plotted in Fig. 48-9 necessarily has its maximum effectiveness for the energy at which the two curves cross each other? Explain your answer.

18. Why does it take so long (about 10^6 y!) for gamma-ray photons generated by nuclear reactions in the sun's central core to diffuse to the surface? What kinds of interactions do they have with the protons, α particles, and electrons that make up the core?

19. The primordial matter of the early universe is thought to have been largely hydrogen. From where did all the silicon in the Earth come? All the gold?

20. Do conditions at the core of the sun satisfy Lawson's criterion for a sustained thermonuclear fusion reaction? Explain.

21. To achieve ignition in a tokamak, why do you need a high plasma temperature? A high density of plasma particles? A long confinement time?

22. Which would generate more radioactive waste products, a fission reactor or a fusion reactor?

EXERCISES & PROBLEMS

SECTION 48-2 NUCLEAR FISSION: THE BASIC PROCESS

1E. (a) How many atoms are contained in 1.0 kg of pure ^{235}U? (b) How much energy, in joules, is released by the complete fissioning of 1.0 kg of ^{235}U? Assume $Q = 200$ MeV. (c) For how long would this energy light a 100-W lamp?

2E. The fission properties of the plutonium isotope ^{239}Pu are very similar to those of ^{235}U. The average energy released per fission is 180 MeV. How much energy, in MeV, is liberated if all the atoms in 1.00 kg of pure ^{239}Pu undergo fission?

3E. At what rate must ^{235}U nuclei undergo fission by neutrons to generate energy at the rate of 1.0 W? Assume that $Q = 200$ MeV.

4E. Fill in the following table, which refers to the generalized fission reaction

$$^{235}U + n \rightarrow X + Y + bn.$$

X	Y	b
^{140}Xe	—	1
^{139}I	—	2
—	^{100}Zr	2
^{141}Cs	^{92}Rb	—

5E. Verify that, as stated in Section 48-2, neutrons in equilibrium with matter at room temperature, 300 K, have an average kinetic energy of about 0.04 eV.

6E. Calculate the disintegration energy Q for the fission of ^{52}Cr into two equal fragments. The masses you will need are ^{52}Cr, 51.94051 u; and ^{26}Mg, 25.98259 u. Discuss your result.

7E. Calculate the disintegration energy Q for the fission of ^{98}Mo into two equal parts. The masses you will need are ^{98}Mo, 97.90541 u; and ^{49}Sc, 48.95002 u. If Q turns out to be positive, discuss why this process does not occur spontaneously.

8E. Calculate the energy released in the fission reaction

$$^{235}U + n \rightarrow {}^{141}Cs + {}^{93}Rb + 2n.$$

Needed atomic and particle masses are

^{235}U	235.04392 u	^{93}Rb	92.92157 u
^{141}Cs	140.91963 u	n	1.00867 u

9E. ^{235}U decays by alpha emission with a half-life of 7.0×10^8 y. It also decays (rarely) by spontaneous fission, and if the alpha decay did not occur, its half-life due to this process alone would be 3.0×10^{17} y. (a) At what rate do spontaneous fission decays occur in 1.0 g of ^{235}U? (b) How many alpha-decay events are there for every spontaneous fission event?

10P. Verify that, as reported in Table 48-1, the fission of the ^{235}U in 1.0 kg of UO$_2$ (enriched so that ^{235}U is 3.0% of the total uranium) could keep a 100-W lamp burning for 690 y.

11P. Consider the fission of ^{238}U by fast neutrons. In one fission event no neutrons were emitted and the final stable end products, after the beta decay of the primary fission fragments, were ^{140}Ce and ^{99}Ru. (a) How many beta-decay events were there in the two beta-decay chains, considered together? (b) Calculate Q. The relevant atomic masses are

^{238}U	238.05079 u	^{140}Ce	139.90543 u
n	1.00867 u	^{99}Ru	98.90594 u

12P. In a particular fission event in which ^{235}U is fissioned by slow neutrons, no neutron is emitted and one of

the primary fission fragments is ^{83}Ge. (a) What is the other fragment? (b) How is the disintegration energy $Q = 170$ MeV split between the two fragments? (c) Calculate the initial speed of each fragment.

13P. Assume that just after the fission of ^{236}U* according to Eq. 48-1, the resulting ^{140}Xe and ^{94}Sr nuclei are just touching at their surfaces. (a) Assuming the nuclei to be spherical, calculate the Coulomb potential energy (in MeV) of repulsion between the two fragments. (*Hint:* Use Eq. 47-3 to calculate the radii of the fragments.) (b) Compare this energy with the energy released in a typical fission event. In what form will the Coulomb potential energy ultimately appear in the laboratory?

14P. A ^{236}U* nucleus undergoes fission and breaks up into two middle-mass fragments, ^{140}Xe and ^{96}Sr. (a) By what percentage does the surface area of the ^{236}U nucleus change during this process? (b) By what percentage does its volume change? (c) By what percentage does its electrostatic potential energy change? The potential energy of a uniformly charged sphere of radius r and charge Q is given by

$$U = \frac{3}{5}\left(\frac{Q^2}{4\pi\epsilon_0 r}\right).$$

SECTION 48-4 THE NUCLEAR REACTOR

15E. A 200-MW fission reactor consumes half its fuel in 3.00 y. How much ^{235}U did it contain initially? Assume that all the energy generated arises from the fission of ^{235}U and that this nuclide is consumed only by the fission process.

16E. Repeat Exercise 15 taking into account nonfission neutron capture by the ^{235}U.

17E. ^{238}Np requires 4.2 MeV for fission. To remove a neutron from this nuclide requires an energy expenditure of 5.0 MeV. Is ^{237}Np fissionable by thermal neutrons?

18P. The thermal energy generated when radiations from radionuclides are absorbed in matter can be used as the basis for a small power source for use in satellites, remote weather stations, and so on. Such radionuclides are manufactured in abundance in nuclear power reactors and may be separated chemically from the spent fuel. One suitable radionuclide is ^{238}Pu ($\tau = 87.7$ y), which is an alpha emitter with $Q = 5.50$ MeV. At what rate is thermal energy generated in 1.00 kg of this material?

19P. (See Problem 18.) Among the many fission products that may be extracted chemically from the spent fuel of a nuclear power reactor is ^{90}Sr ($\tau = 29$ y). It is produced in typical large reactors at the rate of about 18 kg/y. By its radioactivity it generates thermal energy at the rate of 0.93 W/g. (a) Calculate the effective disintegration energy Q_{eff} associated with the decay of a ^{90}Sr nucleus. (Q_{eff} includes contributions from the decay of the ^{90}Sr daughter products in its decay chain but not from neutrinos, which escape totally from the sample.) (b) It is desired to construct a power source generating 150 W (electric) to use in operating electronic equipment in an underwater acoustic beacon. If the source is based on the thermal energy generated by ^{90}Sr and if the efficiency of the thermal–electric conversion process is 5.0%, how much ^{90}Sr is needed?

20P. Many fear that helping additional nations develop nuclear power reactor technology will increase the likelihood of nuclear war because reactors can be used not only to produce energy but, as a by-product through neutron capture with inexpensive ^{238}U, to make ^{239}Pu, which is a "fuel" for nuclear bombs. What simple series of reactions involving neutron capture and beta decay would yield this plutonium isotope?

21P. In an atomic bomb, energy release is due to the uncontrolled fission of plutonium ^{239}Pu (or ^{235}U). The bomb's rating is the magnitude of the released energy, specified in terms of the mass of TNT required to produce the same energy release. One megaton (10^6 tons) of TNT releases 2.6×10^{28} MeV of energy. (a) Calculate the rating, in tons of TNT, of an atomic bomb containing 95 kg of ^{239}Pu, of which 2.5 kg actually undergoes fission. (See Exercise 2.) (b) Why is the other 92.5 kg of ^{239}Pu needed if it does not fission?

22P. A 66-kiloton atomic bomb (see Problem 21) is fueled with pure ^{235}U (Fig. 48-15), 4.0% of which actually undergoes fission. (a) How much uranium is in the bomb? (b) How many primary fission fragments are produced? (c) How many neutrons generated in the fissions are released to the environment? (On the average, each fission produces 2.5 neutrons.)

23P. The neutron generation time t_{gen} in a reactor is the average time needed for a fast neutron emitted in one fission to be slowed down to thermal energies by the moder-

ator and to initiate another fission. Suppose that the power output of a reactor at time $t = 0$ is P_0. Show that the power output a time t later is $P(t)$, where

$$P(t) = P_0 k^{t/t_{gen}},$$

where k is the multiplication factor. For constant power output $k = 1$.

24P. The neutron generation time (see Problem 23) of a particular power reactor is 1.3 ms. It is generating energy at the rate of 1200 MW. To perform certain maintenance checks, the power level must temporarily be reduced to 350 MW. It is desired that the transition to the reduced power level take 2.6 s. To what (constant) value should the multiplication factor be set to effect the transition in the desired time?

25P. The neutron generation time t_{gen} (see Problem 23) in a particular reactor is 1.0 ms. If the reactor is operating at a power level of 500 MW, about how many free neutrons are present in the reactor at any moment?

26P. A reactor operates at 400 MW with a neutron generation time (see Problem 23) of 30.0 ms. If its power increases for 5.00 min with a multiplication factor of 1.0003, find the power output at the end of the 5.00 min.

27P. (a) A neutron of mass m_n and kinetic energy K makes a head-on elastic collision with a stationary atom of mass m. Show that the fractional kinetic energy loss of the neutron is given by

$$\frac{\Delta K}{K} = \frac{4 m_n m}{(m + m_n)^2},$$

in which m_n is the neutron mass. (b) Find $\Delta K/K$ for each of following as the stationary atom: hydrogen, deuterium, carbon, and lead. (c) If $K = 1.00$ MeV initially, how many such collisions would it take to reduce the neutron energy to thermal values (0.025 eV) if the material (consisting of the stationary atoms) is deuterium, a commonly used moderator? (*Note:* In actual moderators, most collisions are not "head-on.")

SECTION 48-5 A NATURAL NUCLEAR REACTOR

28E. How long ago was the ratio ^{235}U/^{238}U in natural uranium deposits equal to 0.15?

29E. The natural fission reactor discussed in Section 48-5 is estimated to have generated 15 gigawatt-years of energy during its lifetime. (a) If the reactor lasted for 200,000 y, at what average power level did it operate? (b) How much ^{235}U did it consume during its lifetime?

30P. In addition to ^{238}U, uranium mined today contains 0.72% of fissionable ^{235}U, too little to make reactor fuel for thermal-neutron fission. For this reason, the natural uranium must be enriched or concentrated in ^{235}U. Both ^{235}U ($\tau = 7.0 \times 10^8$ y) and ^{238}U ($\tau = 4.5 \times 10^9$ y) are radioactive. How far back in time would natural uranium have been a practical reactor fuel, with a ^{235}U/^{238}U ratio of 3.0%?

FIGURE 48-15 Problem 22. A "button" of ^{235}U, ready to be recast and machined for a warhead.

31P. Some uranium samples from the natural reactor site described in Section 48-5 were found to be slightly *enriched* in ^{235}U, rather than depleted. Account for this in terms of neutron absorption by the abundant isotope ^{238}U and the subsequent beta and alpha decay of its products.

SECTION 48-6 THERMONUCLEAR FUSION: THE BASIC PROCESS

32E. Calculate the height of the Coulomb barrier for the head-on collision of two protons. The effective radius of a proton may be taken to be 0.80 fm.

33E. From information given in the text, collect and write down the approximate heights of the Coulomb barriers for (a) the alpha decay of ^{238}U, (b) the fission of ^{235}U by thermal neutrons, and (c) the head-on collision of two deuterons.

34E. Verify that the fusion of 1.0 kg of deuterium by the reaction

$$^2H + {}^2H \rightarrow {}^3He + n, \qquad Q = +3.27 \text{ MeV},$$

could keep a 100-W lamp burning for 2.5×10^4 y.

35E. Methods other than heating the material have been suggested for overcoming the Coulomb barrier for fusion. For example, one might consider using particle accelerators. If you were to use two of them to accelerate two beams of deuterons directly toward each other so as to collide "head-on," (a) what voltage would each accelerator require for the colliding deuterons to overcome the Coulomb barrier? (b) Would this voltage be difficult to achieve? (c) Why do you suppose this method is not presently used?

36P. The equation of the curve $n(K)$ in Fig. 48-9 is

$$n(K) = 1.13n \frac{K^{1/2}}{(kT)^{3/2}} e^{-K/kT},$$

where n is the total density of particles. At the center of the sun the temperature is 1.50×10^7 K and the mean proton energy $\overline{K}$ is 1.94 keV. Find the ratio of the density of protons at 5.00 keV to that at the mean proton energy.

37P. Calculate the Coulomb barrier height for two ^7Li nuclei, fired at each other with the same initial kinetic energy K. (*Hint:* Use Eq. 47-3 to calculate the radii of the nuclei.)

38P. Expressions for the Maxwell speed and energy distributions for the molecules in a gas are given in Chapter 21. (a) Show that the *most probable energy* is given by

$$K_p = \tfrac{1}{2}kT.$$

Verify this result with the energy distribution curve of Fig. 48-9, for which $T = 1.5 \times 10^7$ K. (b) Show that the *most probable speed* is given by

$$v_p = \sqrt{\frac{2kT}{m}}.$$

Find its value for protons at $T = 1.5 \times 10^7$ K. (c) Show that *the energy corresponding to the most probable speed* (which is not the same as the most probable energy) is

$$K_{v,p} = kT.$$

Locate this quantity on the energy-distribution curve of Fig. 48-9.

SECTION 48-7 THERMONUCLEAR FUSION IN THE SUN AND OTHER STARS

39E. We have seen that Q for the overall proton–proton cycle is 26.7 MeV. How can you relate this number to the Q values for the three reactions that make up this cycle, as displayed in Fig. 48-10?

40E. Show that the energy released when three α particles fuse to form ^{12}C is 7.27 MeV. The atomic mass of ^4He is 4.0026 u, and that of ^{12}C is 12.0000 u.

41E. At the center of the sun the density is 1.5×10^5 kg/m^3 and the composition is essentially 35% hydrogen by mass and 65% helium. (a) What is the density of protons at the sun's center? (b) How much larger is this than the density of particles in an ideal gas at standard conditions of temperature and pressure?

42P. Verify the values of Q_1, Q_2, and Q_3 reported for the three reactions in Fig. 48-10. The needed atomic and particle masses are

^1H	1.007825 u	^4He	4.002603 u
^2H	2.014102 u	$e^{\pm}$	0.0005486 u
^3He	3.016029 u		

(*Hint:* Distinguish carefully between atomic and nuclear masses, and take the positrons properly into account.)

43P. Calculate and compare the energy released by (a) the fusion of 1.0 kg of hydrogen deep within the sun and (b) the fission of 1.0 kg of ^{235}U in a fission reactor.

44P. The sun has a mass of 2.0×10^{30} kg and radiates energy at the rate of 3.9×10^{26} W. (a) At what rate does the sun transfer its mass into other forms of energy? (b) What fraction of its original mass has the sun lost in this way since it began to burn hydrogen, about 4.5×10^9 y ago?

45P. (a) Calculate the rate at which the sun is generating neutrinos. Assume that energy production is entirely by the proton–proton cycle. (b) At what rate do solar neutrinos impinge on the Earth?

46P. Coal burns according to

$$C + O_2 \rightarrow CO_2.$$

The heat of combustion is 3.3×10^7 J/kg of atomic carbon consumed. (a) Express this in terms of energy per carbon atom. (b) Express it in terms of energy per kilogram of the initial reactants, carbon and oxygen. (c) Suppose that the sun (mass = 2.0×10^{30} kg) were made of carbon and oxygen in combustible proportions and that it continued to radiate energy at its present rate of 3.9×10^{26} W. How long would it last?

47P. In certain stars the *carbon cycle* is more likely than the proton–proton cycle to be effective in generating energy. This cycle is

$$^{12}\text{C} + {}^{1}\text{H} \rightarrow {}^{13}\text{N} + \gamma, \qquad Q_1 = 1.95 \text{ MeV},$$
$$^{13}\text{N} \rightarrow {}^{13}\text{C} + e^+ + \nu, \qquad Q_2 = 1.19,$$
$$^{13}\text{C} + {}^{1}\text{H} \rightarrow {}^{14}\text{N} + \gamma, \qquad Q_3 = 7.55,$$
$$^{14}\text{N} + {}^{1}\text{H} \rightarrow {}^{15}\text{O} + \gamma, \qquad Q_4 = 7.30,$$
$$^{15}\text{O} \rightarrow {}^{15}\text{N} + e^+ + \nu, \qquad Q_5 = 1.73,$$
$$^{15}\text{N} + {}^{1}\text{H} \rightarrow {}^{12}\text{C} + {}^{4}\text{He}, \qquad Q_6 = 4.97.$$

(a) Show that this cycle of reactions is exactly equivalent in its overall effects to the proton–proton cycle of Fig. 48-10. (b) Verify that the two cycles, as expected, have the same Q.

48P. Let us assume that the core of the sun comprises one-eighth of the sun's mass and is compressed within a sphere whose radius is one-fourth of the solar radius. Assume further that the composition of the core is 35% hydrogen by mass and that essentially all of the sun's energy is generated there. If the sun continues to burn hydrogen at the rate calculated in Sample Problem 48-5, how long will it be before the hydrogen is entirely consumed? The sun's mass is 2.0×10^{30} kg.

49P. The effective Q for the proton–proton cycle of Fig. 48-10 is 26.2 MeV. (a) Express this as energy per kilogram of hydrogen consumed. (b) The power of the sun is 3.9×10^{26} W. If its energy derives from the proton–proton cycle, at what rate is it losing hydrogen? (c) At what rate is it losing mass? Account for the difference in the results for (b) and (c). (d) The sun's mass is 2.0×10^{30} kg. If it loses mass at the constant rate calculated in (c), how long will it take before it loses 0.10% of its mass?

50P. After converting all its hydrogen to helium, a particular star is 100% helium in composition. It now proceeds to convert the helium to carbon via the triple-alpha process,

$$^{4}\text{He} + {}^{4}\text{He} + {}^{4}\text{He} \rightarrow {}^{12}\text{C} + 7.27 \text{ MeV}$$

(see Exercise 40). The mass of the star is 4.6×10^{32} kg, and it generates energy at the rate of 5.3×10^{30} W. How long will it take to convert all the helium to carbon?

51P. Figure 48-16 shows an idealized schematic of a hydrogen bomb. The fusion fuel is deuterium, ^{2}H. The high temperature and particle density needed for fusion are provided by an atomic-bomb "trigger," arranged so as to impress an imploding, compressive shock wave upon the deuterium. The operative fusion reaction is

$$5\,{}^{2}\text{H} \rightarrow {}^{3}\text{He} + {}^{4}\text{He} + {}^{1}\text{H} + 2\text{n}.$$

(a) Calculate Q for the fusion reaction. For needed atomic masses see Problem 42. (b) Calculate the "rating" (see

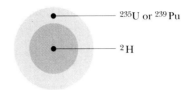

FIGURE 48-16 Problem 51.

Problem 21) of the fusion part of the bomb if it contains 500 kg of deuterium, 30.0% of which undergoes fusion.

SECTION 48-8 CONTROLLED THERMONUCLEAR FUSION

52E. Verify the Q values reported in Eqs. 48-9, 48-10, and 48-11. The needed masses are

^{1}H	1.007825 u	^{4}He	4.002603 u
^{2}H	2.014102 u	n	1.008665 u
^{3}H	3.016049 u		

53P. In the deuteron–triton fusion reaction of Eq. 48-11, how is the reaction energy Q shared between the α particle and the neutron? Neglect the relatively small kinetic energies of the two combining particles.

54P. Ordinary water consists of roughly 0.0150% by mass of "heavy water," in which one of the two hydrogens is replaced with deuterium, ^{2}H. How much average fusion power could be obtained if we "burned" all the ^{2}H in 1 liter of water in 1 day through the reaction $^{2}\text{H} + {}^{2}\text{H} \rightarrow {}^{3}\text{He} + \text{n}$?

SECTION 48-9 THE TOKAMAK

55E. By the summer of 1985, the TFTR tokamak at the Princeton Plasma Physics Laboratory could be run consistently with plasma number densities of 3×10^{13} cm^{-3}, confinement times of 400 ms, and ion temperatures of 3 keV. Under special experimental conditions these same parameters are 6×10^{12} cm^{-3}, 100 ms, and 10 keV. Plot the two points representing these data on Fig. 48-13. Realizing that these data represent just one machine while the figure represents many, what is your impression of the progress being made in this type of fusion research?

SECTION 48-10 LASER FUSION

56E. Assume that a plasma temperature of 1×10^8 K is reached in a laser-fusion device. (a) What is the most probable speed of a deuteron at this temperature? (b) How far would such a deuteron move in the confinement time calculated in Sample Problem 48-7?

57P. The uncompressed radius of the fuel pellet of Sample Problem 48-7 is 200 μm. Suppose that the compressed fuel pellet "burns" with an efficiency of 10%. That is, only

10% of the deuterons and 10% of the tritons participate in the fusion reaction of Eq. 48-11. (a) How much energy is released in each such microexplosion of a pellet? (b) To how much TNT is each such pellet equivalent? (The heat of combustion of TNT is 4.6 MJ/kg.) (c) If a fusion reactor were constructed on the basis of 100 microexplosions per second, what would be its power? (Note that part of this power must be used to operate the lasers.)

ADDITIONAL PROBLEMS

58. Electrons emitted by a radioactive source undergoing beta decay are absorbed and their kinetic energy is converted to electricity. The efficiency of the conversion is 10%, the initial power produced is 10 W, there is initially 0.050 mol of the emitter, and the half-life of the decay is 88 days. What is the average kinetic energy of the emitted electrons?

59. Assume that the charge of a nucleus $^A M_Z$ is uniformly distributed in a sphere of radius $R = R_0 A^{1/3}$. Derive an expression for the contribution to the energy of the fission fragments (in a spherically symmetrical fission) due to the Coulomb energy difference between the initial nucleus and the fission fragments.

60. The power emitted by the sun is 3.9×10^{26} W, which corresponds to a loss in mass of 6.2×10^{11} kg/s. Assume that the sun's mass (1.99×10^{30} kg) is uniformly distributed in a sphere of radius 6.96×10^8 m. At what rate does the sun lose gravitational potential energy due to its power emission?

61. A nucleus of mass M_1 and kinetic energy K_1 moves along the x axis and is incident on a stationary nucleus of mass M_2. The collision results in two new nuclei. One of them, of mass M_3 and kinetic energy K_3, moves at angle ϕ to the x axis. The other one, of mass M_4 and kinetic energy K_4, moves at angle θ to the x axis. Assuming that the particles are moving at nonrelativistic speeds, show that the Q of the reaction is

$$Q = K_3 \left(1 + \frac{M_3}{M_4} \right) - K_1 \left(1 - \frac{M_1}{M_4} \right) - \frac{2\sqrt{M_1 K_1 M_3 K_3}}{M_4} \cos \phi.$$

QUARKS, LEPTONS, AND THE BIG BANG

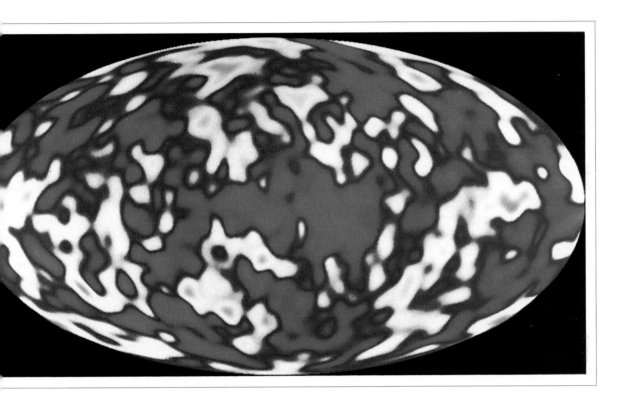

This color-coded image is effectively a photograph of the universe when it was only 300,000 y old, which was about 15×10^9 y ago. This is what you would have seen then as you looked away in all directions (the view has been condensed to this oval). Patches of light from atoms stretch across the "sky," but galaxies, stars, and planets have not yet formed. How can a photograph of the early universe be taken?

49-1 LIFE AT THE CUTTING EDGE

Physicists often refer to the theory of relativity and the quantum theory as "modern physics," to distinguish them from the theories of Newtonian mechanics and Maxwellian electromagnetism, which are lumped together as part of "classical physics." As the years go by, the word "modern" seems less and less appropriate for theories whose foundations were laid down in the opening years of this century. Nevertheless, the label hangs on.

In this closing chapter we consider two lines of investigation that are at once truly "modern" but at the same time have the most ancient of roots. They center around the questions:

> Of what is the universe made?

and

> How did the universe come to be the way it is?

Progress in answering these questions has been rapid in the last few decades, and there are those who think (perhaps unwisely) that definitive answers lie not far beyond our present horizons.

These two questions are not independent. As physicists bang particles together at higher and higher energies, using larger and larger accelerators, they come to realize that no conceivable Earth-bound accelerator can generate particles with energies in the region toward which their theories are tending. There is only one source of particles with these energies and that is the universe itself within the first minute of its existence. The "quark soup" that constituted the universe at these early times on the cosmic clock is the ultimate testing ground for the theories of particle physics!

This chapter is different from the others in that the plan is to bring you close to the frontier of current research and to identify some of the physicists whose efforts have brought us to this point. You will encounter a host of new terms and a veritable flood of particles with names that you should not try to remember. If you are temporarily bewildered, you are sharing the bewilderment of the physicists who lived through these developments and who at times saw nothing but increasing complexity with little hope of understanding. If you stick with it, however, you will come to share the excitement physicists have felt as marvelous new accelerators poured out new results, as the theorists put forth ideas each more daring than the last, and as clarity finally sprang from obscurity.

You may wish to begin by rereading Section 2-9, in which we first discussed the basic particles of physics.

49-2 PARTICLES, PARTICLES, PARTICLES

In the 1930s, there were many who thought that the problem of the ultimate structure of matter was well on the way to being solved. The atom could be understood in terms of only three particles—the electron, the proton, and the neutron. Quantum theory accounted well for the structure of the atom and for radioactive alpha decay. The neutrino had been postulated and, although not yet observed, had been incorporated by Fermi into a successful theory of beta decay. There was hope that quantum theory, applied to protons and neutrons, would soon account for the structure of the nucleus. What else was there?

FIGURE 49-1 The OPAL (Omni-Purpose Apparatus) detector at CERN. It is designed to measure the energies of particles produced in electron–positron collisions at energies of 50 GeV. Although the detector is huge (weighing over 3000 tons), it is small compared with the collider itself, which is a ring with a circumference of 27 km.

The euphoria did not last. The end of that same decade saw the beginning of a period of discovery of new particles that continues to this day. The new particles have names and symbols such as the *muon* (μ), the *pion* (π), the *kaon* (K), the *sigma* (Σ), and so on. All the new particles are unstable, their half-lives ranging from about 10^{-6} to 10^{-23} s. This last value is so small that the very existence of such particles can be established only by indirect methods.

The new particles were first found in reactions triggered by the high-energy protons (the *cosmic rays*) that stream in from space and collide with nuclei of atmospheric atoms. Increasingly, however, the new particles were produced in head-on collisions between protons or electrons accelerated to high energies in accelerators at places like Fermilab (near Chicago), CERN (near Geneva), and SLAC (at Stanford). The particle detectors grew in sophistication until (see Fig. 49-1) they rivaled in size and complexity the accelerators themselves of only a decade or so ago.

Today there are several hundred known particles. Naming them has strained the resources of the Greek alphabet, and most are known only by an assigned number in a periodically issued compilation. Fermi is said to have remarked that if he had known that there were so many particles whose properties he was expected to memorize, he would have taken up botany!

To make sense of this array of particles, we look for simple physical criteria, each of which will allow us to place each particle into either of two categories. Each time we apply a new criterion, we refine both the classification scheme and our understanding. We can make a first "rough cut" among the particles in at least three ways:

1. *Spin.* All particles have an intrinsic angular momentum (as if they are spinning) given by

$$L = s\hbar, \qquad (49\text{-}1)$$

in which $\hbar = h/2\pi$ and s, the *spin quantum number*, can have either half-integer ($\frac{1}{2}, \frac{3}{2}, \ldots$) or integer ($0, 1, \ldots$) values.

Particles with half-integer spins are called **fermions,** after Fermi, who (simultaneously with Dirac) discovered the statistical rules that govern their behavior. Electrons, protons, and neutrons, all of which have $s = \frac{1}{2}$, are fermions.

Particles with integer spins are called **bosons,** after Indian physicist Satyendra Nath Bose, who (simultaneously with Einstein) discovered the governing statistical rules for *these* particles. Photons, which have $s = 1$, are bosons; you will soon meet other important particles in this class.

This may seem a trivial way to classify particles but it is very important for this reason:

Fermions obey the Pauli exclusion principle, which asserts that only a single particle can be assigned to a given quantum state. Bosons *do not* obey this principle. Any number of bosons can be assigned to a given state. Since particles prefer to be in states of lowest energy, bosons tend to cluster together in the lowest possible states.

You have seen how important the Pauli principle is in the process by which electrons (fermions) assume quantum states of the atom.

2. *Forces.* We can also classify particles in terms of the forces that act on them. In Section 6-5 (which you may wish to reread) we outlined the four known fundamental forces. The *gravitational force* acts on *all* particles, but its effects at the level of subatomic particles are so weak that we need not (yet!) consider them. The *electromagnetic force* acts on all *charged* particles; its effects are well known and we can take them into account when we need to; we largely ignore this force in the rest of this chapter.

We are left with the *strong force*, which is the force that binds the nucleus together, and the *weak force*, which is involved in beta decay and similar processes. The weak force acts on all particles, the strong force only on some particles.

We can roughly classify particles on the basis of whether or not the strong force acts on them.

Particles on which the *strong force* acts are called **hadrons.** Particles on which the strong force does *not* act, leaving the weak force as the dominant force, are called **leptons.** Protons, neutrons, and pions are hadrons; electrons and neutrinos are leptons. You will soon meet other members of each class.

We can make a further distinction among the hadrons because some of them (we call them **mesons**) are bosons; the pion is an example. The

THE FORCES THAT ACT ON EACH CATEGORY OF PARTICLE[a]

| | LEPTONS | HADRONS | |
		MESONS	BARYONS
Fermions	WEAK		STRONG WEAK
Bosons		STRONG WEAK	

[a]No particles exist in categories corresponding to the empty boxes. Thus all leptons and all baryons are fermions and all mesons are bosons.

other hadrons (we call them **baryons**) are fermions; the proton is an example. Table 49-1 summarizes these two criteria for classifying particles.

3. Particles and Antiparticles. In 1928, Dirac predicted that the electron should have a positively charged counterpart. This particle, the *positron*, was discovered in the cosmic radiation in 1932 by Carl Anderson. It gradually became clear that *every* particle has a corresponding **antiparticle** with the same mass and spin but (if it is a charged particle) with a charge of the opposite sign. (Particles and antiparticles also differ in the signs of other quantum numbers that we have not yet discussed.) We often represent an antiparticle by putting a bar over the symbol for the particle. Thus p is the symbol for the proton, and $\bar{\text{p}}$ (pronounced "p bar") for the antiproton.

When a particle meets its antiparticle, they can annihilate each other. That is, the particles can disappear, their combined rest energies becoming available to appear in other forms. For an electron annihilating with its antiparticle, this energy appears as two gamma-ray photons:

$$e^- + e^+ \rightarrow \gamma + \gamma \qquad (Q = 1.02 \text{ MeV}). \quad (49\text{-}2)$$

If the two particles are stationary when they annihilate, the photons share the energy equally between them and—to conserve momentum and because photons cannot be stationary—they fly off in opposite directions.

One of the current puzzles in particle physics is the fact that the world we live in is a world of *particles,* not of antiparticles. This dominance of matter over antimatter certainly extends throughout our own galaxy. One can speculate that there are distant anti-

matter galaxies in which the atoms have nuclei with antiprotons, surrounded by clouds of positrons. One can contemplate the disaster that would occur if an antiphysicist from such a galaxy met a physicist from our galaxy in deep space and shook hands! The present view, however, is that the dominance of matter over antimatter extends throughout the universe and that there are no antiphysicists.

49-3 AN INTERLUDE

Before pressing on with the task of classifying the particles, let us step aside for a moment and capture some of the spirit of particle research by analyzing a typical particle event, that shown in the bubble-chamber photograph of Fig. 49-2*a*.

The tracks in this figure are streams of bubbles formed in the wake of energetic charged particles as they move through a chamber filled with liquid hydrogen. We can identify the particle that leaves a particular track—among other ways—by measuring the relative spacing between the bubbles. A magnetic field permeates the chamber, deflecting the tracks of positively charged particles counterclockwise and those of negatively charged particles clockwise. By measuring the radius of curvature of a track, we can calculate the momentum of the particle that made it. Table 49-2 shows some properties of the particles that participated in the event of Fig. 49-2*a*.

Our tools for analysis are the laws of conservation of energy, of linear momentum, of angular momentum, and of charge, along with other conservation laws that we have not yet discussed. Figure 49-2*a* is one member only of a stereo pair of photographs so that, in practice, these analyses can be carried out in three dimensions.

The event of Fig. 49-2*a* is triggered by an energetic antiproton ($\bar{\text{p}}$) that, generated in an accelerator at the Lawrence Berkeley Laboratory, enters the chamber from the left. There are three separate subevents, occurring at points 1, 2, and 3 in Fig. 49-2*b*.

1. Proton–Antiproton Annihilation. At point 1 in Fig. 49-2*b*, the initiating antiproton slams into a proton of the chamber fluid and they annihilate each other. We can tell that the annihilation process occurred while the incoming antiproton was in flight because most of the particles generated in the encounter move in the forward direction, that is, toward the right in Fig. 49-2. From the principle of conservation

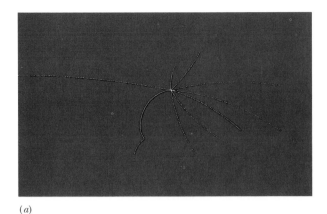

(a)

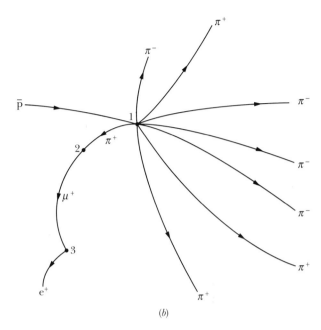

(b)

FIGURE 49-2 (a) A bubble chamber photograph of a series of events initiated by an antiproton that enters the chamber from the left. (b) The tracks redrawn and labeled for clarity. The dots at points 1, 2, and 3 indicate the sites of a sequence of specific secondary events that are described in the text. The tracks are curved because a magnetic field is present that acts on the moving charged particles.

of linear momentum then, the incoming antiproton must have had a forward momentum.

The total energy involved in the collision of the antiproton and proton is the sum of the antiproton's kinetic energy and the two (identical) rest energies of the particles (2×938.3 MeV or 1876.6 MeV). This is enough energy to create a number of lighter particles and to endow them with kinetic energy. In this case, the annihilation produces four positive pions and four negative pions. (For simplicity, we assume that no gamma-ray photons or neutral particles, which would leave no tracks, are produced.) The process is

$$p + \overline{p} \rightarrow 4\pi^+ + 4\pi^-. \qquad (49\text{-}3)$$

The reaction of Eq. 49-3 is a *strong interaction* (it involves the strong force), because all the particles involved are hadrons.

Note that charge is conserved. We can write the charge of a particle as Qe in which Q is a *charge quantum number*. These numbers for the interaction of Eq. 49-3 are

$$(+1) + (-1) = 4 \times (+1) + 4 \times (-1),$$

which tells us that the net charge is zero before the interaction and zero afterward.

For the energy balance, note from above that the energy available from the p–$\overline{p}$ annihilation process is at least the sum of the rest energies,

TABLE 49-2
THE PARTICLES OR ANTIPARTICLES INVOLVED IN THE EVENT OF FIG. 49-2

PARTICLE	SYMBOL	CHARGE	REST ENERGY (MeV)	SPIN	IDENTITY	MEAN LIFE[a] (s)	ANTIPARTICLE
Neutrino	ν	0	0	$\frac{1}{2}$	Lepton	Stable	$\overline{\nu}$
Electron	e^-	-1	0.511	$\frac{1}{2}$	Lepton	Stable	e^+
Muon	μ^-	-1	105.7	$\frac{1}{2}$	Lepton	2.2×10^{-6}	μ^+
Pion	π^+	$+1$	139.6	0	Meson	2.6×10^{-8}	π^-
Proton	p	$+1$	938.3	$\frac{1}{2}$	Baryon	Stable	$\overline{p}$

[a]The *mean life* ($= 1/\lambda$) differs from the *half-life* [$= (\ln 2)/\lambda$]; see Section 47-3.

1876.6 MeV. The rest energy of a pion is 139.6 MeV, so the rest energies of the eight pions amount to 8×139.6 MeV or 1116.8 MeV. This leaves a substantial amount of energy (at least about 760 MeV) to distribute among the eight pions as kinetic energy.

2. *Pion Decay.* Pions are unstable particles, charged pions decaying with a mean life of 2.6×10^{-8} s. At point 2 in Fig. 49-2b one of the positive pions comes to rest in the chamber and decays spontaneously into a (anti)muon and a neutrino:

$$\pi^+ \rightarrow \mu^+ + \nu. \qquad (49\text{-}4)$$

The latter particle, being uncharged, leaves no track. Both the muon and the neutrino are leptons; that is, they are particles on which the strong force does not act. The muon rest energy is 105.7 MeV, so an energy of 139.6 MeV − 105.7 MeV or 33.9 MeV is available to share between the muon and the neutrino as kinetic energy.

The spin of the pion is zero and those of the muon and the neutrino are both one-half; therefore angular momentum is conserved if the spins of the muon and neutrino are opposite each other.

3. *Muon Decay.* Muons are also unstable, decaying with a mean life of 2.2×10^{-6} s. At point 3 in Fig. 49-2b, the muon produced in the reaction of Eq. 49-4 comes to rest in the chamber and decays spontaneously according to

$$\mu^+ \rightarrow e^+ + \nu + \bar{\nu}. \qquad (49\text{-}5)$$

The track of the positron is clearly visible; again, we cannot see the neutrinos because they leave no tracks. The rest energy of the muon is 105.7 MeV and that of the electron is only 0.511 MeV, leaving 105.2 MeV to be shared as kinetic energy among the three particles produced in Eq. 49-5.

You may wonder: Why *two* neutrinos in Eq. 49-5? Why not just one, as in the pion decay in Eq. 49-4? One answer is that the spins of the muon, the electron, and the neutrino are each one-half; with only one neutrino, angular momentum could not be conserved in the muon decay of Eq. 49-5.

SAMPLE PROBLEM 49-1

In 1964 some experiments at the Brookhaven National Laboratory employed a focused beam of kaons (K^-).

The kaons, whose kinetic energy was 5000 MeV, were generated in the Brookhaven synchrotron and traveled a distance of 140 m through a highly evacuated beam tube to a bubble chamber, where the experiments took place.

The rest energy mc^2 of a kaon is 494 MeV and its half-life τ_0 for decay is 8.6×10^{-9} s. By what factor had the intensity of the kaon beam diminished while the particles were traveling from the synchrotron to the bubble chamber?

SOLUTION   The kinetic energy of a kaon is related to its rest energy mc^2 by (see Eq. 42-36)

$$K = mc^2(\gamma - 1),$$

so that the Lorentz factor γ is

$$\gamma = \frac{K}{mc^2} + 1$$

$$= \frac{5000 \text{ MeV}}{494 \text{ MeV}} + 1 = 11.1.$$

The half-life of these kaons in the reference frame of the laboratory is related to their half-life at rest by the time dilation factor (see Eq. 42-8):

$$\tau = \gamma\tau_0 = (11.1)(8.6 \times 10^{-9} \text{ s}) = 9.55 \times 10^{-8} \text{ s}.$$

These energetic kaons travel at approximately the speed of light. At that speed and in a time τ, a kaon beam could cover a distance of

$$L = c\tau = (3.00 \times 10^8 \text{ m/s})(9.55 \times 10^{-8} \text{ s}) = 28.7 \text{ m},$$

after which its intensity would have fallen to half the initial value. Over the full travel distance of 140 m, the beam intensity would drop to

$$\left(\frac{1}{2}\right)^{(140/28.7)} = 0.034 \text{ or } 3.4\% \qquad \text{(Answer)}$$

of its initial value, due to particle decay alone.

Such a beam loss—though unwelcome—is acceptable. Note, however, that *if it had not been for the time dilation effect,* the beam would have weakened to

$$\left(\frac{1}{2}\right)^{(140/28.7)(11.1)} \approx 5 \times 10^{-17}$$

of its initial value. Thus time dilation increased the beam intensity by a factor of nearly a million billion!

SAMPLE PROBLEM 49-2

A stationary pion decays as described by Eq. 49-4:

$$\pi^+ \rightarrow \mu^+ + \nu.$$

What is the kinetic energy of the muon? What is the kinetic energy of the neutrino?

SOLUTION From Table 49-2 the rest energies of the pion and the muon are 139.6 MeV and 105.7 MeV, respectively. The difference between these quantities must appear as kinetic energy of the muon and the neutrino, or

$$139.6 \text{ MeV} - 105.7 \text{ MeV} = 33.9 \text{ MeV}$$

$$= K_\mu + K_\nu. \quad (49\text{-}6)$$

To conserve linear momentum, we must have

$$p_\mu = p_\nu$$

in which p_μ is the magnitude of the linear momentum of the muon and p_ν that of the neutrino. For convenience, we cast this in the form

$$(p_\mu c)^2 = (p_\nu c)^2. \quad (49\text{-}7)$$

Equation 42-40,

$$(pc)^2 = K^2 + 2Kmc^2, \quad (49\text{-}8)$$

gives the relativistic relation between the kinetic energy K of a particle and its momentum p. If we apply this relation to Eq. 49-7, we find that

$$K_\mu^2 + 2K_\mu m_\mu c^2 = K_\nu^2 \quad (49\text{-}9)$$

because $mc^2 = 0$ for the neutrino. Combining this result with Eq. 49-6 and solving for K_μ, we find

$$K_\mu = \frac{(33.9 \text{ MeV})^2}{(2)(33.9 \text{ MeV} + m_\mu c^2)}$$

$$= \frac{(33.9 \text{ MeV})^2}{(2)(33.9 \text{ MeV} + 105.7 \text{ MeV})}$$

$$= 4.12 \text{ MeV}. \quad \text{(Answer)}$$

The kinetic energy of the neutrino is then, from Eq. 49-6,

$$K_\nu = 33.9 \text{ MeV} - K_\mu = 33.9 \text{ MeV} - 4.12 \text{ MeV}$$

$$= 29.8 \text{ MeV}. \quad \text{(Answer)}$$

We see that, although the magnitude of the momentum of the two recoiling particles is the same, the neutrino gets the larger share (88%) of the kinetic energy.

SAMPLE PROBLEM 49-3

Protons in a bubble chamber are bombarded by energetic negative pions, and the following reaction occurs:

$$\pi^- + p \rightarrow K^- + \Sigma^+.$$

The rest energies of the particles involved are

π^-	139.6 MeV	K^-	493.7 MeV
p	938.3 MeV	Σ^+	1189.4 MeV

What is the disintegration energy of the reaction?

SOLUTION The disintegration energy is given by

$$Q = (m_\pi c^2 + m_p c^2) - (m_K c^2 + m_\Sigma c^2)$$

$$= (139.6 \text{ MeV} + 938.3 \text{ MeV})$$

$$- (493.7 \text{ MeV} + 1189.4 \text{ MeV})$$

$$= -605 \text{ MeV}. \quad \text{(Answer)}$$

The minus sign means that the reaction is *endothermic.* That is, if the proton is at rest, the incoming pion (π^-) must have a kinetic energy larger than a certain threshold value to make the reaction go. The threshold energy is larger than 605 MeV because linear momentum must be conserved, which means that the kaon (K^-) and the sigma (Σ^+) must be not only created but also endowed with some kinetic energy. It can be shown by a relativistic calculation whose details are beyond our scope that the threshold energy for the incident pion is 907 MeV.

49-4 THE LEPTONS

Now let us press on with our classification program for the particles. We turn first to leptons, those particles on which the strong force does *not* act.

So far, we have encountered, as leptons, the familiar electron and the neutrino that accompanies it in beta decay. The muon, whose decay is described in Eq. 49-5, is another member of this family. Physicists gradually learned that the neutrino that appears in Eq. 49-4, associated with the production of a muon, is *not the same particle* as the neutrino produced in beta decay, associated with the appearance of an electron. We call the former the **muon neutrino** (symbol ν_μ) and the latter the **electron neutrino** (symbol ν_e) when it is necessary to distinguish between them. In Eq. 49-5, for example, one of the two neutrinos is a muon neutrino and the other is an electron neutrino.

These two neutrinos are known to be different particles because, if a beam of muon neutrinos (produced from pion decay as in Eq. 49-4) is allowed to strike a solid target, *only muons*—and never electrons—are produced. On the other hand, if electron neutrinos (produced by the beta decay of

TABLE 49-3
THE LEPTONS[a]

PARTICLE	SYMBOL	REST ENERGY (MeV)	CHARGE	ANTIPARTICLE
Electron	e^-	0.511	-1	e^+
Electron neutrino[b]	ν_e	0	0	$\bar{\nu}_e$
Muon	μ^-	105.7	-1	μ^+
Muon neutrino[b]	ν_μ	0	0	$\bar{\nu}_\mu$
Tauon	τ^-	1784	-1	τ^+
Tauon neutrino[b]	ν_τ	0	0	$\bar{\nu}_\tau$

[a]All leptons have spin $\frac{1}{2}$ and are thus fermions.

[b]If the neutrino masses are not zero, they are at least very small. This is still an open question.

fission products in a nuclear reactor) are allowed to strike a solid target, *only electrons*—and never muons—are produced.

Another lepton, the **tauon,** was discovered in 1975. It has its own associated neutrino, different still from the other two. Table 49-3 lists the known leptons. There are reasons for dividing the leptons into three "generations," each consisting of a particle (electron, muon, or tauon) and its associated neutrino. Furthermore, there are reasons to believe that there are *only* the three generations of leptons shown in Table 49-3. Leptons have no discernible internal structure, no measurable dimensions, and are believed to be truly point-like fundamental particles.

49-5 A NEW CONSERVATION LAW

We are now ready to consider hadrons (baryons and mesons), those particles whose interactions are governed by the *strong* force. We start by adding another conservation law to the list of conservation laws that are more familiar to us, such as conservation of charge, energy, linear momentum, and angular momentum. It is the *conservation of baryon number*.

To develop this conservation law, let us consider the decay process

$$p \rightarrow e^+ + \gamma \quad (Q = 937.8 \text{ MeV}). \quad (49\text{-}10)$$

This process *never* happens. We should be glad that it does not because otherwise all protons in the universe would gradually change into positrons, with di-

sastrous consequences. Yet this decay process violates none of the conservation laws that we have so far discussed.

We account for the apparent stability* of the proton—and for the absence of many other processes that might otherwise occur—by introducing a new quantum number, the *baryon number B*, and a new conservation law, the conservation of baryon number.

▬▬▬

To every baryon we assign $B = +1$. To every antibaryon we assign $B = -1$. To mesons and leptons we assign $B = 0$. A process or reaction of fundamental particles cannot occur if it changes the net baryon number.

The proton is a baryon, whereas the positron and (gamma-ray) photon are not. Thus the process of Eq. 49-10 cannot occur because it violates the law of conservation of baryon number:

$$(+1) \neq (0) + (0).$$

Baryon number conservation will prove useful in accounting for the many particle decays and reactions that—although not otherwise forbidden—simply do not occur.

▬▬▬

*The proton may yet prove to be unstable, as some current theories predict. Its predicted half-life, however, is at least 10^{30} y, many orders of magnitude greater than the age of the universe. Attempts to detect proton decay have so far proved unsuccessful.

SAMPLE PROBLEM 49-4

Analyze the proposed decay of a stationary proton according to the scheme

$$p \to \pi^0 + \pi^+ \quad \text{(doesn't happen!)}$$

by testing it against the various conservation laws. (Both pions are mesons, with spin and baryon number both equal to zero. The rest energy of the π^0 meson is 135.0 MeV.)

SOLUTION We see at once that charge is conserved and that linear momentum can also be readily conserved. All that is necessary is that the two pions move in opposite directions from the site of the stationary proton, with momenta of equal magnitude.

The disintegration energy is found by subtracting the rest energies of the particles. Thus, using Table 49-2, we have

$$Q = (m_\text{p} c^2) - (m_0 c^2 + m_+ c^2)$$

$$= (938.3 \text{ MeV}) - (135.0 \text{ MeV} + 139.6 \text{ MeV})$$

$$= 663.7 \text{ MeV}.$$

The fact that Q is positive shows that the process cannot be ruled out on energy conservation grounds; the energy is there.

We have noted that both pions have zero spin. The proton, however, has a spin of one-half. Thus angular momentum is *not* conserved and—for that reason alone—the process cannot occur.

Moreover, baryon number is not conserved. For the proton we have $B = +1$, and for the two pions we have $B = 0$. The process is thus doubly forbidden, violating two of our five conservation laws.

SAMPLE PROBLEM 49-5

A particle identified as Ξ^- decays as follows:

$$\Xi^- \to \Lambda^0 + \pi^-.$$

Both decay products are unstable. The following additional reactions occur in cascade until, ultimately, only stable products remain:

$$\Lambda^0 \to \eta + \pi^0, \qquad \eta \to \text{p} + \text{e}^- + \nu,$$

$$\pi^0 \to \gamma + \gamma, \qquad \pi^- \to \mu^- + \nu,$$

$$\mu^- \to \text{e}^- + \nu + \nu.$$

a. Write the overall decay scheme for the Ξ^- particle.

SOLUTION Study of the six given decay equations shows that the overall decay scheme is

$$\Xi^- \to \text{p} + 4\nu + 2\text{e}^- + 2\gamma. \quad \text{(Answer)}$$

All the products on the right side are stable. Note that charge is conserved, the net charge quantum number being -1 on each side.

b. Is the Ξ^- particle a meson or a baryon?

SOLUTION The proton in the overall equation is a baryon (baryon number $= +1$). All the other particles on the right side of the equation have $B = 0$. Thus, for conservation of baryon number, the baryon number of the Ξ^- particle must be $+1$. Hence the particle is a *baryon*. If it were a meson, its baryon number would be zero.

c. What can you say about the spin of the Ξ^- particle?

SOLUTION All particles on the right side of the overall equation except the gamma photons have a spin of $\frac{1}{2}$; the photon has a spin of 1. All these spins can add up to only a half-integer spin for the Ξ^- particle. This is additional evidence that this particle is a baryon. If it were a meson, its spin number would be an integer. (Actually, the spin of the Ξ^- particle is $\frac{1}{2}$; this particle is listed with other spin-$\frac{1}{2}$ baryons in Table 49-4.)

49-6 ANOTHER NEW CONSERVATION LAW!

Particles have more intrinsic properties than mass, charge, spin, and baryon number that we have listed so far. The first of these additional properties emerged when researchers observed that certain new particles, such as the kaon (K) and the sigma (Σ), always seemed to be produced in pairs. It seemed impossible to produce only one of them at a time. Thus if a beam of energetic pions interacts with the protons in a bubble chamber, the reaction

$$\pi^+ + \text{p} \to \text{K}^+ + \Sigma^+ \quad (49\text{-}11)$$

often occurs. The reaction

$$\pi^+ + \text{p} \to \pi^+ + \Sigma^+, \quad (49\text{-}12)$$

which violates no conservation law known at the time, never occurs.

It was eventually proposed (by Murray Gell-Mann in the United States and independently by K.

Nishijima in Japan) that certain particles possess a new property, called *strangeness,* with its own quantum number S and its own conservation law. The name arises from the fact that, before the identities of these certain particles were pinned down, they were known as "strange particles," and the label stuck.

The proton, neutron, and pion have $S = 0$; that is, they are not "strange." It was proposed, however, that the K^+ particle has a strangeness denoted by $S = +1$ and that Σ^+ has $S = -1$. Thus strangeness is conserved in Eq. 49-11,

$$(0) + (0) = (+1) + (-1) \qquad \text{(values of } S),$$

but is *not* conserved in Eq. 49-12,

$$(0) + (0) \neq (0) + (-1) \qquad \text{(values of } S).$$

The reaction of Eq. 49-12 does not occur because it violates the law of *conservation of strangeness:*

> Strangeness is conserved in interactions involving strong forces.

It may seem heavy-handed to invent a new property of particles just to account for a little puzzle like that posed by Eqs. 49-11 and 49-12. However, strangeness and its quantum number soon revealed themselves in many other areas in particle physics, and strangeness is now fully accepted as a legitimate particle attribute, on a par with charge and spin. To those who know and love particles, "strangeness" is no longer "strange."

Do not be misled by the whimsical character of the name. "Strangeness" is no more mysterious a property of particles than is "charge." Both are properties that particles may (or may not) have; each is described by an appropriate quantum number. Each obeys a conservation law. Still other properties of particles have been discovered and given even more whimsical names, such as *charm* and *bottomness,* but all are perfectly legitimate properties. Let us see, as an example, how the new property of strangeness "earns its keep" by leading us to uncover important regularities in the properties of the particles.

49-7 THE EIGHTFOLD WAY

There are eight baryons—the neutron and the proton among them—that have a spin quantum num-

TABLE 49-4
THE EIGHT SPIN-$\frac{1}{2}$ BARYONS

PARTICLE	SYM-BOL	REST ENERGY (MeV)	QUANTUM NUMBERS	
			CHARGE	STRANGENESS
Proton	p	938.3	+1	0
Neutron	n	939.6	0	0
Lambda	Λ^0	1115.6	0	-1
Sigma	Σ^+	1189.4	+1	-1
Sigma	Σ^0	1192.5	0	-1
Sigma	Σ^-	1197.3	-1	-1
Xi	Ξ^0	1314.9	0	-2
Xi	Ξ^-	1321.3	-1	-2

ber of $\frac{1}{2}$. Table 49-4 shows some of their properties. Figure 49-3a shows the fascinating pattern that emerges if we plot the strangeness of these baryons against their charge, using a sloping axis for the

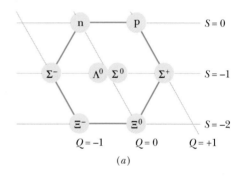

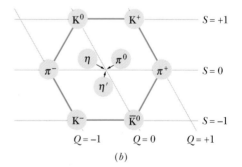

FIGURE 49-3 (*a*) The Eightfold Way pattern for the eight spin-$\frac{1}{2}$ baryons listed in Table 49-4. The particles are represented as points on a strangeness–charge plot, using a sloping axis for the charge quantum number. (*b*) A similar pattern for the nine spin-0 mesons listed in Table 49-5. Note in this case that particles lie opposite their antiparticles; the three mesons in the center form their own antiparticles.

TABLE 49-5
THE NINE SPIN-0 MESONS [a]

| PARTICLE | SYMBOL | REST ENERGY (MeV) | QUANTUM NUMBERS | | ANTIPARTICLE |
			CHARGE	STRANGENESS	
Pion	π^0	135.0	0	0	π^0
Pion	π^+	139.6	$+1$	0	π^-
Kaon	K^+	493.7	$+1$	$+1$	K^-
Kaon	K^0	497.7	0	$+1$	$\overline{K}^0$
Eta	η	548.8	0	0	η
Eta prime	η'	957.6	0	0	η'

[a] Counting both particles and antiparticles, there are a total of nine spin-0 mesons. Note that the π^0, η, and η' mesons are their own antiparticles.

charge quantum numbers. Six of the eight form a hexagon with the two remaining baryons at its center.

Let us turn now from the hadrons called baryons to the hadrons called mesons. There are nine of these (listed in Table 49-5), each with a spin of zero. If we plot them on a strangeness–charge diagram, as in Fig. 49-3b, the same fascinating pattern emerges! These and related plots, called the *Eightfold Way* patterns,* were proposed independently in 1961 by Murray Gell-Mann at the California Institute of Technology and by Yuval Ne'eman at Imperial College, London. The two patterns of Fig. 49-3 are representative of a larger number of symmetrical patterns in which various groups of baryons and mesons can be displayed.

The symmetry of the Eightfold Way pattern for the spin-$\frac{3}{2}$ baryons (not shown here) calls for *ten* particles arranged in a pattern like that of the tenpins in a bowling alley. However, when the pattern was first proposed, only *nine* such particles were known; the "headpin" was missing. In 1962, guided by theory and the symmetry of the pattern, Gell-Mann made a prediction in which he essentially said:

There exists a spin-$\frac{3}{2}$ baryon with a charge of -1, a strangeness of -3, and a rest energy of about 1680 MeV. If you look for this *omega minus* particle (as I propose to call it), I think you will find it.

A team of physicists headed by Nicholas Samios of the Brookhaven National Laboratory took up the challenge and promptly found the "missing" particle, confirming all of its predicted properties. There is nothing like the prompt experimental confirmation of a prediction to build confidence in a theory!

The Eightfold Way patterns bear the same relationship to particle physics that the periodic table does to chemistry. In each case, there is a pattern of organization in which vacancies (missing particles or missing elements) stick out like sore thumbs, guiding experimenters in their searches. In the case of the periodic table, its very existence strongly suggests that the atoms of the elements are not fundamental particles but have a common underlying structure. In the same way, the Eightfold Way patterns strongly suggest that the mesons and the baryons must have an underlying structure, in terms of which their properties can be understood. That structure is the *quark model,* which we now discuss.

49-8 THE QUARK MODEL

In 1964 Murray Gell-Mann and George Zweig independently pointed out that the Eightfold Way patterns can be understood in a simple way if the mesons and the baryons are built up out of subunits that Gell-Mann called **quarks.**† We deal first with

---

*A borrowing from Eastern mysticism. The "Eight" refers to the eight quantum numbers (only a few of which we have defined here) that are involved in the symmetry-based theory that predicts the existence of the patterns.

†The name comes from James Joyce's *Finnegans Wake* ("Three quarks for Muster Mark!"). Joyce is said to have coined this word after listening to the squawks of seagulls.

TABLE 49-6
THE QUARKS

| | | | QUANTUM NUMBERS | | | | |
PARTICLE	SYMBOL	MASS[a]	CHARGE	STRANGENESS	BARYON NUMBER	SPIN	ANTIPARTICLE
Up	u	10	$+\frac{2}{3}$	0	$+\frac{1}{3}$	$\frac{1}{2}$	$\bar{u}$
Down	d	20	$-\frac{1}{3}$	0	$+\frac{1}{3}$	$\frac{1}{2}$	$\bar{d}$
Strange	s	200	$-\frac{1}{3}$	-1	$+\frac{1}{3}$	$\frac{1}{2}$	$\bar{s}$
Charm	c	3,000	$+\frac{2}{3}$	0	$+\frac{1}{3}$	$\frac{1}{2}$	$\bar{c}$
Bottom	b	9,000	$-\frac{1}{3}$	0	$+\frac{1}{3}$	$\frac{1}{2}$	$\bar{b}$
Top[b]	t	60,000	$+\frac{2}{3}$	0	$+\frac{1}{3}$	$\frac{1}{2}$	$\bar{t}$

[a]Masses are reported in terms of the electron mass, which is taken as unity.

[b]Evidence for the top quark has not yet been observed as of 1992.

three of them, called the *up quark* (symbol *u*), the *down quark* (symbol *d*), and the *strange quark* (symbol *s*), and we assign to them the properties displayed in the first three rows of Table 49-6. (The names of the quarks, along with those assigned to three other quarks that we shall meet later, have no meanings other than as convenient labels. Collectively, these names are called the *quark flavors*. We could, for all the difference it would make, have called them vanilla, chocolate, and pistachio instead of up, down, and strange.)

The fractional charges of the quarks may jar you a little. However, withhold judgment until you see how neatly these fractional charges account for the observed integral charges of the mesons and the baryons. Quarks have not (yet) been convincingly observed in the laboratory as free particles, and theorists have put forward plausible reasons why this should be the case. In any event, the quark model is so useful that the failure to see free quarks is not regarded as a hindrance to their acceptance.

We have seen how we can put atoms together by combining electrons and nuclei. Now let us see how we can put mesons and baryons together by combining quarks. We state in advance that success will be complete. That is, for particles formed from the up, down, and the strange quark:

There is no known meson or baryon whose properties cannot be understood in terms of an appropriate combination of quarks. Conversely, there is no possible quark combination to which there does not correspond an observed meson or baryon.

Let us look first at the baryons.

Quarks and Baryons

Each baryon is a combination of three quarks; the combinations are given in Fig. 49-4a. With regard to baryon number, we see that any three quarks (each with $B = +\frac{1}{3}$) yield a proper baryon (with $B = +1$). The spins work out also. With three spins of $\frac{1}{2}$ to work with, we can arrange them so that two spins are parallel and one antiparallel. This leads to $s = \frac{1}{2}$, which is the spin of all the baryons displayed in Table 49-4 and Fig. 49-3a. If all three spins are parallel, we have $s = \frac{3}{2}$; there are 10 baryons with this spin.

Charges also work out, as we can see from three examples. The proton has a quark composition of *uud* so that its charge is

$$Q(uud) = (+\tfrac{2}{3}) + (+\tfrac{2}{3}) + (-\tfrac{1}{3}) = +1.$$

The neutron has a quark composition of *udd* and its charge is

$$Q(udd) = (+\tfrac{2}{3}) + (-\tfrac{1}{3}) + (-\tfrac{1}{3}) = 0.$$

The Σ^- particle has a quark composition of *dds* and its charge is

$$Q(dds) = (-\tfrac{1}{3}) + (-\tfrac{1}{3}) + (-\tfrac{1}{3}) = -1.$$

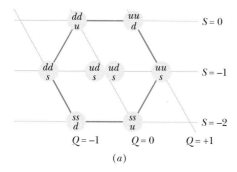

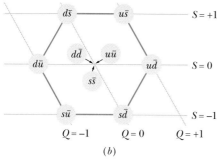

FIGURE 49-4 (a) The quark compositions of the baryons plotted in Fig. 49-3a. (Although the two central baryons share the same quark structure, the sigma is an excited state of the lambda, decaying into it by emission of a gamma-ray photon.) (b) The quark compositions of the mesons plotted in Fig. 49-3b.

The charge and strangeness quantum numbers of all the baryons displayed in Fig. 49-3a and Table 49-4 work out in exact agreement with those of the quark composites in Fig. 49-4a.

Quarks and Mesons

Mesons are quark–antiquark pairs; their compositions are given in Fig. 49-4b and are consistent with the fact that the spins of all the mesons displayed in Fig. 49-3b and Table 49-5 are zero. Both quarks and antiquarks have $s = \frac{1}{2}$, so the two particles that comprise a meson must have opposite spins to give spin-0 for the meson.

The quark–antiquark model is also consistent with the fact that mesons are not baryons; that is, mesons have a baryon number $B = 0$. The baryon number for a quark is $+\frac{1}{3}$ and for an antiquark is $-\frac{1}{3}$, the combination adding to zero.

Consider the meson π^+, which is made up of an up quark u and an antidown quark $\bar{d}$. We see from Table 49-6 that the charge of the up quark is $+\frac{2}{3}$ and that of the antidown quark is $+\frac{1}{3}$ (that is, its sign is opposite that of the charge of the down quark). This adds nicely to a charge of $+1$ for the π^+ meson. Thus

$$Q(u\bar{d}) = (+\tfrac{2}{3}) + (+\tfrac{1}{3}) = +1.$$

All the charge and strangeness quantum numbers of Fig. 49-4b agree with those of Table 49-5 and Fig. 49-3b. Convince yourself that all possible up, down, and strange quark–antiquark combinations are used and that all known mesons with $s = 0$ are accounted for. Everything fits perfectly.

A New Look at Beta Decay

Let us see how beta decay appears from the quark point of view. In Eq. 47-10, we presented a typical example of this process:

$$^{32}\text{P} \rightarrow {}^{32}\text{S} + \text{e}^- + \nu.$$

After the neutron was discovered and Fermi had worked out his theory of beta decay, physicists came to view the fundamental beta-decay process as the changing of a neutron into a proton inside the nucleus. Thus

$$\text{n} \rightarrow \text{p} + \text{e}^- + \nu.$$

Today we look deeper and see that a neutron (udd) can change into a proton (uud) by changing a down quark into an up quark. We now view the fundamental beta-decay process as

$$d \rightarrow u + \text{e}^- + \nu.$$

Thus as we come to know more and more about the fundamental nature of matter, we can examine familiar processes at deeper and deeper levels. We see too that the quark model not only helps us to understand the structure of particles but also throws light on their interactions.

Still More Quarks

There are other particles and other Eightfold Way patterns that we have not discussed. To account for them, it turns out that we need to postulate three more quarks, the *charm quark c*, the *top quark t*, and the *bottom quark b*.

We note in Table 49-6 that these three new quarks are exceptionally heavy, the lightest of them (charm) being almost twice as heavy as a proton. To generate particles that contain such quarks, we must go to higher and higher energies, which is the reason that these three new quarks were not discovered earlier and do not enter as fully as their lighter siblings into particle research.

The first observed particle that contains a charmed quark was the J/Ψ meson, whose quark structure is ($c\bar{c}$). It was discovered simultaneously and independently in 1974 by groups headed by Samuel Ting at the Brookhaven National Laboratory and by Burton Richter at Stanford University.

Compare Table 49-6 (the quark family) and Table 49-3 (the lepton family) carefully, noting the neat symmetry of these two "six packs" of particles, each dividing naturally into three corresponding two-particle "generations." In terms of what we know today, these two families, the quarks and the leptons, seem to be truly fundamental particles.

SAMPLE PROBLEM 49-6

The Ξ^- particle has a spin of $\frac{1}{2}$ and quantum numbers $Q = -1$ and $S = -2$. It is known to be a three-quark combination involving only up, down, and strange quarks. What must this combination be?

SOLUTION   Because its strangeness is -2, this particle must contain two strange quarks, each of which (see Table 49-6) has $S = -1$. The third quark must then be either an up quark or a down quark (both having $S = 0$). The two strange quarks have a combined charge of $(-\frac{1}{3}) + (-\frac{1}{3})$ or $-\frac{2}{3}$. We require a charge of -1 for the Ξ^- particle so that we must add a third quark whose charge is $-\frac{1}{3}$; that is the down quark. Thus the quark composition of Ξ^- is *dss*.

49-9 FORCES AND MESSENGER PARTICLES (OPTIONAL)

We turn now from cataloging the particles to considering the forces that act between them.

The Electromagnetic Force

Two electrons exert electromagnetic forces on each other according to Coulomb's law. At a deeper level,

this interaction is described by a highly successful theory called *quantum electrodynamics* (QED). From this point of view we say that each electron detects the presence of the other by exchanging photons with it, the photon being the quantum of the electromagnetic field.

We cannot detect these photons because they are emitted by one electron and absorbed by the other a very short time later. Because of their transitory existence, we call them **virtual photons.** Because of their role in communicating between the two interacting charges, we sometimes call these photons *messenger particles.*

If a stationary electron emits a photon and remains itself unchanged, energy is not conserved. The principle of conservation of energy is saved, however, by the uncertainty principle, written in the form

$$\Delta E \cdot \Delta t \approx h. \qquad (49\text{-}13)$$

We interpret this relation to mean that you can "overdraw" an amount of energy ΔE, violating conservation of energy, *provided* that you "return" it within an interval Δt given by $h/\Delta E$. The virtual photons do just that. When an electron emits a virtual photon, the overdraw in energy is quickly set right when that electron receives a virtual photon from the second electron, and the violation of the principle of conservation of energy is hidden by the inherent uncertainty.

The Weak Force

A field theory of the weak force was developed by analogy with the field theory of the electromagnetic force. The messenger particles that transmit the weak force between leptons, however, are not (massless) photons but massive particles, identified by the symbols W and Z. The theory was so successful that it revealed the electromagnetic force and the weak force as different aspects of a single *electroweak force.* This accomplishment is a logical extension of the work of Maxwell, who revealed the electric and the magnetic force as different aspects of a single *electromagnetic* force.

The electroweak theory was specific in predicting the properties of the messenger particles. Their charges and rest energies, for example, were predicted to be

PARTICLE	CHARGE	REST ENERGY
W	$\pm e$	82 ± 2 GeV
Z	0	92 ± 2 GeV

Recall that the proton rest energy is only 0.938 GeV; these are massive particles! The Nobel prize for 1979 was awarded to Sheldon Glashow, Steven Weinberg, and Abdus Salam for their development of the electroweak theory.

The theory was confirmed in 1983 by Carlo Rubbia and his group at CERN. Both messenger particles were observed, their rest energies agreeing with the predicted values. The Nobel prize for 1984 went to Rubbia and Simon van der Meer for this brilliant experimental work.

Some notion of the complexity of particle physics in this day and age can be found by looking at an earlier Nobel prize particle physics experiment— the discovery of the neutron in 1932. This vitally important discovery was a "tabletop" experiment, employing particles emitted by naturally occurring radioactive materials as projectiles; it was reported under the title "Possible Existence of a Neutron," the single author being James Chadwick.

The discovery of the W and Z messenger particles in 1983, by contrast, was carried out at a large particle accelerator, about 7 km in circumference and operating in the several-hundred-GeV range. The principal detector alone weighed 2000 tons. The experiment employed more than 130 physicists from 12 institutions in 8 countries, along with a large support staff.

The Strong Force

A theory of the strong force, that is, the force that acts between quarks, has also been developed. The messenger particles in this case are called *gluons* and, like the photon, they are predicted to be massless. The theory assumes that each "flavor" of quark comes in three varieties that, for convenience, have been labeled *red, yellow,* and *blue.* Thus there are three up quarks, one of each color, and so on. The antiquarks also come in three colors, which we call *antired, antiyellow,* and *antiblue.* You must not think that quarks are actually colored, like tiny jelly beans. The names are labels of convenience but (for once!)

they do have a certain formal justification, as we shall see.

The force acting between quarks is called a **color force** and the underlying theory, by analogy with quantum electrodynamics (QED), is called **quantum chromodynamics** (QCD). An important prediction of the theory is that quarks can be assembled only in *color-neutral* combinations.

There are two ways to bring this about. In the theory of actual colors, red + yellow + blue yields white, which is color-neutral; thus we can assemble three quarks to form a baryon. Antired + antiyellow + antiblue is also white, so that we can assemble three antiquarks to form an antibaryon. Finally, red + antired, or yellow + antiyellow, or blue + antiblue also yield white. Thus we can assemble quark–antiquark combinations to form a meson. The color-neutral rule does not permit any other combinations of quarks, and none are observed.

Unification—Einstein's Dream

The attempt to unify the fundamental forces of nature—which occupied Einstein's attention for much of his later life—is very much a current problem. Table 6-2 summarizes the current status. We have seen that the weak force has been successfully combined with electromagnetism so that they may be jointly viewed as aspects of a single *electroweak force.* Theories that attempt to add the strong force to this combination—called *Grand Unification Theories* (GUTs)—are being pursued actively. Theories that seek to complete the job by adding gravity— sometimes called *Theories of Everything* (TOE)—are at an encouraging but speculative stage at this time.

49-10 A PAUSE FOR REFLECTION

Let us put what you have just learned in perspective. If all we are interested in is the structure of the world around us, we can get along nicely with the electron, the neutrino, the neutron, and the proton. As someone has said, we can operate "Spaceship Earth" quite well with just these particles. We can see a few of the more exotic particles by looking for them in the cosmic rays but, to see most of them, we must build massive accelerators and look for them carefully at great effort and expense.

The reason we must go to such lengths is that—measured in energy terms—we live in a world of very low temperatures. Even at the center of the sun, the value of kT is only about 1 keV. To produce the exotic particles, we must be able to accelerate protons or electrons to energies in the GeV and TeV range and higher. Once upon a time, however, the temperature *was* high enough to provide just such energies, and far beyond. That was when the universe began. So let us now turn our attention to that time.

The most distant objects that we can "see" are quasars* whose distances may be up to 13×10^9 light-years. As we look out in space we are also looking back in time. Thus if we see a quasar 13×10^9 ly away, we are seeing the quasar as it was 13×10^9 y ago, when the photons that are reaching us now left it. Since the Big Bang that represents the creation of the universe occurred about 15×10^9 y ago, we are looking back to its earliest days. In the remainder of this chapter we shall examine and analyze the results of such "looking back."

49-11 THE UNIVERSE IS EXPANDING

As we have seen in Section 18-8, it is possible to measure the relative speeds at which galaxies are approaching us or receding from us by measuring the Doppler shift of the light that they emit. If we look only at distant galaxies, beyond our immediate galactic neighbors, we find an astonishing fact. They are all moving away from us!

In 1929, Edwin P. Hubble (Fig. 49-5) established a connection between the speed v of recession of a galaxy and its distance r from us, namely, that they are proportional. That is,

$$v = Hr \quad \text{(Hubble's law)}, \quad (49\text{-}14)$$

in which H, the **Hubble parameter,** has the value

$$H \approx 17 \times 10^{-3} \text{ m/(s·ly)}. \quad (49\text{-}15)$$

The value of the Hubble parameter is somewhat uncertain because of the difficulty of measuring the distances of remote galaxies. These distances are established by an involved chain of measurements and

---

Quasars (quasistellar objects) are extremely luminous objects whose nature is not yet fully understood.

assumptions, its starting point (once the dimensions of our solar system have been established) being the measurement of the distances of our closest neighboring stars by parallax methods.

We interpret Hubble's law to mean that the universe is expanding, much as the raisins in what is to be a loaf of raisin bread grow farther apart as the dough rises. Observers on all other galaxies would find that distant galaxies were rushing away from them also, in accordance with Hubble's law. In keeping with our analogy, all raisins are alike.

Hubble's law fits in well with the Big Bang hypothesis. What we are seeing are the outward-flying fragments of that primordial explosion.

SAMPLE PROBLEM 49-7

As judged by Doppler shift measurements of the light that it emits, the most distant object that we can see from Earth is a quasar that is receding from us at 2.2×10^8 m/s. (Note that this is 73% of the speed of light!) How far away is that object?

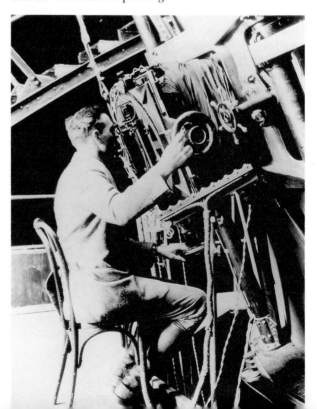

FIGURE 49-5 Edwin Hubble (1889–1953) shown at the controls of the 100-in. telescope at Mount Wilson, where he carried out much of the work that led him to propose that the universe is expanding.

SOLUTION From Hubble's law (Eq. 49-14)

$$r = \frac{v}{H} = \frac{2.2 \times 10^8 \text{ m/s}}{17 \times 10^{-3} \text{ m/(s} \cdot \text{ly)}}$$

$$= 13 \times 10^9 \text{ ly}. \qquad \text{(Answer)}$$

This result is only approximate because the quasar has not always been receding from us at the same speed.

SAMPLE PROBLEM 49-8

Assume that the quasar in Sample Problem 49-7 has been moving at its calculated speed with respect to us ever since the Big Bang. What minimum limit does this impose on how long ago the Big Bang occurred? That is, what is the minimum age of the universe based on that speed?

SOLUTION We can find the time from

$$t = \frac{r}{v} = \frac{1}{H}$$

$$= \frac{1}{17 \times 10^{-3} \text{ m/(s} \cdot \text{ly)}} = 58.8 \text{ s} \cdot \text{ly/m}$$

$$= (58.8 \text{ s} \cdot \text{ly/m})(9.46 \times 10^{15} \text{ m/ly})$$

$$\times (1 \text{ y}/3.16 \times 10^7 \text{ s})$$

$$\approx 18 \times 10^9 \text{ y}. \qquad \text{(Answer)}$$

This is in reasonable agreement with the age of the universe as computed by other methods.

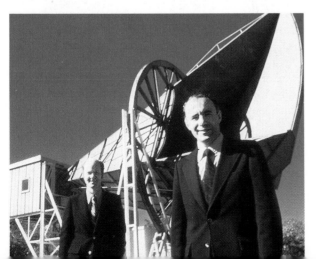

FIGURE 49-6 Arno Penzias (right) and Robert Wilson, standing in front of the large horn antenna with which they first detected the microwave background radiation. This radiation, left over from the early universe, continuously arrives from all directions; indeed, the whole universe is filled with it.

49-12 THE MICROWAVE BACKGROUND RADIATION

In 1965, Arno Penzias and Robert Wilson (Fig. 49-6), of what is now AT&T Bell Laboratories, were testing a sensitive microwave receiver used for communications research. They discovered a faint background "hiss" that remained unchanged in intensity no matter in what direction their antenna was pointed. It soon became clear that Penzias and Wilson were observing a *microwave background radiation,* generated in the early universe and filling all space almost uniformly. This background radiation, whose maximum intensity occurs at a wavelength of 1.1 mm, has the same distribution in wavelength as does cavity radiation (discussed in Section 43-5) at a temperature of 2.7 K, the "cavity" in this situation being the entire universe. Penzias and Wilson were awarded the 1978 Nobel prize for their discovery.

This radiation originated about 300,000 years after the Big Bang, when the universe suddenly became transparent to electromagnetic waves. The radiation at that time corresponded to cavity radiation at a temperature of perhaps 10^5 K. As the universe expanded, however, the temperature dropped to its present value of 2.7 K, much as the temperature of a gas expanding under adiabatic conditions will fall.

49-13 THE MYSTERY OF THE DARK MATTER

Vera Rubin of the Carnegie Institution of Washington and her associates have been measuring the rotation rates of distant galaxies for a number of years. They do so by measuring the Doppler shifts of bright clusters of stars located within each galaxy at various distances from the galactic center. Her conclusion, based on these measurements, is surprising: the orbital speed of objects at the outer visible edge of a rotating galaxy is about the same as that of objects close to the galactic center.

That is *not* what we find in the solar system. The orbital speed of Pluto (the planet most distant from the sun) is only about one-tenth that of Mercury (the planet closest to the sun).

The only explanation for Rubin's findings that is consistent with Newtonian mechanics is that there is much more matter in a typical rotating galaxy than what we can actually see. (See Problems 37 and 38.)

In fact, the visible portion of a galaxy represents only about 5–10% of the total galactic mass.

What then is this other matter that we cannot see—this *dark matter* that permeates and surrounds a typical galaxy? If neutrinos have even a small mass, they might account for it. If not, theoretical physicists have an abundant supply of predicted new particles that might do the job, many with exotic names such as axions, winos, and wimps. At present, however, there is no experimental evidence to support any of these predictions. Philip Morrison, after describing the evidence for the existence of dark matter on a television series ("The Ring of Truth," Episode 6) put it well:

> If you ask me what the universe as a whole is made of, I must admit that as of now I remain in great doubt. I just do not know. But one thing I do know: we will try very hard to find out.

49-14 THE BIG BANG

In 1985, a physicist remarked at a scientific meeting:

> It is as certain that the universe started with a Big Bang about 15 billion years ago as it is that the Earth goes around the Sun.

This strong statement suggests the level of confidence in which the Big Bang theory, first advanced by George Gamow and his colleagues, is held by those who study these matters.

You must not imagine that the Big Bang was like the explosion of some giant firecracker and that, in principle at least, you could have stood to one side and watched. There was no "one side" because the Big Bang represents the beginning of spacetime itself. From the point of view of our present universe, there is no position in space to which you can point and say, "The Big Bang happened there." It happened everywhere.

Moreover, there was no "before the Big Bang," because time *began* with that creation event. Let us see what went on during succeeding intervals of time after time began.

Big Bang zero to 10^{-43} s. Of this tiny but important period* we know little, because the laws of physics as we know them do not hold. (As one author has written: "How you do physics in a situation like this, when space and time are disconnected, is *not* described in Halliday and Resnick.") At 10^{-43} s, the temperature of the universe is about 10^{23} K and the universe is expanding rapidly.

10^{-43} s to 10^{-35} s. During this period, the strong, the weak, and the electromagnetic forces act as a single force, described by a Grand Unified Theory (GUT). Gravitation acts separately, as it does today.

10^{-35} s to 10^{-10} s. The strong force "freezes out," leaving the electroweak force to continue to act as a single force.

10^{-10} s to 10^{-5} s. All four forces appear separately, just as they do today. The universe consists of a hot "soup" of quarks, leptons, and photons.

10^{-5} s to 3 min. Quarks join to form mesons and baryons. Matter and antimatter annihilate, wiping out the antimatter and leaving the slight excess of matter from which our present universe is formed.

3 min to 10^5 y. Protons and neutrons join to form the light nuclides, such as ^{4}He, ^{3}He, ^{2}H, and ^{7}Li, in just the abundances that we find today. The universe consists of a plasma made up of nuclei and electrons.

10^5 y to the present. At the beginning of this period, electrons were finally able to orbit protons to form hydrogen atoms without being immediately knocked out of their orbits by photons. The light that was emitted during these atom formations is now the microwave background radiation that was first detected by Penzias and Wilson in 1965. Thus the radiation released by the atoms during this early time gives us a picture of what the universe looked like when it was about 10^5 y old.

Measurements made after 1965 suggested that the microwave background radiation is uniform in all directions, implying that all the matter (particles and atoms) in the universe, at age 10^5 y, was uniformly distributed. This finding was most puzzling, because the matter in the present universe is not uniformly distributed, but instead is collected in galaxies, galactic clusters, and galactic superclusters.

---

*The approximate time 10^{-43} s is known as the Planck time; see Sample Problem 23-7.

There are also vast *voids* in which there is relatively little matter, and there are regions so crowded with matter that they are called *walls*. If the Big Bang theory of the beginning of the universe is even approximately correct, the seeds for this nonuniform distribution of matter must have been in place before the universe was 10^5 y old and thus must now show up as a nonuniform distribution of the microwave background radiation.

In 1992, measurements made by NASA's Cosmic Background Explorer satellite revealed that the background radiation is, in fact, not perfectly uniform. The image shown in this chapter's opening photograph was made from those measurements and shows us the universe when the universe was only 300,000 y old. As you can see, large-scale collecting of matter had already begun; thus, the Big Bang theory is, in principle, on the right track.

49-15 A SUMMING UP

Let us, in these closing paragraphs, step aside for a moment and consider where our rapidly accumulating store of knowledge about the universe is leading us. That it provides satisfaction to a host of curiosity-motivated physicists is beyond dispute.

However, some view it as a humbling experience in that each increase in knowledge seems to reveal more clearly our own relative insignificance in the grand scheme of things. Thus in rough chronological order, we came to realize that:

Our Earth is not the center of the solar system. Our sun is but one star among many.

Our galaxy is but one of many, and our sun is an insignificant star near its outer edge.

Our Earth has existed for perhaps only a third of the age of the universe and will surely disappear when our sun burns up its fuel and becomes a red giant.

We have lived on the Earth, as a species, for less than a million years, a blink in cosmological time.

The last crushing blow: The neutrons and protons of which we are made are not the predominant form of matter in the universe. As someone has said, we are not even made of the right stuff!

However, the bright side is that it is we ourselves who discovered all these facts. Although our position in the universe may be insignificant, the laws of physics that we have discovered (uncovered?) seem to hold throughout the universe and—as far as we know—for all past and future time. At least, there is no evidence that other laws hold in other parts of the universe. Thus until someone complains, we are entitled to stamp the laws of physics "Discovered on Earth." There remains much more to be discovered, and so we close this text with the forward-looking words of a philosopher:

> The universe is full of magical things
> patiently waiting for our wits to grow sharper.

REFERENCES

GENERAL REFERENCES ON PARTICLE PHYSICS

"Particle Physics for Everybody," by Paul Davies, *Sky & Telescope,* December 1987. Any of the many books by Davies is a fascinating read.

James S. Trefil, *From Atoms to Quarks* (Charles Scribner's Sons, New York, 1980).

Frank Close, *The Cosmic Onion* (The American Institute of Physics, New York, 1983).

Yuval Ne'eman and Yoram Kirsh, *The Particle Hunters* (Cambridge University Press, Cambridge, 1987).

Jeremy Bernstein, *The Tenth Dimension—An Informal History of High Energy Physics* (McGraw-Hill, Inc., New York, 1989).

SPECIAL TOPICS IN PARTICLE PHYSICS

"Heavy Leptons," by Martin L. Perl and William T. Kirk, *Scientific American,* March 1978.

George L. Trigg, *Landmark Experiments in Twentieth Century Physics* (Crane, Russak & Company, New York, 1975), Chapter 15.

"Quarks with Color and Flavor," by Sheldon Glashow, *Scientific American,* October 1975.

"The Discovery of the Intermediate Vector Bosons," by Anne Kernan, *American Scientist,* January–February 1986. Discusses the discovery of the *W* and *Z* messenger particles.

Michael Riordan, *The Hunting of the Quark* (Simon and Schuster, Inc., New York, 1987).

GENERAL REFERENCES ON
BIG BANG PHYSICS

"The Early Universe and High-Energy Physics," by David N. Schramm, *Physics Today*, April 1983.

James S. Trefil, *The Moment of Creation* (Macmillan Publishing Company, New York, 1983).

Steven Weinberg, *The First Three Minutes*, 2nd ed. (Basic Books, New York, 1988).

Dennis Overbye, *Lonely Hearts of the Cosmos* (HarperCollins Publishers, New York, 1991).

SPECIAL TOPICS IN BIG BANG PHYSICS

"Dark Matter in the Spiral Galaxies," by Vera C. Rubin, *Scientific American*, June 1983. Dr. Rubin's findings in her own words.

"The Inflationary Universe," by Alan H. Guth and Paul J. Steinhardt, *Scientific American*, May 1984.

"The Cosmic Background Radiation and the New Aether Drift," by Richard A. Muller, *Scientific American*, May 1978.

"COBE Measures Anisotropy in Cosmic Microwave Background Radiation," by Barbara Levi, *Physics Today*, June 1992.

REVIEW & SUMMARY

Leptons and Quarks

Current research supports the view that all matter is made of six kinds of **leptons** (Table 49-3) and six kinds of **quarks** (Table 49-6). All these have spin quantum numbers equal to $\frac{1}{2}$ and are thus **fermions** (particles with half-integer spin). There are also 12 **antiparticles,** one corresponding to each of the leptons and quarks.

The Interactions

Particles with electric charge interact by the electromagnetic force by exchanging "messenger particle" **virtual photons.** Leptons can interact with each other and with quarks through the **weak force,** massive W and Z particles as messengers. In addition, quarks interact with each other by the **color force.** The electromagnetic and weak forces are different manifestations of the same force, called the **electroweak force.**

Leptons

Three of the leptons (the **electron, muon,** and **tauon**) have electric charge equal to $-1e$; these also have non-zero rest energy. There are uncharged **neutrinos** (also leptons), one corresponding to each of the charged leptons. The neutrinos have very small, possibly zero, rest energy. The antiparticles for the charged leptons have positive charge.

Quarks

The six quarks (up, down, strange, charm, bottom, and top, in order of increasing mass) each have baryon number $+\frac{1}{3}$ and charge equal to either $+(\frac{2}{3})e$ or $-(\frac{1}{3})e$. The strange quark has strangeness -1 while the others all have strangeness 0. These algebraic signs are reversed for the antiparticles.

Hadrons: Baryons and Mesons

Quarks combine into strongly interacting particles called **hadrons. Baryons** are hadrons with half-integer spin quantum numbers ($\frac{1}{2}$ or $\frac{3}{2}$). **Mesons** are hadrons with integer spin quantum numbers (0 or 1). Baryons are fermions and mesons are **bosons** (particles with integer spin). There are nine spin-0 mesons, each of which can be understood in terms of a quark–antiquark combination involving up, down, and strange quarks; their properties and compositions are given in Table 49-5 and in the Eightfold Way patterns of Figs. 49-3*b* and 49-4*b*. All these mesons have baryon number equal to zero. Quarks also combine into the eight three-quark spin-$\frac{1}{2}$ hadrons listed in Table 49-4 (among which are the familiar proton and neutron) and arranged into the Eightfold Way patterns of Figs. 49-3*a* and 49-4*a*; these have baryon number equal to $+1$. **Quantum chromodynamics** predicts that the possible combinations of quarks are either a quark with an antiquark, three quarks, or three antiquarks (this prediction is consistent with experiment). All hadrons, except for protons, are unstable.

Particle Interactions and Decay

Subatomic particles are studied by observing their decays and their interactions with each other. These reactions are governed by conservation laws: energy, angular momentum, electric charge, baryon number, lepton number, and strangeness.

Expansion of the Universe

Current evidence strongly suggests that the universe is expanding, with the distant galaxies moving away from us at a rate given by **Hubble's law:**

$$v = Hr \qquad \text{(Hubble's law)}, \qquad (49\text{-}14)$$

in which H, the **Hubble parameter,** has the value

$$H \approx 17 \times 10^{-3} \text{ m/(s·ly)}. \qquad (49\text{-}15)$$

History of the Universe

The expansion described by Hubble's law and the presence of ubiquitous background microwave radiation suggest that the universe began in a "big bang" about 15 billion years ago. A general outline of its history, as we now understand it, is given in Section 49-14.

QUESTIONS

1. What is really meant by an elementary particle? In arriving at an answer, consider such properties as lifetime, mass, size, decays into other particles, fusion to make other particles, and reactions.

2. Why do particle physicists want to accelerate particles to higher and higher energies?

3. The words *chemistry, Mendeleev, periodic table, missing elements,* and *wave mechanics* suggest a line of development in our understanding of the structure of atoms. What words suggest a corresponding line of development in our understanding of the particles of physics?

4. Name two particles that have neither mass nor charge. What properties do these particles have?

5. Why do neutrinos leave no tracks in detecting chambers?

6. Neutrinos have (presumably) no mass and travel with the speed of light. How then can they carry varying amounts of energy?

7. Do all particles, photons included, have antiparticles?

8. Photons and neutrinos are alike in that they have zero charge, zero mass (presumably), and travel with the speed of light. What are the differences between these two particles? How would you produce them? How would you detect them?

9. Explain why physicists say that the π^0 meson is its own antiparticle.

10. Why can't an electron decay by disintegrating into two neutrinos?

11. Why is the electron stable? That is, why does it not decay spontaneously into other particles?

12. Why cannot a stationary electron emit a single gamma-ray photon and disappear? Could an electron that is moving do so?

13. A neutron is massive enough to decay by the emission of a proton and two neutrinos. Why does it not do so?

14. A positron invariably finds an electron and they annihilate each other. How then can we call the positron a stable particle?

15. What is the mechanism by which two electrons exert forces on each other?

16. Do the eight pions whose tracks appear in Fig. 49-2*a* all have the same initial kinetic energy? If so, what is that energy? If not, which of them have the greatest energy?

17. Is the magnetic field that is present in Fig. 49-2 directed into the page or out of the page?

18. A particle that responds to the strong force is either a meson or a baryon. You can tell which it is by allowing the particle to decay until only stable end products remain. If there is a proton among these products, the original particle was a baryon. If there is no proton, the original particle was a meson. Explain this classification rule.

19. How many kinds of stable leptons are there? Stable mesons? Stable baryons? In each case, name them.

20. Most particle reactions are endothermic, rather than exothermic. Why?

21. What is the lightest strongly interacting particle? What is the heaviest particle that is unaffected by the strong interaction?

22. For each of the following particles, state which of the four types of interaction are applicable: (a) electron; (b) neutrino; (c) neutron; (d) pion.

23. Just as x rays are used to discover internal imperfections in a metal casting caused by gas bubbles, so cosmic-ray muons have been used in an attempt to discover hidden burial chambers in Egyptian pyramids. Why were muons used?

24. Are strongly interacting particles affected by the weak interaction?

25. Do all weak-interaction decays produce neutrinos?

26. The messengers for quantum electrodynamics are photons and they are virtual. The messengers for the weak force are the *W* and *Z* particles and they are observed. Comment on this difference.

27. What is the difference between a boson and a fermion? A hadron and a lepton?

28. Baryons and leptons are both fermions. In what ways are they different?

29. Mesons and baryons are each sensitive to the strong force. In what way are they different?

30. By comparing Tables 49-3 and 49-6, point out as many similarities between leptons and quarks as you can, and as many differences.

31. Quarks are not observed directly. What is the indirect evidence for them?

32. We can explain the "ordinary" world around us with two leptons and two quarks. Name them.

33. The neutral pion has a quark structure of $u\bar{u}$ and decays with a mean life of only 8.3×10^{-17} s. The positively charged pion, on the other hand, has a quark structure of $u\bar{d}$ and decays with a mean life of 2.6×10^{-8} s. Explain, in terms of their quark structures, why the mean life of the neutral pion should be so much shorter (by a factor of 3×10^8) than that of the charged pion. (*Hint:* Think of the annihilation process.)

34. Do leptons contain quarks? Do mesons? Do photons? Do baryons?

35. The ratio of the magnitude of the gravitational force between the electron and the proton in the hydrogen atom to the magnitude of the electromagnetic force of attraction between them is about 10^{-40}. If the gravitational force is so very much weaker than the electromagnetic force, how was it that the gravitational force was discovered first and is so much more apparent to us?

36. Why can't we find the center of the expanding universe? Are we looking for it? Is there such a thing?

37. Due to the effect of the gravitational force, the rate of expansion of the universe must have decreased in time following the Big Bang. Show that this slowdown implies that the age of the universe is less than $1/H$.

38. It is not possible, using telescopes that are sensitive in any part of the electromagnetic spectrum, to "look back" any farther than about 300,000 y from the Big Bang. Why?

39. How does one arrive at the conclusion that the visible portion of a galaxy represents about only 10% of the galactic mass?

40. Are we *always* looking back into time as we observe the distant universe? Does the direction in which we look make a difference?

EXERCISES & PROBLEMS

SECTION 49-3 AN INTERLUDE

1E. Calculate the difference in mass, in kilograms, between the muon and pion of Sample Problem 49-2.

2E. A neutral pion decays into two gamma rays: $\pi^0 \rightarrow \gamma + \gamma$. Calculate the wavelengths of the gamma rays produced by the decay of a neutral pion at rest.

3E. An electron and a positron are separated by a distance r. Find the ratio of the gravitational force to the electrostatic force between them. From the result, what can you conclude concerning the forces acting between particles detected in a bubble chamber.

4E. The positively charged pion decays by Eq. 49-4: $\pi^+ \rightarrow \mu^+ + \nu$. What then must be the decay scheme of the negatively charged pion? (*Hint:* The π^- is the antiparticle of the π^+.)

5E. How much energy would be "created" if our Earth were annihilated by collision with an anti-Earth?

6P. A neutral pion has a rest energy of 135 MeV and a mean life of 8.3×10^{-17} s. If it is produced with an initial kinetic energy of 80 MeV and it decays after one mean lifetime, what is the longest possible track that this particle could leave in a bubble chamber? Take relativistic time dilation into account. (*Hint:* See Sample Problem 49-1.)

7P. Observations of neutrinos emitted by the supernova SN1987a in the Large Magellanic Cloud (see Fig. 48-11) place an upper limit of 20 eV on the rest energy of the electron neutrino. Suppose that the rest energy of this neutrino, rather than being zero, is in fact equal to 20 eV. How much slower than the speed of light would a 1.5-MeV neutrino emitted in a beta decay move?

8P. Certain theories predict that the proton is unstable, with a half-life of about 10^{32} years. Assuming that this is true, calculate the number of proton decays you would expect to occur in one year in the water of an Olympic-sized swimming pool holding 114,000 gallons of water.

9P. A positive tauon (τ^+, rest energy = 1784 MeV) is moving with 2200 MeV of kinetic energy in a circular path perpendicular to a uniform 1.20-T magnetic field. (a) Calculate the momentum of the tauon in kg·m/s. Relativistic effects must be considered. (b) Find the radius of the circular path. (*Hint:* See Problem 57 in Chapter 42.)

10P. The rest energies of many short-lived particles cannot be measured directly but must be inferred from the measured momenta and known rest energies of the decay products. Consider the ρ^0 meson, which decays by the reaction $\rho^0 \rightarrow \pi^+ + \pi^-$. Calculate the rest energy of the ρ^0 meson given that the oppositely directed momenta of the created pions each have magnitude 358.3 MeV/c. See Table 49-5 for the rest energies of the pions.

11P. (a) A stationary particle m_0 decays into two particles m_1 and m_2, which move off with equal but oppositely directed momenta. Show that the kinetic energy K_1 of m_1 is given by

$$K_1 = \frac{1}{2E_0} [(E_0 - E_1)^2 - E_2^2],$$

where m_0, m_1, and m_2 are masses and E_0, E_1, and E_2 are the corresponding rest energies. (*Hint:* Follow the arguments of Sample Problem 49-2 except that, in this case, neither of the created particles has zero mass.) (b) Show that the result in (a) yields the kinetic energy of the muon as calculated in Sample Problem 49-2.

SECTION 49-5 A NEW CONSERVATION LAW

12E. Verify that the hypothetical proton decay scheme given in Eq. 49-10 does not violate the conservation laws of (a) charge, (b) energy, (c) linear momentum, and (d) angular momentum.

13E. What conservation law is violated in each of these proposed decays? (Assume that the decay products have zero orbital angular momentum.) (a) $\mu^- \to e^- + \nu$; (b) $\mu^- \to e^+ + \nu + \bar{\nu}$; (c) $\mu^+ \to \pi^+ + \nu$.

14P. The A_2^+ particle and its products decay according to the following schemes:

$$A_2^+ \to \rho^0 + \pi^+, \qquad \mu^+ \to e^+ + \nu + \bar{\nu},$$

$$\rho^0 \to \pi^+ + \pi^-, \qquad \pi^- \to \mu^- + \bar{\nu},$$

$$\pi^+ \to \mu^+ + \nu, \qquad \mu^- \to e^- + \nu + \bar{\nu}.$$

(a) What are the final stable decay products? (b) From the evidence, is the A_2^+ particle a fermion or a boson? Is it a meson or a baryon? What is its baryon number? (*Hint:* See Sample Problem 49-5.)

SECTION 49-7   THE EIGHTFOLD WAY

15E. The reaction $\pi^+ + p \to p + p + \bar{n}$ proceeds by the strong interaction. By applying the conservation laws, deduce the charge, baryon number, and strangeness of the antineutron.

16E. By examining strangeness, determine which of the following decays or reactions proceed via the strong interaction: (a) $K^0 \to \pi^+ + \pi^-$; (b) $\Lambda^0 + p \to \Sigma^+ + n$; (c) $\Lambda^0 \to p + \pi^-$; (d) $K^- + p \to \Lambda^0 + \pi^0$.

17E. What conservation law is violated in each of these proposed reactions and decays? (Assume that the products have zero orbital angular momentum.) (a) $\Lambda^0 \to p + K^-$; (b) $\Omega^- \to \Sigma^- + \pi^0$ ($S = -3$, $Q = -1$ for Ω^-); (c) $K^- + p \to \Lambda^0 + \pi^+$.

18E. Calculate the disintegration energy of the reactions (a) $\pi^+ + p \to \Sigma^+ + K^+$ and (b) $K^- + p \to \Lambda^0 + \pi^0$.

19E. A Σ^- particle moving with 220 MeV of kinetic energy decays according to $\Sigma^- \to \pi^- + n$. Calculate the total kinetic energy of the decay products.

20P. Use the conservation laws to identify the particle labeled x in each of the following reactions, which proceed by means of the strong interaction: (a) $p + p \to p + \Lambda^0 + x$; (b) $p + \bar{p} \to n + x$; (c) $\pi^- + p \to \Xi^0 + K^0 + x$.

21P. Show that if, instead of plotting S versus Q for the spin-$\frac{1}{2}$ baryons in Fig. 49-3a and for the spin-0 mesons in Fig. 49-3b, the quantity $Y = B + S$ is plotted against the quantity $T_z = Q - \frac{1}{2}B$, then the hexagonal patterns emerge with the use of nonsloping (perpendicular) axes. (The quantity Y is called *hypercharge* and T_z is related to a quantity called *isospin*.)

22P. Consider the decay $\Lambda^0 \to p + \pi^-$ with the Λ^0 at rest. (a) Calculate the disintegration energy. (b) Find the kinetic energy of the proton. (c) What is the kinetic energy of the pion? (*Hint:* See Problem 11.)

SECTION 49-8   THE QUARK MODEL

23E. The quark makeups of the proton and neutron are *uud* and *udd*, respectively. What are the quark makeups of (a) the antiproton and (b) the antineutron?

24E. From Tables 49-4 and 49-6, determine the identities of the baryons formed from the following combinations of quarks. Check you answers with the baryon octet shown in Fig. 49-3a: (a) *ddu*; (b) *uus*; (c) *ssd*.

25E. What quark combinations are needed to form (a) a Λ^0 and (b) a Ξ^0?

26E. Using the up, down, and strange quarks only, construct, if possible, a baryon (a) with $Q = +1$ and $S = -2$ and (b) with $Q = +2$ and $S = 0$.

27E. There are 10 baryons with spin $\frac{3}{2}$. Their symbols and quantum numbers are as follows:

	Q	S		Q	S
Δ^-	-1	0	Σ^{*0}	0	-1
Δ^0	0	0	Σ^{*+}	$+1$	-1
Δ^+	$+1$	0	Ξ^{*-}	-1	-2
Δ^{++}	$+2$	0	Ξ^{*0}	0	-2
Σ^{*-}	-1	-1	Ω^-	-1	-3

Make a charge–strangeness plot for these baryons, using the sloping coordinate system of Fig. 49-3. Compare your plot with this figure.

28P. There is no known meson with $Q = +1$ and $S = -1$ or with $Q = -1$ and $S = +1$. Explain why, in terms of the quark model.

29P. The spin-$\frac{3}{2}$ Σ^{*0} baryon (see Exercise 27) has a rest energy of 1385 MeV (with an intrinsic uncertainty ignored here); the spin-$\frac{1}{2}$ Σ^0 baryon has a rest energy of 1192.5 MeV. Suppose that each of these particles has a kinetic energy of 1000 MeV. Which, if either, is moving faster and by how much?

SECTION 49-11   THE UNIVERSE IS EXPANDING

30E. If Hubble's law can be extrapolated to very large distances, at what distance would the recessional speed become equal to the speed of light?

31E. What is the observed wavelength of the 656.3-nm H_α line of hydrogen emitted by a galaxy at a distance of 2.40×10^8 ly?

32E. In the laboratory, one of the lines of sodium is emitted at a wavelength of 590.0 nm. In the light from a particular galaxy, however, this line is seen at a wavelength of 602.0 nm. Calculate the distance to the galaxy, assuming that Hubble's law holds.

33P. The recessional speeds of galaxies and quasars at great distances are close to the speed of light, so that the relativistic Doppler shift formula (see Eq. 42-26) must be used. The red shift is reported as fractional red shift $z = \Delta\lambda/\lambda_0$. (a) Show that, in terms of z, the recessional speed parameter $\beta = v/c$ is given by

$$\beta = \frac{z^2 + 2z}{z^2 + 2z + 2}.$$

(b) A quasar detected in 1987 has $z = 4.43$. Calculate its speed parameter. (c) Find the distance to the quasar, assuming that Hubble's law is valid to these distances.

34P. Will the universe continue to expand forever? To attack this question, make the (reasonable?) assumption that the recessional speed v of a galaxy a distance r from us is determined only by the matter that lies inside a sphere of radius r centered on us. If the total mass inside this sphere is M, the escape speed v_e from the sphere is given by $v_e = \sqrt{2GM/r}$ (see Sample Problem 15-8). (a) Show that the average density ρ inside the sphere must be at least equal to the value

$$\rho = 3H^2/8\pi G$$

to prevent unlimited expansion. (b) Evaluate this "critical density" numerically; express your answer in terms of H-atoms/cm^3. Measurements of the actual density are difficult and complicated by the presence of dark matter.

SECTION 49-12   THE MICROWAVE BACKGROUND RADIATION

35P. Due to the presence everywhere of the microwave background radiation, the minimum possible temperature of a gas in interstellar or intergalactic space is not 0 K but 2.7 K. This implies that a significant fraction of the molecules in space that can occupy excited states of low excitation energy may, in fact, be in those excited states. Subsequent de-excitation would lead to the emission of radiation that could be detected. Consider a (hypothetical) molecule with just one excited state. (a) What would the excitation energy have to be in order that 25% of the molecules be in the excited state? (*Hint:* See Eq. 45-24.) (b) What would be the wavelength of the photon emitted in a transition back to the ground state?

SECTION 49-13   THE MYSTERY OF THE DARK MATTER

36E. What would the mass of the sun have to be if Pluto (the outermost planet most of the time) were to have the same orbital speed that Mercury (the innermost planet) has now? Use data from Appendix C, and express your answer in terms of the sun's current mass M. (Assume circular orbits.)

37P. Suppose that the radius of the sun were increased to 5.90×10^{12} m (the average radius of the orbit of the planet Pluto, the outermost planet), that the density of this expanded sun were uniform, and that the planets re-

volved within this tenuous object. (a) Calculate the Earth's orbital speed in this new configuration and compare this with its present orbital speed of 29.8 km/s. Assume that the radius of the Earth's orbit remains unchanged. (b) What would be the new period of revolution of the Earth? (The sun's mass remains unchanged.)

38P. Suppose that the matter (stars, gas, dust) of a particular galaxy, of total mass M, is distributed uniformly throughout a sphere of radius R. A star of mass m is revolving about the center of the galaxy in a circular orbit of radius $r < R$. (a) Show that the orbital speed v of the star is given by

$$v = r\sqrt{GM/R^3},$$

and therefore that the period T of revolution is

$$T = 2\pi\sqrt{R^3/GM},$$

independent of r. Ignore any resistive forces. (b) What is the corresponding formula for the orbital period assuming that the mass of the galaxy is strongly concentrated toward the center of the galaxy, so that essentially all the mass is at distances from the center less than r?

SECTION 49-14   THE BIG BANG

39E. From Planck's radiation law it is possible to derive the following relation between the temperature T of a cavity radiator and the wavelength λ_{max} at which it radiates most strongly:

$$\lambda_{max} T = 2898 \ \mu\text{m} \cdot \text{K}.$$

(This is Wien's law; see Problem 48 in Chapter 43.) (a) The microwave background radiation peaks in intensity at a wavelength of 1.1 mm. To what temperature does this correspond? (b) About 10^5 years after the Big Bang, the universe became transparent to electromagnetic radiation. Its temperature then was about 10^5 K. What was the wavelength at which the background radiation was most intense at that time?

40E. The wavelength of the photons at which a radiation field of temperature T radiates most intensely is given by $\lambda_{max} = (2898 \ \mu\text{m} \cdot \text{K})/T$ (see Exercise 39). (a) Show that the energy E in MeV of such a photon can be computed from

$$E = (4.28 \times 10^{-10} \ \text{MeV/K}) \, T.$$

(b) At what minimum temperature can this photon create an electron–positron pair in a pair production process (see Section 23-6)?

ANSWERS TO ODD-NUMBERED EXERCISES AND PROBLEMS

Chapter 43

3. 2.1 μm; infrared. **5.** (a) 35.4 keV.
(b) 8.57×10^{18} Hz.
(c) 35.4 keV/$c = 1.89 \times 10^{-23}$ kg·m/s.
7. (a) 1.24×10^{20} Hz. (b) 2.43 pm.
(c) 2.73×10^{-22} kg·m/s = 0.511 MeV/c.
9. (a) The infrared bulb. (b) 6.0×10^{20}.
11. 4.7×10^{26}. **13.** (a) 2.96×10^{20} photons/s.
(b) 48,600 km. (c) 280 m.
(d) 5.89×10^{18} m^{-2}·s^{-1}; 1.96×10^{10} m^{-3}.
15. 233 nm. **17.** 10 eV. **19.** 676 km/s. **21.** (a) 1.3 V.
(b) 680 km/s. **23.** (a) 382 nm. (b) 1.82 eV.
27. (a) 3.1 keV. (b) 14 keV. **29.** (a) 2.7 pm.
(b) 6.05 pm. **31.** (a) $+4.8$ pm. (b) -41 keV.
(c) 41 keV. **33.** (a) 8.1×10^{-9}%. (b) 4.9×10^{-4}%.
(c) 8.8%. (d) 66%. **35.** 2.65 fm. **38.** 44°.
43. 9.99 μm. **45.** 91 K. **47.** (a) 0.97 mm; microwave.
(b) 9.9 μm; infrared. (c) 1.6 μm; infrared.
(d) 0.26 nm; x ray.
(e) 2.9×10^{-41} m; very high energy gamma ray.
53. $4\Delta T/T$; 0.0130. **55.** 2.6 eV. **57.** -80.7 keV.
59. (a) 121.5 nm. (b) 91.2 nm.
61. 656 nm; 486 nm; 434 nm. **63.** (a) 12.7 eV.
(b) 12.7 eV $(4 \to 1)$; 2.55 eV $(4 \to 2)$; 0.66 eV $(4 \to 3)$;
12.1 eV $(3 \to 1)$; 1.89 eV $(3 \to 2)$; 10.2 eV $(2 \to 1)$.
65. (a) 30.5 nm; 291 nm; 1050 nm.
(b) 8.25×10^{14} Hz; 3.65×10^{14} Hz; 2.06×10^{14} Hz.
69. 4.1 m/s. **75.** (a) 3×10^{74}. (b) No. **77.** (b) n^2.
(c) n. (d) $1/n$. (e) $1/n^3$. (f) $1/n$. (g) $1/n^4$.
(h) $1/n^4$. (i) $1/n^2$. (j) $1/n^2$. (k) $1/n^2$.

Chapter 44

1. (a) 1.7×10^{-35} m. **3.** 7.75 pm.
5. (a) 3.3×10^{-24} kg·m/s for each.
(b) 38 eV for the electron; 6.2 keV for the photon.
7. (a) 38.8 meV. (b) 146 pm. **9.** (a) 73 pm; 3.4 nm.
(b) Yes. **11.** (a) 1.24 keV; 1.50 eV.
(b) 1.24 GeV; 1.24 GeV. **13.** 0.025 fm. **15.** A neutron.
17. 9.70 kV (relativistic calculation); 9.79 kV (classical
calculation). **19.** 11.5°, 23.6°, 36.9°, 53.1°.
21. (a) 20.5 meV. (b) 37.7 eV. **23.** (a) 1900 MeV.
(b) No. **25.** 90.5 eV.
27. 18.1, 36.2, 54.3, 66.3, 72.4 μeV. **29.** (a) 0.196.
(b) 0.608. (c) 0.196. **31.** 0.323. **33.** 0.439.

37. Proton: 9.2×10^{-6}; deuteron: 7.6×10^{-8}.
39. 10^{104} y. **41.** 7×10^{-23} kg·m/s. **43.** 1.2 m.
45. 0.41 μeV; $E_2 = -3.4$ eV. **47.** (a) 124 keV.
(b) 40.5 keV.

Chapter 45

3. 3.64×10^{-34} J·s. **5.** 24.1°.
7. $n > 3$; $m_\ell = +3, +2, +1, 0, -1, -2, -3$;
$m_s = +\frac{1}{2}, -\frac{1}{2}$. **9.** 50. **17.** 5.54 nm^{-1}. **19.** 0.0054.
21. (b) 16.4 nm$^{-3/2}$. **25.** 0.981 nm^{-1}; 3.61 nm^{-1}.
27. 0.0019. **29.** 54.7°; 125°. **31.** (a) 58 μeV.
(b) 14 GHz. (c) 2.1 cm; short radio wave region.
33. 5.35 cm. **35.** 19 mT. **37.** All statements are true.
39. (a) $(2, 0, 0, \pm\frac{1}{2})$.
(b) $n = 2$, $\ell = 1$; $m_\ell = 1, 0, -1$; $m_s = \pm\frac{1}{2}$. **45.** 12.4 kV.
47. 49.6 pm; 99.2 pm. **49.** (a) 24.8 pm.
(b) and (c) remain unchanged. **51.** 6.44 keV; 10.2 eV.
53. $\frac{9}{16}$. **55.** (a) 69.5 kV. (b) 17.9 pm. (c) 21.4 pm;
18.5 pm. **57.** 282 pm.
59. (b) 24%; 15%; 11%; 7.9%; 6.5%; 4.7%; 3.5%; 2.5%;
2.0%; 1.5%. **61.** 10,000 K. **63.** 4.4×10^{17}.
65. 2.0×10^{16} s^{-1}. **67.** 4.8 km. **69.** 1.8 pm.
71. (a) 7.33 μm. (b) 707 kW/m^2. (c) 24.9 GW/m^2.
73. (a) 53 GPa. (b) 1.2×10^8 K.

Chapter 46

1. 3460 atm. **3.** (a) 2.7×10^{25} m^{-3}.
(b) 8.43×10^{28} m^{-3}. (c) 3100.
(d) 3.3 nm for oxygen, 0.228 nm for the electrons.
7. 1.92×10^{28} m^{-3}·eV^{-1}. **9.** (a) 6.81 eV.
(b) 1.77×10^{28} m^{-3}·eV^{-1}. (c) 1.59×10^{28} m^{-3}·eV^{-1}.
11. 5.53 eV. **13.** $T \gg 10^5$ K. **15.** 3.
19. (a) 5.86×10^{28} m^{-3}. (b) 5.52 eV. (c) 1390 km/s.
(d) 0.522 nm. **21.** 137 MeV. **25.** 200°C.
29. (a) 19.8 kJ. (b) 3 min 18 s. **31.** (a) n-type.
(b) 5×10^{21} m^{-3}. (c) 2.5×10^5.
33. (a) Pure: 4.78×10^{-10}; doped: 0.0141. (b) 0.824.
37. (b) 2.49×10^8. **39.** 4.20 eV.

Chapter 47

1. 15.8 fm. **3.** (a) 0.39 MeV. (b) 4.61 MeV.
5. (a) Six. (b) Eight. **9.** (a) 1150 MeV.
(b) 4.8 MeV/nucleon; 12 MeV/proton.
12. (a) 3.0×10^{17} kg/m^3 for each.

(b) 1.3×10^{25} C/m³ for ^{55}Mn; 1.1×10^{25} C/m³ for ^{209}Bi. **15.** 4×10^{-22} s. **17.** $K \approx 30$ MeV.
21. (a) 19.8 MeV, 6.26 MeV, 2.22 MeV. (b) 28.3 MeV.
(c) 7.07 MeV. **23.** 1.6×10^{25} MeV. **25.** 7.92 MeV.
27. 280 d. **29.** (a) 7.6×10^{16} s^{-1}. (b) 4.9×10^{16} s^{-1}.
31. (a) 4.8×10^{-18} s^{-1}. (b) 4.6×10^{9} y.
33. 5.3×10^{22}. **35.** 265 mg. **37.** 209 d. **39.** 87.8 mg.
41. 730 cm². **45.** (a) 3.66×10^{7} s^{-1}. (b) $t \geqslant 3.82$ d.
(c) 3.66×10^{7} s^{-1}. (d) 6.42 ng.
47. Pu: 5.5×10^{-9}; Cm: zero. **49.** (a) 4.25 MeV.
(b) -24.1 MeV. (c) 28.3 MeV.
51. $Q_3 = -9.50$ MeV; $Q_4 = 4.66$ MeV;
$Q_5 = -1.30$ MeV. **53.** 1.40 MeV. **55.** 0.782 MeV.
59. (b) 0.961 MeV. **61.** 78.4 eV. **63.** 1600 y.
65. 1.7 mg. **67.** 1.02 mg. **69.** (a) 18 mJ.
(b) 0.29 rem. **71.** (a) 6.3×10^{18}. (b) 2.5×10^{11}.
(c) 0.20 J. (d) 0.23 rad. (e) 3.0 rem.
73. 3.87×10^{10} K. **75.** (a) 25.35 MeV. (b) 12.8 MeV.
(c) 25.0 MeV. **77.** (a) 3.85 MeV, 7.95 MeV.
(b) 3.98 MeV, 7.33 MeV.

Chapter 48
1. (a) 2.6×10^{24}. (b) 8.2×10^{13} J. (c) 2.6×10^{4} y.
3. 3.1×10^{10} s^{-1}. **7.** $+5.00$ MeV.
9. (a) 16 fissions/day. (b) 4.3×10^{8}. **11.** (a) 10.

(b) 231 MeV. **13.** (a) 252 MeV.
(b) Typical fission energy $= 200$ MeV. **15.** 463 kg.
17. Yes. **19.** (a) 1.15 MeV. (b) 3.2 kg.
21. (a) 44 kton. **25.** 1.7×10^{k}.
27. (b) 1.0, 0.89, 0.28, 0.019. (c) 8. **29.** (a) 75 kW.
(b) 5800 kg. **33.** (a) 30 MeV. (b) 6 MeV.
(c) 170 keV. **35.** (a) 170 kV. **37.** 1.41 MeV.
41. (a) 3.1×10^{31} photons/m³. (b) 1.2×10^{6} times.
43. (a) 4.0×10^{27} MeV. (b) 5.1×10^{26} MeV.
45. (a) 1.83×10^{38} s^{-1}. (b) 8.25×10^{28} s^{-1}.
49. (a) 6.3×10^{14} J/kg. (b) 6.2×10^{11} kg/s.
(c) 4.3×10^{9} kg/s. (d) 15×10^{9} y. **51.** (a) 24.9 MeV.
(b) 8.65 Mton. **53.** $K_\alpha = 3.5$ MeV; $K_n = 14.1$ MeV.
57. (a) 230 kJ. (b) 0.11 lb. (c) 23 MW.

Chapter 49
1. 6.03×10^{-29} kg. **3.** 2.4×10^{-43}. **5.** 1.08×10^{42} J.
7. 27 cm/s. **9.** (a) 1.90×10^{-18} kg·m/s. (b) 9.90 m.
13. (a) Strangeness. (b) Charge. (c) Energy.
15. $Q = 0$; $B = -1$; $S = 0$. **17.** (a) Energy.
(b) Angular momentum. (c) Charge. **19.** 338 MeV.
23. (a) $\overline{uu}d$. (b) $\overline{u}dd$. **25.** (a) sud. (b) uss.
29. Σ^0; 7530 km/s. **31.** 665 nm. **33.** (b) 0.934.
(c) 1.65×10^{10} ly. **35.** (a) 256 μeV. (b) 4.84 nm.
37. (a) 121 m/s. (b) 246 y. **39.** (a) 2.6 K. (b) 29 nm.

PHOTO CREDITS

Chapter 11

Opener: Guido Alberto Rossi/Image Bank. Page 286: Rick Rickman/Duomo. Page 293: Art Tilley/FPG International. Page 295: Roger Ressmeyer/Starlight. Page 305: Fabricius-Taylor/AllStock. Page 306: Zimmerman/FPG International. Page 312: Courtesy Lick Observatory. Page 316: Courtesy Lawrence Livermore Laboratory, University of California.

Chapter 12

Opener: Courtesy Ringling Brothers and Barnum & Bailey Circus. Page 320: Richard Megna/Fundamental Photographs. Page 321: Courtesy Alice Halliday. Page 322: Elisabeth Weiland/Photo Researchers. Page 327: Doug Lee/Peter Arnold. Page 332: Kerstgens/Sipa Press. Page 335: Steven E. Sutton/Duomo. Page 336: (top) Courtesy NASA; (bottom) E. Sander/Gamma-Liaison. Page 339: copyright © Estate of Harold Edgerton, courtesy of Palm Press, Inc.

ESSAY 2

Page E2-1: (top left and bottom) Susan Cook: (top right) courtesy Kenneth Laws. Pages E2-2–E2-4: Martha Swope.

Chapter 13

Opener: Jose Azel/Woodfin Camp & Associates. Page 354: (left) Fred Hirschmann/Allstock; (right) Andy Levin/Photo Researchers. Page 355: Randy G. Taylor/Leo de Wys. Page 356: Richard Negri/Gamma-Liaison. Page 358: Chris Dalmas/Gamma-Liaison. Page 364: David Bitters/The Picture Cube. Page 368: Courtesy Micro-Measurements Division, Measurements Group, Inc., Raleigh, North Carolina. Page 372: Hideo Kurihara/Tony Stone Worldwide.

Chapter 14

Opener: Tom van Dyke/Sygma. Page 382: Kent Knudson/FPG International. Page 398: Bettmann Archive. Page 404: Courtesy NASA.

Chapter 15

Opener: Courtesy Lund Observatory. Page 412: Courtesy NASA. Page 421: Kim Gordon/AstroStock. Page 422: Courtesy Lockheed Missile and Space Co., Inc. Pages 425 and 438: Courtesy Finley-Holiday Film Corporation. Page 429: Courtesy NASA.

ESSAY 3

Pages E3-1 and E3-3: Courtesy NASA.

Chapter 16

Opener: Steven Frink. Page 453: T. Orban/Sygma. Page 454: John Amos/Photo Researchers. Page 455: (left) Will McIntyre/Photo Researchers; (right) courtesy D. H. Peregrine, University of Bristol, England. Page 456: Courtesy Volvo North America Corporation. Page 463: Courtesy San Francisco Maritime National Historical Park Museum. Page 466: T. Campion/Sygma. Page 470: Kick Stewart/Allsport.

ESSAY 4

Page E4-1: Courtesy Peter Brancazio.

Chapter 17

Opener: John Visser/Bruce Coleman. Page 493: Richard Megna/Fundamental Photographs. Page 494: Courtesy T. D. Rossing, Northern Illinois University.

Chapter 18

Opener: Stephen Dalton/Animals Animals. Page 504: (left) Howard Sochurack/Stock Market; (right) courtesy John Foster, IBM Corporation. Page 510: Ben Rose/Image Bank. Page 513: Bob Gruen/Star File Photos. Page 514: John Eastcott/Yva Momatiuk/DRK Photo. Page 521: Philippe Plailly/Science Photo Library/Photo Researchers. Page 532: Courtesy Mt. Wilson and Palomar Observatories.

ESSAY 5

Page E5-1: (top) Courtesy John Rigden; (center) courtesy Lincoln Center. Page E5-2: Lincoln Russell/Stock, Boston. Pages E5-4 and E5-5: Courtesy Lincoln Center, photos by Norman McGrath.

Chapter 19

Opener: Jim Brandenburg/Minden Pictures. Page 540: AP/Wide World Photos. Page 543: Richard Choy/Peter Arnold.

Chapter 20

Opener: Tom Owen Edmunds/Image Bank. Page 554: Obremski/Image Bank. Page 564: (left) Fritz Goro; (right) Peter Arnold/Peter Arnold. Page 565: Courtesy Daedalus Enterprises. Page 569: Mark Newman.

ESSAY 6

Page E6-1: Jeff R. Werner.

Chapter 21

Opener: Bryan and Cherry Alexander Photography. Page 581: Courtesy NASA. Page 593: Tom Branch/Photo Researchers.

ESSAY 7

Page E7-1: Courtesy Barbara Levi. Page E7-2: Courtesy NASA/Langley Research Center.

Chapter 22

Opener: Steven Dalton/Photo Researchers. Page 608: (top) Richard Ustinich/Image Bank; (bottom) V. Kiselev/Sovfoto. Page 625: (left) Carry Wolinski/Stock, Boston; (right) Courtesy Professor Hallet, University of Washington, Seattle.

Chapter 23

Opener: Michael Watson. Page 636: Fundamental Photographs. Page 637: Courtesy Saran Wrap, Dow Brands, Household Products Division. Page 638: (left) Courtesy Xerox Corporation; (right) Johann Gabriel Doppelmayr, Neuentdeckte Phaenomena von bewündernswurdigen Würckungen der Natur, Nuremberg 1744. Page 646: Courtesy Lawrence Berkeley Laboratory, University of California.

Chapter 24
Opener: Quesada/Burke, NY. Page 655: Stephen Frink/All-Stock. Page 666: Russ Kinne/Comstock. Page 667: Courtesy Environmental Elements Corporation.

Chapter 25
Opener: E. R. Degginger/Bruce Coleman. Page 692: C. Johnny Autery. Page 693: E. Philip Krider, Institute of Atmospheric Physics, University of Arizona, Tucson.

Chapter 26
Opener: Courtesy NOAA. Page 726: (top) Courtesy NASA; (bottom) courtesy Westinghouse Corporation. Page 731: Courtesy NASA.

Chapter 27
Opener: C. Goivaux Communication/PHOTOTAKE. Page 740: Paul Silverman/Fundamental Photographs. Page 749: copyright © The Harold E. Edgerton 1992 Trust, courtesy of Palm Press, Inc. Page 751: Courtesy Royal Institution, England.

Chapter 28
Opener: UPI/Bettmann Newsphotos. Page 772: The Image Works. Page 778: Laurie Rubie/Tony Stone Worldwide. Page 781: (left) Courtesy AT&T Bell Laboratories; (right) courtesy Shoji Tonaka, International Superconductivity Technology Center, Tokyo, Japan.

Chapter 29
Opener: Norbert Wu. Page 790: Courtesy Southern California Edison Co. Page 804: Courtesy Simpson Electric Co.

Chapter 30
Opener: Johnny Johnson/Earth Scenes. Page 818: (left) Schneps/Image Bank; (right) Science Photo Library/Photo Researchers. Page 819: Courtesy Dr. Richard Cannon, Southeast Missouri State University, Cape Girardeau. Page 820: Richard Megna/Fundamental Photographs. Page 821: Courtesy Lawrence Berkeley Laboratory, University of California. Page 822: Richard Megna/Fundamental Photographs. Page 826: Courtesy John Le P. Webb, Sussex University, England. Page 828: Courtesy Dr. L. A. Frank, University of Iowa. Page 830: Courtesy Siemmens Gammasonics. Page 831: (top left and center) Courtesy Fermi National Accelerator Laboratory; (top right) Spaceshots.

Chapter 31
Opener: Courtesy Rafael Testing Unit, Israel. Page 850: Courtesy Educational Services, Inc.

Chapter 32
Opener: Dan McCoy/Black Star. Page 880: Courtesy Fender Musical Instruments Corp. Page 882: Courtesy Jenn-Air Co.

ESSAY 8
Page E8-1: (top) Courtesy Peter Lindenfield; (bottom) From "Superconductivity" by D. Shoenberg, Ph.D., Cambridge University Press, 1952. Page E8-2: (top) Hank Morgan/Photo Researchers; (bottom) T. Matsumoto/Sygma. Page E8-3: David Parker/University of Birmingham High TC Consortium/Science Photo Library/Photo Researchers.

Chapter 33
Opener: Photo by Dan Blodget, courtesy Fisher Research Laboratory. Page 900: Courtesy Royal Institution, England.

ESSAY 9
Page E9-1: Courtesy Thomas Rossing.

Chapter 34
Opener: Bob Zehring. Page 920: Runk/Schoenberger/Grant Heilman. Page 923: Courtesy Dr. Henry Guckel, University of Wisconsin. Page 925: copyright © Colchester & Essex Museum. Page 927: Peter Lerman. Page 931: (top) Courtesy Ralph W. de Blois; (bottom) courtesy R. E. Rosensweig, Research and Science Laboratory, Exxon Corp. Page 932: CNRI/Science Photo Library/Photo Researchers.

ESSAY 10
Page E10-1: Courtesy Charles Bean. Page E10-2: Courtesy R. P. Blakemore and R. B. Frankel, *Scientific American,* December 1981.

Chapter 35
Opener: A. Glauberman/Photo Researchers. Page 942: Courtesy Hewlett-Packard. Page 949: John Chiasson/Gamma-Liaison.

Chapter 36
Opener: Courtesy Haverfield Helicopter Co. Page 964: (left) Steve Kagan/Gamma-Liaison; (right) Ted Cowell/Black Star.

Chapter 37
Opener: Keystone/Sygma; (inset) courtesy Jack Nissenthall.

Chapter 38
Opener: Courtesy Hansen Publications. Page 998: copyright © 1992 Ben and Miriam Rose, from the collection of the Center for Creative Photography, Tucson. Page 1002: Diane Schiumo/Fundamental Photographs. Page 1007: Roger Ressmeyer/Starlight.

ESSAY 11
Pages E11-1–3: Courtesy Raymond C. Turner.

Chapter 39
Opener: Courtesy Courtauld Institute Galleries, London. Page 1012: (center) *PSSC Physics,* 2nd. ed., copyright © 1975 D. C. Heath & Co. with Education Development Center, Newton, Massachusetts; (bottom) courtesy Lockheed Advanced Development Co. Page 1014: Courtesy Bausch & Lomb. Page 1015: Tony Stone Worldwide. Page 1016: Will & Deni McIntyre/Photo Researchers. Page

1019: (left) Frans Lanting/Minden Pictures; (right) Wayne Sorce. Page 1027: Courtesy Matthew G. Wheeler. Page 1042: Piergiorgio Scharandis/Black Star.

ESSAY 12
Page E12-1: Courtesy Suzanne Nagel.

Chapter 40
Opener: E. R. Degginger. Page 1053: Runk/Schoenberger/ Grant Heilman. Page 1054: From *Atlas of Optical Phenomena* by M. Cagnet et al., Springer-Verlag, Prentice Hall, 1962. Pages 1055 and 1062: Richard Megna/Fundamental Photographs. Page 1064: Courtesy Dr. Helen Ghiradella, Department of Biological Sciences, SUNY, Albany. Page 1073: Courtesy Bausch & Lomb.

ESSAY 13
Page E13-1: Courtesy Elsa Garmire.

Chapter 41
Opener: Courtesy The Art Institute of Chicago. Page 1076: Ken Kay/Fundamental Photographs. Pages 1077 and 1083: From *Atlas of Optical Phenomena* by Cagnet, Francon, Thierr, Springer-Verlag, Berlin, 1962. Page 1084: (left) AP/Wide World Photos; (right) Cath Ellis/Science Photo Library/Photo Researchers. Pages 1086 and 1089: From *Atlas of Optical Phenomena* by Cagnet, Francon, Thierr, Springer-Verlag, Berlin, 1962. Page 1091: Department of Physics, Imperial College/Science Photo Library/Photo Researchers. Page 1092: Peter L. Chapman/Stock, Boston. Page 1101: Courtesy Robert Greenler.

ESSAY 14
Pages E14-1–4: Courtesy Tung Jeong.

Chapter 42
Opener: T. Tracy/FPG International. Page 1106: Courtesy Hebrew University of Jerusalem, Israel.

ESSAY 15
Page E15-1: (top) Photo by Vittorio Giannella, courtesy Joseph Ford; (bottom) Reproduced with permission from ONERA, courtesy Joseph Ford.

Chapter 43
Opener: Courtesy Jearl Walker. Page 1143: Courtesy AIP Neils Bohr Library.

Chapter 44
Opener: Lawrence Berkeley Laboratory/Science Photo Library/Photo Researchers. Page 1159: (top left) Courtesy Riber Division of Instruments, Inc.; (top right and bottom) From PSSC film "Matter Waves," courtesy Education Development Center, Newton, Massachusetts. Page 1167: Courtesy IBM Corporation, Research Division, Almadon

Research Center, San Jose, CA. Page 1173: Courtesy A. Tonomura, J. Endo, T. Matsuda, and T. Kawasaki/ Advanced Research Laboratory, Hitachi, Ltd., Kokubinju, Tokyo; H. Ezawa, Department of Physics, Gakushuin University, Mejiro, Tokyo.

ESSAY 16
Pages E16-1 and E16-4: Courtesy Ivar Giaever.

Chapter 45
Opener: Kurt Coste/Tony Stone Worldwide. Page 1180: Courtesy Professor Hatsu Uyeda, Kyoto University. Page 1190: CNRI/Science Photo Library/Photo Researchers. Page 1195: (left) Courtesy AIP Neils Bohr Library; (right) Scott Camazine/Photo Researchers. Page 1197: (left) Will & Deni McIntyre/Photo Researchers; (right) Tony Stone Worldwide.

Chapter 46
Opener: Courtesy IBM Corporation, Research Division. Page 1217: Alfred Pasieka/Peter Arnold. Page 1224: Courtesy AT&T Bell Laboratories.

ESSAY 17
Pages E17-1–2: Courtesy Patricia Cladis.

Chapter 47
Opener: Elscint/Science Photo Library/Photo Researchers. Page 1245: (inset) R. Perry/Sygma; (bottom) George Rockwin/Bruce Coleman.

Chapter 48
Opener: Courtesy U.S. Department of Energy. Page 1266: Ivleva/Magnum. Page 1267: Courtesy Chicago Historical Society. Page 1272: Courtesy Anglo-Australian Telescope Board. Page 1275: (left) Courtesy Los Alamos National Laboratory, New Mexico; (right) Roger Ressmeyer/Starlight. Page 1279: Courtesy Martin Marietta Energy Systems, Inc., for the Department of Energy.

Chapter 49
Opener: Courtesy NASA. Page 1284: David Parker/Science Photo Library/Photo Researchers. Page 1287: Courtesy Lawrence Berkeley Laboratory. Page 1298: Courtesy AIP Neils Bohr Library. Page 1299: Courtesy AT&T Bell Laboratories.

Table of Contents
Jerry Yulsman/Image Bank
Lois Greenfield/Bruce Coleman
Stephen Dalton/Animals Animals
Stephen Dalton/Photo Researchers
Johnny Johnson/Earth Scenes
Courtesy The Art Institute of Chicago

INDEX